Lecture Notes
in Control and Information Sciences 202

Editor: M. Thoma

Bruce A. Francis and Allen R. Tannenbaum

Feedback Control, Nonlinear Systems, and Complexity

Springer-Verlag London Ltd.

Editors

Bruce Allen Francis
Department of Electrical and Computing Engineering, University of Toronto,
Toronto, Ontario M5S 1A4, Canada

Allen Robert Tannenbaum
Department of Electrical Engineering, University of Minnesota, Minneapolis,
MN 55455, USA

ISBN 978-3-540-19943-4   ISBN 978-3-540-39364-1 (eBook)
DOI 10.1007/978-3-540-39364-1

British Library Cataloguing in Publication Data
A catalogue record for this book is available from the British Library

© Springer-Verlag London 1995
Originally published by Springer-Verlag Berlin Heidelberg New York in 1995

Typesetting: Camera ready by contributors

69/3830-543210 Printed on acid-free paper

Preface

This volume is the proceedings of a conference held May 6 and 7, 1994, at McGill University in Montreal to honour Professor George Zames on the occasion of his 60th birthday (January 7, 1994).

The general chair of the conference was Dean Pierre Bélanger, the local organizer Peter Caines, and the technical organizers Bruce Francis and Allen Tannenbaum. The speakers, all invited, were

John Baras	Raymond Kwong
Roger Brockett	Sanjoy Mitter
Peter E. Caines	A. Stephen Morse
Munther A. Dahleh	Kameshwar Poolla
Edward J. Davison	Michael G. Safonov
John C. Doyle	Malcolm C. Smith
Tryphon T. Georgiou	Eduardo D. Sontag
Keith Glover	Le Yi Wang
Pramod Khargonekar	Jan C. Willems

Looking over George's list of publications, one sees the pioneering insights that he has had in the subject of feedback control. Indeed, we think it is fair to say that no other systems theorist has thought so deeply about feedback systems and has shaped the subject in so profound a manner.

Those of us who have had the privilege of working with George Zames and becoming his friend must be impressed with his generosity, his depth, and his devotion to his subject. There is a Jewish saying that a truly full life is reached only at one hundred and twenty. In that case, we can only look forward to the surprises that George will have for us in the next sixty years!

Happy Birthday George!

Bruce Francis and Allen Tannenbaum

Acknowledgements Financial support for the conference was provided by the Natural Sciences and Engineering Research Council of Canada and by the Canadian Institute for Advanced Research. We thank Stephen Zucker of the CIAR for this support, Pierre Bélanger for eloquently opening the conference, Peter Caines for his superb job of organizing the conference venue and banquet, and Jennifer Quinn and Mindle Levitt for all their help with correspondence and registration.

CONTENTS

Biographical Sketch of George Zames

George Zames was born on January 7, 1934. He was a child living with his parents in Warsaw, Poland, when the bombing of that city on September 1, 1939, marked the start of World War II. His family escaped Europe in an odyssey through Lithuania, whose occupation by Soviet tanks they witnessed, then through Russia, Siberia, followed by a triple crossing of the Sea of Japan, eventually reaching Kobe, Japan, early in 1941. (This episode, which involved the extraordinary help of the Japanese consul to Lithuania, Senpo Sugihara, is the subject of the book *The Fugu Plan* by M. Tokayer). Later that year they moved to the Anglo-French International Settlement in Shanghai, China, where they were stranded by the outbreak of the war in the Pacific. Despite the war, their sojourn in Shanghai was a happy one, and George was able to attend school without interruption. The family moved to Canada in 1948. In spite of losing a year in the move, George entered McGill University at age 15. He graduated at the top of the Engineering Physics class, and won an Athlone Fellowship for study in England. He gravitated to Imperial College of London University, where his advisors included J.H. Westcott, Colin Cherry, and Denis Gabor.

In 1956, George began his doctoral studies at the Massachusetts Institute of Technology. He was briefly associated with Doug Ross at the Servomechanisms Laboratory, where he was assigned the task of developing computer graphics for Gordon Brown's recently developed computer-controlled milling machine which, at the time, was the only one in existence. Eventually this program became the APT programming language.

He later switched to the Communications Theory Group of Norbert Wiener, Y.W. Lee, and Amar Bose at MIT's Research Laboratory of Electronics. In 1959, while still working on his thesis, Zames introduced the first global linearization theory for nonlinear feedback systems. Key applications included the earliest versions of the Companding Theorem for the recovery of nonlinearly distorted bandlimited signals, the Small Gain Theorem, and the notion of "physical realizability," later renamed strict causality, for nonlinear systems. These results were described in the 1960 report *Nonlinear Operators for Systems Analysis,* which won wide recognition, and were repeated in his 1963-64 papers. The companding recovery involved the use of contraction–mapping based feedback, and presaged George's continuing interest in nonlinear feedback systems.

In 1957, Niels Bohr arrived for a lecture tour of North America and asked for a "typical American" student to guide him around Cambridge. George was found to be appropriate for this task, and after being asked by Norbert Weiner for an introduction to Bohr, witnessed a remarkable argument between the two men on the merits of research into the natural sciences, such as Physics, versus the sciences which focus on man-made phenomena, notably Cybernetics.

After receiving his doctorate of science degree, George was appointed Assistant Professor at MIT. The following summer he set out for a vacation in Greece. On his way he stopped in Israel and met Eva. He never got to Greece on that trip. They were married two years later. They have two sons, Ethan and Jonathan.

Between 1961 and 1965, he moved back and forth between MIT and Harvard, continuing his work on nonlinear stability. His version of the Circle Criterion, the Positive Operator Theorem and related multiplier methods appeared in a 1964 N.E.C. paper. In 1965, he won a Guggenheim Fellowship, which he spent at the NASA Cambridge Research Center, forming the nucleus of what was to become the Office of Control Theory, with which W.M. Wonham, S. Morse, P.L. Falb, W. Wolovich, and H. Kushner later became associated. At the Center he won the first of a half-dozen outstanding paper awards.

In 1970, the NASA Center closed and was replaced by a Center for Transportation Systems. For a year he worked on Transportation planning with G. Kovatch, authoring the 1971 twenty-year *Transportation Technology Forecast* and the studies of *Personalized Rapid Transit Systems.*

George then took an extended sabbatical at the Technion in Haifa, Israel. There he worked out a theory of structural stabilization of nonlinear systems by dither, which led to a seminal paper co-authored with his student at the time, N. Schneydor. He also hosted Claude Shannon during Shannon's visit to receive the Harvey Prize. George's interest in metric complexity theory was stimulated by this event as well as interactions with the Technion professors Jacob Ziv and Moshe Zakai.

In 1974, George returned to McGill where he was appointed Professor of Electrical Engineering. At that point he had gotten the idea that the objective of complexity reduction could provide a basis for a unified theory of metric organizations, of which simple feedback is an example and which, more generally, take the form of feedback hierarchies. These powerful new ideas were described in a plenary lecture at the 1976 Conference on Decision and Control, which has been updated and repeated on many occasions. (It is one of George's best talks!)

By then he had concluded that existing indicators of feedback performance were inadequate, and that without a better indicator a satisfactory synthesis theory could not be achieved. It was for this reason (and partly also in an attempt to show that feedback reduces metric complexity) that George introduced the H^∞ feedback synthesis theory in 1979-81. He subsequently collaborated on this subject with B. Francis, W. Helton, A. Tannenbaum, and S.J. Mitter among others. In 1981 together with his student A. El-Sakkary, he also proposed the "gap" as a metric for unstable systems. George's work on H^∞ theory has had an enormous influence on control theory for the past decade. Besides becoming the dominant methodology in robust control, it has also influenced a number of world-class mathematicians to work in problems in feedback control.

George was awarded the Macdonald Chair of Electrical Engineering at McGill in 1983 and the IEEE Field Award for Control Science in 1984. He is a Fellow of the Canadian Institute for Advanced Research and the Royal Society of Canada.

Publications of George Zames

Papers in Journals

G. Zames, L. Lin, and L.Y. Wang, "Fast Identification n-Widths and Uncertainty Principles for LTI and Slowly Varying Systems," *IEEE Trans. Auto. Control*, **39**, Sept. 1994, pp. 1827-1838.

G. Zames and J.G. Owen, "Duality Theory for MIMO Robust Disturbance Rejection," *IEEE Trans. Auto. Control*, **38**, May 1993, pp. 743-751.

G. Zames and J.G. Owen, "Metric Dimension and Feedback in Discrete Time," *IEEE Trans. Auto. Control*, **38**, April 1993, pp. 664-667.

J.G. Owen and G. Zames, "Duality Theory of Robust Disturbance Attenuation," *Automatica*, **29**, 1993, pp. 695-705.

J.G. Owen and G. Zames, "Robust H^∞ Disturbance Minimization by Duality," *Systems and Control Letters*, **19**, 1992, pp. 255-263.

G. Zames and L.Y. Wang, "Local-global Double Algebras for Slow H^∞ Adaptation: Part I - Inversion and Stability," *IEEE Trans. Auto. Control*, **36**, Feb. 1991, pp. 130-142.

L.Y. Wang and G. Zames, "Local-global Double Algebras for Slow H^∞ Adaptation: Part II - Optimization of Stable Plants, *IEEE Trans. Auto. Control*, **36**, Feb. 1991, pp. 143-151.

L.Y. Wang and G. Zames, "Local-global Double Algebras for Slow H^∞ Adaptation: The Case of L^2 Disturbances," *IMA Journal of Control and Optimization*, **2**, 1991, pp. 287-319.

L.Y. Wang and G. Zames, "Lipschitz-continuity of H^∞ Interpolation," *Systems and Control Letters*, **14**, 1990, pp. 381-387.

G. Zames and S.J. Mitter, "On the Essential Spectra and Norms of Hankel* Hankel plus Toeplitz* Toeplitz Operators," *Systems and Control Letters*, **10**, 1988, pp. 159-165.

C. Foias, A.Tannenbaum, and G. Zames, "Sensitivity Minimization for Arbitrary SISO Distributed Plants," *Systems and Control Letters*, **8**, pp. 189-195, 1987.

C. Foias, A.Tannenbaum, and G. Zames, "Some Explicit Formulae for the Singular Values of Certain Hankel Operators with Factorizable Symbols," *SIAM J. Mathematical Analysis*, **19**, 1988, pp. 1081-1089.

C. Foias, A.Tannenbaum, and G. Zames, "On the H^∞-Optimal Sensitivity Problem for Systems with Delays," *SIAM J. Control and Optimization*, **25**, 1987, pp. 686-705.

C. Foias, A.Tannenbaum, and G. Zames, "Weighted Sensitivity Minimization for Delay Systems," *IEEE Trans. Auto. Control*, **31**, August 1986, pp. 763-766.

C. Foias, A.Tannenbaum, and G. Zames, "On Decoupling the H^∞-Optimal Sensitivity Problem for Products of Plants," *Systems and Control Letters*, **7**, 1986, pp. 239-249.

B. Francis, J.W. Helton, and G. Zames, "H^∞ - Optimal Multivariable Feedback Controllers," *IEEE Trans. Auto. Control*, **29**, October 1984, pp. 888-900.

B. Francis and G. Zames, "On Optimal Sensitivity Theory for SISO Feedback Systems," *IEEE Trans. Auto. Control*, **29**, January 1984, pp. 9-16.

G. Zames and D. Bensoussan, "Multivariable Feedback Sensitivity and Decentralized Control," *IEEE Trans. Auto. Control*, **28**, November 1983, pp. 1030-1035.

G. Zames and B.A. Francis, "Feedback, Minimax Sensitivity, and Optimal Robustness," *IEEE Trans. Auto. Control*, **28**, May 1983, pp. 585-601.

G. Zames, "Feedback and Optimal Sensitivity: Model Reference Transformations, Multiplicative Seminorms and Approximate Inverses," *IEEE Trans. Auto. Control*, **26**, April 1981, pp. 301-320.

G. Zames, "On the Metric Complexity of Casual Linear Systems: ε - Entropy and ε - Dimension for Continuous Time," *IEEE Trans. Auto. Control*, **24**, April 1979, pp. 222-230.

G. Zames and N. Schneydor, "Structural Stability, Continuity and Quenching," *IEEE Trans. Auto. Control*, **22**, June 1977, pp. 352-361.

G. Zames and N. Schneydor, "Dither in Nonlinear Systems," *IEEE Trans. Auto. Control*, **21**, Oct. 1976, pp. 660-667.

G. Zames and R.R. Kallmann "On Spectral Mappings, Higher Order Circle Criteria and Periodically Varying Systems," *IEEE Trans. Auto. Control*, **15**, December 1970, pp. 649-652.

P.L. Falb, M. Freedman and G. Zames, "A Hilbert Space Stability Theory Over Locally Compact Abelian Groups," in *SIAM J. on Control*, **7**, August 1969.

G. Zames, "Stability of Systems with Sector Nonlinearities," *IEEE Trans. Auto. Control*, **13**, December 1968, pp. 709-711.

G. Zames and M. Freedman, "Logarithmic Variations Criteria for the Stability of Systems with Time-Varying Gains," *SIAM J. Control*, **6**, 1968, pp. 487-507.

G. Zames and P.L. Falb, "On the Stability of Systems with Monotone and Odd Monotone Nonlinearities," *IEEE Trans. Auto. Control*, **12**, April 1967, pp. 221-223.

G. Zames and P.L. Falb, "On Crosscorrelation Bounds and the Positivity of Certain Nonlinear Operators," *IEEE Trans. Auto. Control*, April 1967.

G. Zames, "On the Input-Output Stability of Time-varying Nonlinear Feedback Systems, Part I: Conditions Derived Using Concepts of Loop Gain, Conicity, and Positivity," *IEEE Trans. Auto. Control*, **11**, April 1966, pp. 228-238.

G. Zames, "On the Input-Output of Time-Varying Nonlinear Feedback Systems, Part II: Conditions Involving Circles in the Frequency Plane and Sector Nonlinearities," *IEEE Trans. Auto. Control*, **11**, July 1966, pp. 465-476.

L. Stark, Y. Takahashi, and G. Zames, "Non-linear Analysis of Human Lens Accommodation," *IEEE Trans. System Science and Cybernetics*, **SSC-1**, November 1965, pp. 75-83.

G. Zames, "Realizability Conditions for Nonlinear Feedback Systems," *IEEE Trans. on Circuit Theory*, **CT-11**, June 1964, pp. 186-194.

G. Zames, "Functional Analysis Applied to Nonlinear Feedback Systems," *IEEE Trans. on Circuit Theory*, **CT-10**, September 1963, pp. 392-404.

D.B. Brick and G. Zames, "Bayes Optimum Filters Derived Using Wiener Canonical Forms," *IRE Trans. on Information Theory*, **IT-8**, September 1962.

Papers in Conference Proceedings

G. Zames, "Adaptive Feedback, Identification and Complexity: An Overview," *Proc. 32nd. IEEE Conf. Dec. Control*, 1993, pp. 2068-2075.

J.G. Owen and G. Zames, "Adaptation in the Presence of Unstructured Uncertainty," *Proc. 32nd. IEEE Conf. Dec. Control*, 1993, pp. 2110-2114.

G. Zames and L.Y. Wang, "Adaptive vs. Robust Control: Information Based Concepts," IFAC Internat. Symposium on *Adaptive Control and Signal Processing*, Grenoble, July 1992, pp. 533-536.

L. Lin, L.Y. Wang, and G. Zames, "Uncertainty Principles and Identification n-Widths for LTI and Slowly Varying Systems," *Proc. 1992 Amer. Control Conf.*, Chicago, pp. 296-300.

J.G. Owen and G. Zames, "Unstructured H^∞: Duality and Hankel Approximations for Robust Disturbance Attenuation," *Proc. MTNS Congress*, Kobe, June 1991, pp. 345-350.

J.G. Owen and G. Zames, "Robust H^∞ Disturbance Minimization by Duality," *Proc. 30th. IEEE Conf. Dec. Control*, Brighton, December 1991, pp. 206-209.

G. Zames and L.Y. Wang, "What is an Adaptive-learning System?," *Proc. 29th. IEEE Conf. Dec. Control*, **5**, Dec. 1990, pp. 2861-2864.

G. Zames, "Feedback Organizations and Learning," *NATO Workshop on Knowledge-based Robot Control*, Bonas, France, 1988.

G. Zames, "Feedback Organizations, Learning and Complexity in H^∞," Plenary Lecture, *American Control Conf.*, Atlanta, 1988.

G. Zames, A. Tannenbaum, and C. Foias, "Optimal H^∞ Interpolation: A New Approach," *Proc. 25th. Conf. Dec. Control*, **I**: pp. 350-355, 1986.

C. Foias, A. Tannenbaum, and G. Zames, "H^∞ Optimization Theory for Distributed Systems," *Proc. 25th. IEEE Conf. on Decision and Control*, **2**: pp. 899-904, 1986.

C. Foias, A. Tannenbaum, and G. Zames, "On Decoupling the H^∞-Optimal Sensitivity Problem for Products of Plants," *Proc. 1986 Amer. Contr. Conf.*, June 18-20, 1986.

C. Foias, A. Tannenbaum, and G. Zames, "Weighted Sensitivity Minimization for Delay Systems," *Proc. IEEE Conf. on Decision and Control*, 1985, **I**, pp. 244-249.

B.A. Francis and G. Zames, "Design of H^∞-Optimal Multivariable Feedback Systems," *22nd. IEEE Conf. on Decision and Control*, Dec. 1983, pp. 103-108.

B. Francis and G. Zames, "On Optimal Sensitivity Theory for SISO Feedback Systems," *Proc. 21st IEEE Conf. on Decision and Control*, Dec. 1982, pp. 623-628.

G. Zames and A. El-Sakkary, "Uncertainty in Unstable Systems: The Gap Metric," *Proc. 8th Congress*, Kyoto, Japan, 1981, No. 5.6.

G. Zames, "Feedback, Optimal Sensitivity, and Plant Uncertainty via Multiplicative Seminorms," *Proc. 9th IFAC Congress*, Kyoto, Japan, 1981, No. 38.7.

G. Zames and B. Francis, "Feedback and Minimax Sensitivity," *Proc. NATO-AGARD*, No. 117, October 1981, pp. 10.1-10.8.

G. Zames and B. Francis, "A New Look at Classical Compensation Methods: Feedback and Minimax Sensitivity," *Proc. 20th IEEE Conf. on Decision and Control*, Dec. 1981, pp. 867-874.

B. Francis and G. Zames, "On Minimax Servomechanisms," *Proc. 20th IEEE Conf. on Decision and Control*, Dec. 1981, pp. 188-193.

G. Zames and D. Bensoussan, "Multivariable Feedback and Decentralized Control," *20th. IEEE Conf. on Decision and Control*, Dec. 1981.

G. Zames, "The Tradeoff Between Identification and Feedback," *19th. IEEE Conf. on Decision and Control*, Albuquerque, Dec. 1980.

G. Zames and A. El-Sakkary, "Unstable Systems and Feedback: The Gap Metric," *Proc. 18th Allerton Conf.*, Oct. 1980, pp. 380-385.

G. Zames, "Optimal Sensitivity and Feedback: Model Reference Transformations, Weighted Seminorms and Approximate Inverses," *Proc. 17th. Allerton Conf.*, Oct. 1979, pp. 744-751.

G. Zames, "On the Metric Complexity of Causal Linear Systems," *Proc. 1977 IEEE Conf. on Decision and Control*, New Orleans, Dec. 1977, pp. 807-810.

N. Schneydor and G. Zames, "On the Stabilization of Nonlinear Systems by Dither," *Proc. 13th. Allerton Conf.*, 1975, pp. 560-579.

G. Zames, "Feedback, Hierarchies and Complexity," *Proc. 1976 IEEE Conf. on Decision and Control*, Dec. 1976, Addenda.

G. Zames, "Modeling Transportation Systems: An Overview," *Proc. 1971 IEEE Decision and Control Conf.*, Miami, Dec. 1971, 294-304.

P.L. Falb, M.J. Freedman, and G. Zames, "Input-Output Stability: A General Viewpoint," *IFAC Congress*, Warsaw, June 1969.

G. Zames and P. Falb, "On the Stability of Systems with Monotone and Odd Monotone Nonlinearities," Conference on *Mathematical Control Theory*, USC, January 1967.

G. Zames and M.I. Freedman, "Logarithmic Variation Criteria for the Stability of Systems with Time-Varying Gains," *Proc. Joint Auto. Contr. Conf.*, 1968, 769-786.

G. Zames, "On the Input-Output Stability of Nonlinear Systems–Conditions for L_∞ - Boundedness Derived Using Conic Operators on Exponentially Weighted Spaces," *Proc. 1965 Allerton Conference*, Urbana, October 1965, pp. 460-471.

G. Zames, "On the Stability of Nonlinear Feedback Systems," *Proc. National Electronics Conference*, Chicago, Illinois, Vol. 20, October 1964, pp. 725-730.

G. Zames, "Contracting Transformations-A Theory of Stability for Nonlinear, Time Varying Systems," *International Conference on Microwaves, Circuit Theory and Information Theory*, Tokyo, September 1964.

Book Chapters

J.G. Owen and G. Zames, "Unstructured Uncertainty in H^∞," *Control of Uncertain Dynamical Systems*, S.P. Bhattacharyya and L.H. Keel, Eds., CRC Press, 1992, pp. 3-21.

M. Verma and G. Zames, "An Energy and 2-port Framework for H^∞ Control," *Control of Uncertain Dynamical Systems*, S.P. Bhattacharya and L.H. Keel, Eds., CRC Press, 1992, pp. 39-51.

J.G. Owen and G. Zames, "Unstructured H^∞: Duality and Hankel Approximations for Robust Disturbance Attenuation". *Recent Advances in Mathematical Theory of Systems, Control, Networks and Signal Processing I*, H. Kimura and S. Kodama, Eds., MITA Press, Tokyo, 1991, pp. 345-350.

L.Y. Wang and G. Zames, "Local-Global Double Algebras". *Advances in Control and Dynamical Systems*, **50**, Academic Press, 1991.

G. Zames and D. Bensoussan, "Multivariable feedback and decentralized control," *Advances in Control and Dynamical System*, **22**, Part I, Academic Press, 1985.

G. Zames, "Biological Control Mechanisms: Human Accommodation as an Example of a Neurological Servomechanisms," *Theoretical and Experimental Biophysics*, Arthur Cole Ed., Marcel Dekker, New York, 1967, with L. Stark and Y. Takahashi, pp. 130-160.

Reports or Monographs

G. Zames, *Transportation Systems Technology: A Twenty-Year Outlook*, with G. Kovatch, F. Casey, and J. Barber. Report for the Secretary of Transportation, DOT, Washington, D.C.; also presented at the Conference on *Personalized Rapid Transit*, University of Minnesota, Nov. 1972.

G. Zames, *Personalized Rapid Transit System: A First Analysis*, with G. Kovatch, (Report No. DOT-TSC-ST-71-11, Transportation Systems Center, Cambridge, Mass.) prepared for the Office of the Secretary of Transportation, DOT, Washington, D.C., *Proc. Second Conference on Personalized Rapid Transit*, University of Minnesota, Minneapolis, Minn., Nov. 1972.

G. Zames, "On the Mapping of Random Processes into Random Variables on the Unit Interval and their Characterization by Probability Distribution Operators," ARM No. 258, July 21, 1961; Applied Research Laboratory, Sylvania Electronic Systems, Waltham, Massachusetts, 02154.

G. Zames, "On the Transmission of Signals through Randomly Time Varying Media; Some Preliminary Studies," ARM No. 213, June 15, 1960; Applied Research Laboratory, Sylvania Electronic Systems, Waltham, Massachusetts, 02154.

G. Zames, "Nonlinear Operators for Systems Analysis," Technical Report No. 370, Research Laboratory of Electronics, MIT, September 1960.

Research Laboratory of Electronics QPR's

G. Zames, "Conservation of Bandwidth in Nonlinear Operations," Quarterly Progress Report No. 55, Research Laboratory of Electronics, MIT, October 15, 1969, pp. 98-109.

G. Zames, "Realizability of Nonlinear Filters, and Feedback Systems," Quarterly Progress Report No. 56, Research Laboratory of Electronics, MIT, January 15, 1960, pp. 137-143.

G. Zames, "Canonical Forms for Nonlinear Statistical Estimators," Quarterly Progress Report No. 52, Research Laboratory of Electronics, MIT, January 15, 1959.

Other Presentations

G. Zames, "On Learning, Adaptation and Complexity," *Workshop on Intelligent Control*, Cortona, Italy, April 1993.

G. Zames, "Unstructured Uncertainty in H^∞," *International Conf. on Robust Control*, San Antonio, March 1991.

G. Zames, "Visions of Tomorrow in Communication and Control," distinguished panel, MIT Symposium "From Servo-loops to Fiber Nets: 50 years of Systems, Communications and Control," Cambridge, Oct. 1990.

G. Zames, "Feedback organizations and complexity". Plenary paper delivered to the *Conference on Operator Theory*, College Park, Maryland, Dec. 1985.

G. Zames, "Feedback organizations and complexity in H^∞, plenary paper delivered to the *International Symposium on the Mathematical Theory of Networks and Systems*, Stockholm, June 1985.

G. Zames, "Feedback and Nonparametric Modelling Uncertainty in H^∞," plenary session *NSF Workshop on PDE's and Flexible Structures*, Tampa, March 1985.

G. Zames, "Feedback Organizations and Complexity," keynote lecture, *ONR Symposium on Multivariable Design*, Minneapolis, Oct. 1984.

G. Zames, "H^∞ Feedback Theory," Workshop on Adaptive systems, Grenoble, France, June 1984.

G. Zames, "H^∞-Methods in Optimal Design," *Amer. Control, Conf.*, San Diego, June 1984.

G. Zames, "Optimally Robust Feedback Theory: Recent Developments," *Internat. Symp. Circuits and Systems*, Montreal, May 1984, S10.

B.A. Francis, G. Zames, J.W. Helton, "H^∞-Optimal Feedback Controllers for Linear Multivariable Systems," *Internat. Symposium Math. Theory of Networks and Systems*, Beersheva, Israel, June, 1983.

G. Zames, "The Optimally Robust Servomechanism Problem," *IFAC Workshop on Singular Perturbations*, Ohrid, Yugoslavia, July 1982.

G. Zames, "Robustness of Feedback Systems," *Internat. Symposium on Circuits and Systems*, Rome, May 1982, S2.2.

G. Zames, "Multivariable Feedback Synthesis," *Bilateral Meeting on Control Systems*, Shanghai, China, August 10-12, 1981, No. 1.3.

G. Zames, "Multivariable Feedback Under Plant Uncertainty: Explicit Input-Output Methods based on Approximate-Inverse Optimization," Parts I and II, *NATO distinguished Lecturer Series*, Ankara, Turkey; Bolkesjo, Norway; and Delft, the Netherlands; October 1981.

G. Zames, "Feedback and Optimal Sensitivity," *Math. Theory of Networks and Systems Workshop*, Virginia Beach, May 1980.

G. Zames, "Feedback, Optimal Sensitivity, and Plant Uncertainty," *IEEE International Symposium Circuits and Systems*, Houston, April 1980.

G. Zames, "Feedback Organizations and Complexity," *Mathematical Theory of Networks and Systems Conf.*, Delft Netherlands, July 1979.

G. Zames, "Decentralized Control of Large Transportation Networks," Transportation Workshop, *1972 IFAC Congress*, Paris, June 1972.

On the Structured Singular Value for Operators on Hilbert Space

Hari Bercovici and Ciprian Foias
Department of Mathematics
Indiana University
Bloomington, Indiana 47405

Allen Tannenbaum
Department of Electrical Engineering
University of Minnesota
Minneapolis, Minnesota 55455

Abstract

In this note, we discuss some new results concerning a lifting method introduced by the authors in order to study the structured singular value applied to input/output operators of control systems. We moreover give a new criterion which guarantees that the structured singular value equals its upper bound defined by D-scalings.

With great admiration, this paper is dedicated to Professor George Zames on the occasion of his 60-th birthday.

1   Introduction

This paper is concerned with the study of the structured singular value for very general linear operators. More precisely, let A be a linear operator on a Hilbert space $\mathcal{E}$, and let Δ be an algebra of operators on $\mathcal{E}$. The structured singular value of A (relative to Δ) is the number

$$\mu_\Delta(A) = 1/\inf\{\|X\| : X \in \Delta,\ -1 \in \sigma(AX)\} .$$

This quantity was introduced by Doyle and Safonov [7, 13] under a more restrictive context, and it has proved to be a powerful tool in robust system

---

*This work was supported in part by grants from the Research Fund of Indiana University, by the National Science Foundation DMS-8811084, ECS-9122106, by the Air Force Office of Scientific Research AFOSR AF/F49620-94-1-00S8DEF, and by the Army Research Office DAAH04-94-G-0054 and DAAH04-93-G-0332.

analysis and design. In system analysis, the structured singular value gives a measure of robust stability with respect to certain perturbation measures. Unfortunately, $\mu_\Delta(A)$ is very difficult to calculate, and in practice an upper bound for it is used. This upper bound is defined by

$$\widehat{\mu}_\Delta(A) := \inf\{\|XAX^{-1}\| : X \in \Delta', X \text{ invertible}\},$$

where Δ' is the commutant of the algebra Δ.

In [1, 6], we formulated a lifting technique for the study of the structured singular value. The basic idea is that $\widehat{\mu}_\Delta(A)$ can be shown to be equal to the structured singular value of an operator on a bigger Hilbert space. (In [1] this was done for finite dimensional Hilbert spaces, and then in [6] this was extended to the infinite dimensional case.) The problem with these results is that the size of the ampliation necessary to get $\widehat{\mu}_\Delta(A)$ equal to a structured singular value, was equal to the dimension of the underlying Hilbert space. Hence in the infinite dimensional case we needed an infinite ampliation. In this work, we will show that in fact, one can always get by with a finite lifting. (Note that in this paper we will be using the terms "ampliation" and "lifting" interchangeably.) For the block diagonal algebras of interest in robust control, the ampliation only depends on the number of blocks of the given perturbation structure. (See Theorem 2 below.) We moreover, give a new result when $\widehat{\mu}_\Delta(A) = \mu_\Delta(A)$, that is, when no lifting is necessary and so $\widehat{\mu}_\Delta(A)$ gives a nonconservative measure of robustness. (See Theorem 3.) This is then used to derive an elegant result of Shamma [14, 15] on Toeplitz operators. See also [8, 10, 11] for related work in this area. Complete details of the results in this note may be found in [5].

2 Background Results

Denote by $\mathcal{L}(\mathcal{E})$ the algebra of all bounded linear operators on the (complex, separable) Hilbert space $\mathcal{E}$. Fix an operator $A \in \mathcal{L}(\mathcal{E})$ and a subalgebra $\Delta \subset \mathcal{L}(\mathcal{E})$. The quantities $\mu_\Delta(A)$ and $\widehat{\mu}_\Delta(A)$ have already been defined in the Introduction. Note that $\Delta \subset \Delta''$ and $\Delta''' = (\Delta'')' = \Delta'$ so that we have the inequalities

$$\mu_\Delta(A) \leq \mu_{\Delta''}(A), \quad \widehat{\mu}_\Delta(A) = \widehat{\mu}_{\Delta''}(A) .$$

Note that the algebras Δ considered in [7] consisted of block diagonal matrices, so our approach is more general in this respect. In the following proposition we summarize some of the elementary properties of μ_Δ; see Doyle [7] or [1] for proofs. We will denote by $\|T\|_{\text{sp}}$ the spectral radius of the operator T.

Lemma 1 *(i)* $\mu_\Delta(A) = \sup\{\|AX\|_{\text{sp}} : X \in \Delta, \ \|X\| \leq 1\}$;

(ii) μ_Δ *is upper semicontinuous;*

(iii) *If $\mathcal{E}$ is finite dimensional, then μ_Δ is continuous;*

(iv) $\mu_\Delta(A) \leq \widehat{\mu}_\Delta(A)$.

In our study we will need further singular values which we now define. For $n \in \{1, 2, \ldots, \infty\}$ we denote by $\mathcal{E}^{(n)}$ the orthogonal sum of n copies of $\mathcal{E}$, and by $T^{(n)}$ the orthogonal of n copies of $T \in \mathcal{L}(\mathcal{E})$. Operators on $\mathcal{E}^{(n)}$ can be represented as $n \times n$ matrices of operators in $\mathcal{L}(\mathcal{E})$, and $T^{(n)}$ is represented by a diagonal matrix, with diagonal entries equal to T.

Denote by Δ_n the algebra of all operators on $\mathcal{E}^{(n)}$ whose matrix entries belong to Δ, and observe that $(\Delta_n)'' = (\Delta'')_n$, and $(\Delta_n)' = (\Delta')^{(n)} = \{T^{(n)} : T \in \Delta'\}$. Therefore we will denote these algebras by Δ_n'' and Δ_n', respectively.

Lemma 2 *For every finite number n we have*

$$\mu_\Delta(A) \leq \mu_{\Delta_n}(A^{(n)}) \leq \mu_{\Delta_{n+1}}(A^{(n+1)}) \leq \mu_{\Delta_\infty}(A^{(\infty)}) \leq \hat{\mu}_\Delta(A) \, ,$$

and

$$\mu_{\Delta''}(A) \leq \mu_{\Delta_n''}(A^{(n)}) \leq \mu_{\Delta_{n+1}''}(A^{(n+1)}) \leq \mu_{\Delta_\infty''}(A^{(\infty)}) \leq \hat{\mu}_\Delta(A) \, .$$

Proof. It clearly suffices to prove the first sequence of inequalities. Observe that for every $X \in \Delta_n$ and for $m > n$ we can define an operator $Y \in \Delta_m$ by $Y = X \oplus 0$. Clearly $\sigma(A^{(n)}X) = \sigma(A^{(m)}Y) \cup \{0\}$ and hence $-1 \in \sigma(A^{(n)}X)$ implies $-1 \in \sigma(A^{(m)}Y)$. Since $\mu_\Delta(A) = \mu_{\Delta_1}(A^{(1)})$, this proves the first three inequalities. The last one follows because $\mu_{\Delta_\infty}(A^{(\infty)}) \leq \hat{\mu}_{\Delta_\infty}(A^{(\infty)}) = \hat{\mu}_\Delta(A)$.
$\square$

We will now state (without proof) several results from [1, 2, 3, 4, 5, 6] which we will need in the sequel.

Lemma 3 ([6]) *Let A be finite dimensional C^*-algebra. Then A has only finitely many equivalence classes of cyclic representations.*

Lemma 4 ([6]) *Let the sequence Y_j of operators on $\mathcal{H}$, and the sequence $h_j \in \mathcal{H}$ satisfy*

(i) $\sup_j \text{rank } Y_j < \infty$, $\sup_j \|Y_j\| < \infty$;

(ii) $\lim_{j \to \infty} \|(Y_j - I)h_j\| = 0$;

(iii) $\lim_{j \to \infty} \|h_j\| = 1$.

Then $\liminf_{n \to \infty} \|Y_j\|_{\text{sp}} \geq 1$.

Lemma 5 ([4]) . *Let $\mathcal{H}$ be a Hilbert space, $T \in \mathcal{L}(\mathcal{H})$, and $D_j \in \mathcal{L}(\mathcal{H})$ invertible so that*

$$T_0 = \lim_{j \to \infty} D_j T D_j^{-1}.$$

If the set $\{D_j, D_j^{-1} : j = 1, 2, \ldots\}$ is contained in a finite dimensional subspace, then $\|T_0\|_{\text{sp}} = \|T\|_{\text{sp}}$.

3 Relative Numerical Range

We will also need some results in what follows about the relative numerical range. Let $\mathcal{H}$ be a complex separable Hilbert space, and let $\mathcal{L}(\mathcal{H})$ denote the set of bounded linear operators on $\mathcal{H}$. Let $T_1, \ldots, T_m, Q \in \mathcal{L}(\mathcal{H})$. Then we define the following *relative numerical ranges*:

$$W_Q(T_1, \ldots, T_m) := \{\lambda \in \mathbf{C}^n, \ \lambda = \lim_{n \to \infty} (\langle T_j h_n, h_n \rangle)_{j=1}^m :$$

$$h_n \in \mathcal{H}, \|h_n\| = 1, \lim_{n \to \infty} \|Q h_n\| = 0\},$$

and

$$W_Q^0(T_1, \ldots, T_m) := \{\lambda \in \mathbf{C}^n, \ \lambda = \lim_{n \to \infty} (\langle T_j h_n, h_n \rangle)_{j=1}^m :$$

$$h_n \in \mathcal{H}, \|h_n\| = 1, \lim_{n \to \infty} \|Q h_n\| = 0, h_n \to 0 \text{ weakly}\}.$$

Lemma 6 ([5]) $W_Q^0(T_1, \ldots, T_m)$ *is a compact convex subset of* $\mathbf{C}^m$.

Lemma 7 ([5]) $W_Q(T_1, \ldots, T_m)$ *is the union of all segments*

$$\{\theta\lambda + (1 - \theta)\mu : 0 \leq \theta \leq 1\},$$

where $\lambda \in W_Q^0(T_1, \ldots, T_m)$ *and* $\mu = (\langle T_j h, h \rangle)_{j=1}^m$ *for some* $h \in \ker Q, \|h\| = 1$.

Corollary 1 ([5]) *For all* $T, Q \in \mathcal{L}(\mathcal{H})$, *the set*

$$W_Q(T) = \{\lambda = \lim_{n \to \infty} \langle T h_n, h_n \rangle : h_n \in H, \|h_n\| = 1, \lim_{n \to \infty} \|Q h_n\| = 0\}$$

is a compact convex set.

Finally, for the proof of our lifting theorem (to be given in Section 4), we will need the following elementary fact:

Lemma 8 ([5]) *Let* $\mathcal{Z}$ *denote a finite dimensional normed space, and let* S *be a set of linear functionals on* $\mathcal{Z}$. *Suppose that for every* $z \in \mathcal{Z}$ *there exists a sequence* $\ell_n \in S$ *such that* $\lim_{n \to \infty} \ell_n(z) = 0$. *Then there exists a sequence* ℓ_n *in the convex hull of* S *such that* $\lim_{n \to \infty} \|\ell_n\| = 0$.

4 Liftings of Perturbations

In this section, we will formulate and prove a new lifting result relating $\mu_\Delta(A)$ and $\widehat{\mu}_\Delta(A)$. For finite dimensional $\mathcal{E}$, a lifting result of this type was first proven in [1]. The result was then generalized to the infinite dimensional case in [6]. (For another proof of this type of lifting result in finite dimensions, see [8].) In these theorems, the lifting or ampliation of the operator A and perturbation structure Δ depends on the dimension of $\mathcal{E}$. Thus if $\mathcal{E}$ is infinite dimensional, we get an infinite lifting. In the new result proven below, we only have to lift up to the dimension of Δ' which in the cases of interest in the control applications of this theory only depends on the number of blocks of the given perturbation structure.

We now state the original *Lifting Theorem* which turns out to be very useful in analyzing the structured singular value. In the finite dimensional case, i.e., $\mathcal{E}$ being finite-dimensional, it was employed in [1] to show that $\mu_\Delta(A) = \widehat{\mu}_\Delta(A)$ when the relevant diagonal algebra has three or fewer blocks, and so gave an alternative proof of a result due to Doyle [7]. We therefore briefly review the set-up of [1, 6]. Let $HS(\mathcal{E})$ denote the space of all Hilbert-Schmidt operators on $\mathcal{E}$ equipped with the Hilbert space structure

$$\langle T_1, T_2 \rangle := Tr(T_2^* T_1),$$

where Tr denotes the trace. Define the operator $L_A : HS(\mathcal{E}) \rightarrow HS(\mathcal{E})$ by $L_A := AX$. Now we set

$$\bar{\mu}_\Delta(A) := \mu_{\tilde{\Delta}}(L_A)$$

where

$$\tilde{\Delta} := \{L_X : X \in \Delta'\}'.$$

In what follows, we will assume that Δ' is finite dimensional $*$-algebra, but that $\mathcal{E}$ is arbitrary. We can now state the following lifting result:

Theorem 1 ([1, 6]) *Let Δ' be a finite dimensional $*$-algebra. Then*

$$\widehat{\mu}_\Delta(A) := \bar{\mu}_\Delta(A).$$

We can now state our new lifting theorem:

Theorem 2 *Assume that Δ' is a $*$-algebra of finite dimension n. Then*

$$\widehat{\mu}_\Delta(A) = \mu_{\Delta_n''}(A^{(n)})$$

for every $A \in \mathcal{L}(\mathcal{E})$.

Proof. The argument starts as in the proof of Theorem 3 in [1], and of Theorem 1 of [6]. Without loss of generality, we may assume that $\widehat{\mu}_\Delta(A) = 1$. We must show that $\mu_{\Delta_n''}(A^{(n)}) \geq 1$. Choose a sequence of invertible operators $X_j \in \Delta'$ such that $\|X_j A X_j^{-1}\| \rightarrow \widehat{\mu}_\Delta(A)$. Since $X_j A X_j^{-1}$ belongs to the finite dimensional space generated by $\Delta' A \Delta'$, we may assume that the sequence $X_j A X_j^{-1}$ converges to some operator A_0 such that $\|A_0\| = 1$. Obviously $\|X A_0 X^{-1}\| \geq \|A_0\|$ for every invertible operator $X \in \Delta'$. In particular we have

$$\|(I - X)A_0(I + X + X^2 + \cdots)\| \geq 1$$

for $X \in \Delta'$ with $\|X\| < 1$. Fix an operator $X \in \Delta'$ and a sequence $\varepsilon_j > 0$ converging to zero. There exist vectors $h_j \in \mathcal{E}$ with $\|h_j\| = 1$, such that

$$\|(I - \varepsilon_j X)A_0(I + \varepsilon_j X + \varepsilon_j^2 X^2 + \cdots)h_j\|^2 \geq 1 - \varepsilon_j^2.$$

This can be rewritten as,

$$\langle A_0^* A_0 h_j, h_j \rangle + 2\varepsilon_j \Re \langle A_0^*(A_0 X - X A_0)h_j, h_j \rangle + O(\varepsilon_j^2) \geq 1 - \varepsilon_j^2$$

or equivalently,

$$2\varepsilon_j \Re \langle A_0^*(A_0 X - X A_0)h_j, h_j \rangle + O(\varepsilon_j^2) \geq \langle (I - A_0^* A_0)h_j, h_j \rangle - \varepsilon_j^2 \geq -\varepsilon_j^2.$$

Dividing by ε_j and letting $\varepsilon_j \to 0$ as $j \to \infty$, we see from the last equation that

$$\langle (I - A_0^* A_0)h_j, h_j \rangle \to 0, \tag{1}$$
$$\liminf_{j \to \infty} \Re \langle A_0^*(A_0 X - X A_0)h_j, h_j \rangle \geq 0. \tag{2}$$

We easily conclude that

$$\liminf_{j \to \infty} \Re \langle (X - A_0^* X A_0)h_j, h_j \rangle \geq 0. \tag{3}$$

Set

$$Q = I - A_0^* A_0, \quad T = X - A_0^* X A_0.$$

Then from (1,3), we see that

$$Q h_j \to 0, \quad \liminf_{j \to \infty} \Re \langle T h_j, h_j \rangle \geq 0. \tag{4}$$

Applying the above argument to ζX for any $\zeta \in \partial \mathbf{D}$ (the unit circle), we see that there exists a sequence $h_j^{(\zeta)}$, $\|h_j^{(\zeta)}\| = 1$ such that

$$Q h_j^{(\zeta)} \to 0, \quad \liminf_{j \to \infty} \Re \zeta \langle T h_j, h_j \rangle \geq 0. \tag{5}$$

We claim that $0 \in W_{Q,0}(T)$. Indeed, if this were not the case, Corollary 1 would imply the existence of $\zeta \in \partial \mathbf{D}$ such that

$$\liminf_{j \to \infty} \Re \zeta \langle T h_j, h_j \rangle < 0$$

for all sequences of unit vectors h_j such that $Q h_j \to 0$ contradicting (5).

Thus, we have shown that for each $X \in \Delta'$, there exists a sequence of unit vectors $h_j \in \mathcal{E}$ such that

$$(I - A_0^* A_0)h_j \to 0, \quad \text{and} \quad \langle (X - A_0^* X A_0)h_j, h_j \rangle \to 0. \tag{6}$$

Let

$$\Delta'_{\mathrm{sa}} := \{X = X^* : X \in \Delta'\}.$$

Consider now a subspace $D \subset \Delta'_{\mathrm{sa}}$ of real dimension $n - 1$ such that $\Delta'_{\mathrm{sa}} = D + \mathbf{R}I$. Set $\mathcal{Z} = \{X - A_0^* X A_0 : X \in D\}$, and for every unit vector $h \in \mathcal{E}$ define a linear functional $\ell(h)$ on $\mathcal{Z}$ by $\ell(h)(T) = \langle T h, h \rangle$, $T \in \mathcal{Z}$. Then Lemma 8 applied to the set $S_k = \{\ell(h) : \|(I - A_0^* A_0)h\| \leq 1/k\}$ implies the existence of linear functionals ℓ_k in the convex hull of S_k such that $\|\ell_k\| \leq 1/k$. Note further that the real dimension of $\mathcal{Z}$ is at most $n - 1$. Then from a standard result (see e.g., [12], page 73), each ℓ_k is a convex combination of at most n functionals $\ell(h)$, say $\ell_k = \sum_{j=1}^n \alpha_j^{(k)} \ell(h_j^{(k)})$, where $\alpha_j^{(k)} \geq 0$, $\sum_{j=1}^n \alpha_j^{(k)} = 1$, and the $h_j^{(k)}$ are unit vectors in $\mathcal{E}$, such that

$$\|(I - A_0^* A_0) h_j^{(k)}\| \leq 1/k.$$

Let us define unit vectors vectors $u_k \in \mathcal{E}^{(n)}$ by

$$u_k = \oplus_{j=1}^n (\alpha_j^{(k)})^{1/2} h_j^{(k)}, \tag{7}$$

and observe that $\lim_{k \to \infty} \langle (X^{(n)} - A_0^{*(n)} X^{(n)} A_0^{(n)}) u_k, u_k \rangle = 0$ for every $X \in \Delta'$. Taking $X = Y^* Y$ we obtain

$$\lim_{k \to \infty} (\|Y^{(n)} A_0^{(n)} u_k\| - \|Y^{(n)} u_k\|) = 0 \tag{8}$$

for every $Y \in \Delta'$.

Consider now the spaces $\mathcal{H}_k = \Delta_n' A_0^{(n)} u_k$ and $\mathcal{K}_k = \Delta_n' u_k$. Lemma 3 implies that, by passing to appropriate subsequences, we may assume that all the representations $X \to X^{(n)}|\mathcal{H}_k$ (resp. $X \to X^{(n)}|\mathcal{K}_k$) are unitarily equivalent. It follows that we can find partial isometries U_k, V_k in Δ_n'' such that $U_k \mathcal{H}_k = \mathcal{H}_1$ and $V_k \mathcal{K}_k = \mathcal{K}_1$. Dropping again to appropriate subsequences, we may assume that the limits $u = \lim_{k \to \infty} U_k A_0^{(n)} u_k$ and $v = \lim_{k \to \infty} V_k u_k$ exist. Then (8) implies that

$$\|Y^{(n)} u\| = \|Y^{(n)} v\|$$

for every $Y \in \Delta'$. Therefore there exists a partial isometry $W \in \Delta_n''$ such that

$$W Y^{(n)} u = Y^{(n)} v$$

for every $Y \in \Delta'$. Of course, W can be chosen equal to zero on the orthogonal complement of $\Delta_n' u$ and thus to have finite rank at most n. The partial isometries $R_k := V_k^* W U_k$ are in Δ_n'', they have uniformly bounded rank, and

$$\lim_{k \to \infty} (R_k A_0^{(n)} - I) u_k = \lim_{k \to \infty} V_k^* (W U_k A_0^{(n)} u_k - V_k u_k) = 0.$$

Therefore Lemma 4 implies that

$$\liminf_{k \to \infty} \|R_k A_0^{(n)}\|_{\mathrm{sp}} \geq 1.$$

Finally, since R_k commutes with $X^{(n)}, X \in \Delta'$, and we have

$$X_j^{(n)} R_k A^{(n)} X_j^{(n)-1} \to R_k A_0^{(n)}$$

in norm as $j \to \infty$. Lemma 5 shows that

$$\|R_k A_0^{(n)}\|_{\mathrm{sp}} = \|R_k A^{(n)}\|_{\mathrm{sp}}.$$

Consequently, we have

$$\liminf_{k \to \infty} \|R_k A^{(n)}\|_{\mathrm{sp}} = \liminf_{k \to \infty} \|R_k A_0^{(n)}\| \geq 1.$$

Thus,

$$\mu_{\Delta_n''}(A^{(n)}) \geq \liminf_{k \to \infty} \|A^{(n)} X_k\|_{\mathrm{sp}} \geq 1 = \hat{\mu}_\Delta(A),$$

which completes the proof of the theorem. $\square$

Remark. In the cases of interest in control,

$$\Delta'' = \Delta,$$

and so one has from Theorem 2 that

$$\mu_{\Delta_n}(A) = \hat{\mu}_\Delta(A).$$

5 Conditions for $\mu = \hat{\mu}$

In this section, we will discuss some new conditions when $\mu = \hat{\mu}$ without any need for lifting. In the finite dimensional case, there have been some results of this kind, the most famous of which is that of Doyle [7], who showed that no lifting is necessary for perturbation structures with three or fewer blocks.

We begin by noting that in the proof of Theorem 2, we established a useful property of the critical operators A_0 in the closed Δ' similarity orbit

$$\overline{\mathcal{O}_{\Delta'}(A)} = \overline{\{X A X^{-1} : X \in \Delta'\}}$$

of A. Namely, if we call *critical* any $A_0 \in \overline{\mathcal{O}_{\Delta'}(A)}$ satisfying

$$\limsup_{\epsilon \downarrow 0} \|(I - \epsilon X) A_0 (I - \epsilon X)^{-1}\| \geq \|A_0\|, \quad \forall X \in \Delta',$$

then the first part of the proof of Theorem 1 establishes the following:

Lemma 9 *If A_0 is a critical operator in $\overline{\mathcal{O}_{\Delta'}(A)}$, then it enjoys the following property $(\mathcal{O})$:*

$$0 \in W_Q(\|A_0\|^2 X - A_0^* X A_0), \quad X \in \Delta',$$

where $Q = \|A_0\|^2 I - A_0^* A_0$.

Indeed, property $(\mathcal{O})$ is a reformulation of equation (6) in the case in which the norm of A_0 may be different from 1.

The next lemma is the key step in adapting the proof of Theorem 2 in order to show that

$$\mu_\Delta(A) = \hat{\mu}_\Delta(A)$$

in several interesting cases.

Lemma 10 *Let A_0 be an operator on $\mathcal{E}$ which satisfies the essential version of property $(\mathcal{O})$, property $(\mathcal{O}^0)$, namely*

$$0 \in W_Q^0(\|A_0\|^2 X - A_0^* X A_0), \ X \in \Delta',$$

where $Q = \|A_0\|^2 I - A_0^* A_0$. Then there exists a sequence $\{h_k\}_{k=1}^\infty \subset \mathcal{E}$, $\|h_k\| = 1$, $k = 1, 2, \ldots$, such that

$$Q h_k \to 0 \ \text{strongly and} \ \langle (\|A_0\|^2 X - A_0^* X A_0) h_k, h_k \rangle \to 0,$$

for all $X \in \Delta'$.

Proof. Without loss of generality we can assume that $\|A_0\| = 1$. Let $X_1, \ldots, X_n$ be an algebraic basis of Δ'. (Note that Δ' is finite dimensional.) Set $T_j := X_j - A_0^* X_j A_0$, $j = 1, \ldots, n$. Then by virtue of Lemma 6, $W_Q^0(T_1, \ldots, T_n)$ is convex and compact. If $0 \notin W_Q^0(T_1, \ldots, T_n)$, there exists $\psi = (\psi_1, \ldots, \psi_n) \in \mathbf{C}^n$ and $\epsilon > 0$ such that

$$\Re \sum_{j=1}^n \psi_j \lambda_j \geq \epsilon, \ \forall \lambda = (\lambda_1, \ldots, \lambda_n) \in W_Q^0(T_1, \ldots, T_n).$$

Set

$$T = \sum_{j=1}^n \psi_j T_j.$$

Property $(\mathcal{O}^0)$ implies that there exists a sequence $\{g_k\}_{k=1}^\infty \subset \mathcal{E}$, $\|g_k\| = 1$, $g_k \to 0$ weakly such that $\langle T g_k, g_k \rangle \to 0$. Without loss of generality (by passing to a subsequence if necessary), we can assume that

$$\langle T_j g_k, g_k \rangle \to \lambda_j, \ (j = 1, \ldots, n),$$

for $k \to \infty$. Thus

$$\lambda = (\lambda_1, \ldots, \lambda_n) \in W_Q^0(T_1, \ldots, T_n).$$

Hence

$$0 \leftarrow \Re \langle T g_k, g_k \rangle = \Re \sum_{j=1}^n \psi_j \langle T_j g_k, g_k \rangle \to \Re \sum_{j=1}^n \psi_j \lambda_j \geq \epsilon,$$

a contradiction. We therefore conclude that $0 \in W_Q^0(T_1, \ldots, T_n)$, i.e., there exists a sequence $\{h_k\}_{k=1}^\infty \subset \mathcal{E}$ satisfying the properties $\|h_k\| = 1$, $k = 1, 2, \ldots$, $\|Q h_k\| \to 0$, $h_k \to 0$ weakly, and

$$\langle (X_j - A_0^* X_j A_0) h_k, h_k \rangle = \langle T_j h_k, h_k \rangle \to 0$$

for all $j = 1, 2, \ldots, n$. This implies that

$$\langle (X - A_0^* X A_0) h_k, h_k \rangle \to 0,$$

for all $X \in \Delta'$. $\square$

We can now state the second main result of this paper:

Theorem 3 *If there exists a critical operator A_0 satisfying property $\mathcal{O}^0$ in the closed Δ'-orbit of A, then*

$$\mu_{\Delta''}(A) = \hat{\mu}_\Delta(A).$$

Proof. We only have to note that because of Lemma 10, in the proof of Theorem 1, we need not take direct sums. More precisely, referring to equation (7) in the proof of Theorem 1, we can take $u_k = h_k$, where

$$\{h_k\}_{k=1}^\infty$$

is the sequence provided by Lemma 10. The proof then proceeds exactly as in Theorem 1 with A_0 replacing $A_0^{(n)}$, X replacing $X^{(n)}$, and Y replacing $Y^{(n)}$.
□

Remark. Under the hypotheses of Theorem 3, when $\Delta'' = \Delta$ (which happens in all cases of interest in control), we have that

$$\mu_\Delta(A) = \hat{\mu}_\Delta(A).$$

Let $L(\Delta'A\Delta')$ denote the linear space generated by

$$\Delta'A\Delta' = \{XAY : X, Y \in \Delta'\}.$$

Obviously $L(\Delta'A\Delta')$ is finite dimensional, and therefore closed. Hence $\overline{\mathcal{O}_{\Delta'}(A)} \subset L(\Delta'A\Delta')$.

Corollary 2 *If for every $B \in L(\Delta'A\Delta')$, $B \neq 0$, the norm of B is not attained (that is, there is no $h \in \mathcal{H}$ such that $\|Bh\| = \|B\|\|h\| \neq 0$), then*

$$\mu_{\Delta''}(A) = \hat{\mu}_\Delta(A).$$

Proof. The critical operator A_0 constructed in the first part of the proof of Theorem 1 belongs to $L(\Delta'A\Delta')$, and therefore its norm is not attained. However in equation (6), we can assume that the sequence $\{h_j\}_{j=1}^\infty$ is weakly convergent, say $h_j \to h$ weakly. Without loss of generality, we may assume that $\|A_0\| = 1$. Then (6) shows that

$$(I - A_0^* A_0)h = 0.$$

Therefore if $h \neq 0$, we would have

$$\|A_0 h\|^2 = \|h\|^2 = \|A_0\|^2 \|h\|^2 \neq 0,$$

and so the norm of A_0 would be attained. We conclude that $h_j \to 0$ weakly, and so A_0 satisfies property $(\mathcal{O}^0)$. The required result now follows by Theorem 3.
□

Remark. Note that Corollary 2 applies only to infinite dimensional Hilbert spaces $\mathcal{E}$.

6 LTI Systems

In this section, we want to use our lifting methodology in order to derive a beautiful result of Shamma [14, 15] on the structured singular value of a linear time invariant system, i.e., a Toeplitz operator.

Accordingly, set $\mathcal{E} = H^2(\mathbf{C}^n)$ and let A denote the multiplication (analytic Toeplitz) operator on $\mathcal{E}$ defined by

$$(Ah)(z) = A(z)h(z), \quad |z| < 1, \ h \in \mathcal{E},$$

where

$$A(z) = [a_{jk}]_{j,k=1}^n, \quad |z| < 1,$$

has H^∞ entries. Let Δ' be any $*$–subalgebra of $\mathcal{L}(\mathbf{C}^n)$, the elements of which are regarded as multiplication operators on $\mathcal{E}$. Note that in this case, $\Delta'' = \Delta$ is the algebra generated by operators of the form

$$(Bh)(z) = B(z)h(z), \quad |z| < 1, \ h \in \mathcal{E}$$

with $B(z)X = XB(z)$, $|z| < 1$, $X \in \Delta'$ as well as of the form

$$B \begin{bmatrix} h_1 \\ h_2 \\ \vdots \\ h_n \end{bmatrix} = \begin{bmatrix} Y h_1 \\ Y h_2 \\ \vdots \\ Y h_n \end{bmatrix},$$

with $Y \in \mathcal{L}(H^2(\mathbf{C}))$ arbitrary. We can now state:

Lemma 11 *Let A_0 be an analytic Toeplitz operator. Then if A_0 has property $(\mathcal{O})$, it also has property $(\mathcal{O}^0)$.*

Proof. Without loss of generality we may assume $\|A_0\| = 1$. Let $X \in \Delta'$ and let $h_j, j = 1, 2, \ldots$ be a sequence of unit vectors satisfying

$$\|(I - A_0^* A_0)h_j\|^2 \to 0, \quad \langle (X - A_0^* X A_0)h_j, h_j \rangle \to 0. \tag{9}$$

Note that since $I - A_0^* A_0 \geq 0$ the first condition in (9) is equivalent to

$$\langle (I - A_0^* A_0)h_j, h_j \rangle \to 0.$$

Let U denote the canonical unilateral shift on $\mathcal{E} = H^2(\mathbf{C}^n)$, that is,

$$(Uh)(z) := zh(z), \quad |z| < 1, \ h \in \mathcal{E}.$$

As is well-known, we can view $H^2(\mathbf{C}^n)$ as a subspace of $L^2(\mathbf{C}^n)$. In particular, in this representation the relations (9) are equivalent to

$$\int_0^{2\pi} (\|h_j(e^{it})\|^2 - \|A_0(e^{it})h_j(e^{it})\|^2)dt \quad \to \quad 0$$

$$\int_0^{2\pi} [\langle Xh_j(e^{it}), h_j(e^{it}) \rangle - \langle XA_0(e^{it})h_j(e^{it}), h_j(e^{it}) \rangle]dt \quad \to \quad 0. \tag{10}$$

Note that X is an $n \times n$ matrix with constant coefficients. Therefore in (10), h_j can be replaced by $U^k h_j$ for any $k \geq 0$ without changing the values of the integrals. We infer that

$$\|(I - A_0^* A_0)U^{k_j}h_j\|^2 \to 0, \quad \langle(X - A_0^* X A_0)U^{k_j}h_j, U^{k_j}h_j\rangle \to 0,$$

for any sequence $\{k_j\}_{j=1}^{\infty}$ of natural numbers. Since for any $g, h \in \mathcal{E}$

$$|\langle U^k g, h\rangle| = |\langle g, U^{*k}h\rangle| \leq \|g\|\|U^{*k}h\| \to 0 \quad (k \to \infty),$$

we can choose k_j sufficiently large in order to guarantee that

$$|\langle U^{k_j}h_j, h\rangle| \leq \frac{1}{2^j},$$

for any h of the form

$$h = (z^m \delta_{pk})_{k=1}^n, \ 0 \leq m \leq j, \ 1 \leq p \leq n, \tag{11}$$

where δ_{pk} is the Kronecker delta. Thus

$$\langle U^{k_j}h_j, h\rangle \to 0, \quad \text{as } j \to \infty$$

for all vectors of the form (11). Since these vectors form an orthonormal basis of $\mathcal{E}$, we see that $U^{k_j}h_j \to 0$ weakly, which concludes the proof of the lemma. □

Corollary 3 ([14, 15]) *For A and Δ' as above, we have that*

$$\mu_{\Delta}(A) = \hat{\mu}_{\Delta}(A).$$

Proof. First, note that any operator B in $L(\Delta' A \Delta')$ is also an analytic Toeplitz operator. In particular, the critical operator A_0 obtained in the proof of Theorem 2 is a multiplication operator given by

$$A_0(z) = [a_{jk}^0(z)]_{j,k=1}^n, \ |z| < 1.$$

By Lemma 9, the operator A_0 has property $(\mathcal{O})$, and thus also property $(\mathcal{O}^0)$, by virtue of Lemma 11. The conclusion now follows from Theorem 3. □

7 Concluding Remarks

There are a number of interesting questions which still must be solved before we have a complete lifting theory for the structured singular value. First of all, we would like to extend these results to infinite dimensional algebras Δ'. Moreover, we would like to remove the hypothesis that Δ' be a *-algebra. This would allow us to consider algebras of Toeplitz operators as our perturbation structure.

Finally, we would like to know the minimal possible lift as well as the most general conditions when $\mu = \hat{\mu}$. These are all interesting problems which demand further investigation.

References

[1] H. Bercovici, C. Foias, and A. Tannenbaum, "Structured interpolation theory," *Operator Theory: Advances and Applications* **47** (1990), pp. 195–220.

[2] H. Bercovici, C. Foias, and A. Tannenbaum, "A spectral commutant lifting theorem," *Trans. AMS* **325** (1991), pp. 741–763.

[3] H. Bercovici, C. Foias, and A. Tannenbaum, "A relative Toeplitz-Hausdorff theorem," to appear in *Operator Theory: Advances and Applications.*

[4] H. Bercovici, C. Foias, and A. Tannenbaum, "Continuity of the spectrum on closed similarity orbits," *Integral Equations and Operator Theory* **18** (1994), 242–246.

[5] H. Bercovici, C. Foias, and A. Tannenbaum, "The structured singular value for linear input/output operators," submitted to *SIAM J. Control and Optimization.*

[6] H. Bercovici, C. Foias, P. Khargonekar, and A. Tannenbaum, "On a lifting theorem for the structured singular value," to appear in *Journal of Math. Analysis and Applications.*

[7] J. C. Doyle, "Analysis of feedback systems with structured uncertainties," *IEE Proc.* **129** (1982), pp. 242–250.

[8] M. Fan, "A lifting result on structured singular values," Technical Report, Georgia Institute of Technology, Atlanta, Georgia, November 1992.

[9] P. Halmos, *A Hilbert Space Problem Book*, Springer-Verlag, New York, 1982.

[10] M. Khammash and J. B. Pearson, "Performance robustness of discrete-time systems with structured uncertainty," *IEEE Trans. Aut. Control* **AC-36** (1991), pp. 398–412.

[11] A. Magretski, "Power distribution approach in robust control," Technical Report, Royal Institute of Technology, Stockholm, Sweden, 1992.

[12] W. Rudin, *Functional Analysis*, 2nd Edition, McGraw-Hill, 1991.

[13] M. G. Safonov, *Stability Robustness of Multivariable Feedback Systems*, MIT Press, Cambridge, Mass., 1980.

[14] J. Shamma, "Robust stability with time-varying structured uncertainty," to appear in *IEEE Trans. Aut. Control.*

[15] J. Shamma, "Robust stability for time-varying systems," *31st Proc. IEEE Conference on Decision and Control*, Tucson, Arizona, 1992, pp. 3163–3168.

Pulses, Periods, and Cohomological Terms in Functional Expansions

Roger Brockett*
Division of Applied Sciences
Harvard University

Dedicated to George Zames on the occasion of his 60th birthday.

Abstract

We introduce a new device, based on a geometrical characterization of pulses, which allows us to extend the applicability of functional expansions as a tool for dealing with pulse driven systems. Basic to our method is the use of certain closed forms whose periods coincide with the spacing between stable equilibrium points of a nonlinear system. By incorporating the integrals of such forms in a functional expansion it is possible to shape the domain of convergence, adapting it to specific situations.

1 Introduction

The efforts made by various groups in the late 1950's to create a theory of nonlinear systems using the language and methods of functional analysis are well known. The idea of making the black box point of view rigorous and computationally effective had enormous appeal. The 1963 paper of George Zames [1] embodies this point of view and, even when read thirty years later, allows one to sense the excitement that surrounded this activity. Much of the appeal of this work lies in the generality of its formulation but quantitative methods were also developed. Norbert Wiener and Y. W. Lee emphasized the use of Volterra expansions as a means to carry out computations; Zames worked more with Picard's method of successive approximations. Both approaches have difficulty dealing with systems displaying hysteresis or other manifestations of nondecaying memory [2]. We describe here a method of dealing with some problems involving infinite memory, remaining faithful to the functional analysis point of view.

*This work was supported in part by the National Science Foundation under Engineering Research Center Program, NSF D CDR-8803012, by the US Army Research Office under grant DAAL03-86-K-0171(Center for Intelligent Control Systems), and by the Office of Naval Research under Grant N00014-90-J-1887

The reason for considering the subject anew is because of its potential for shedding light on some of the challenging questions relating to how neurobiological systems make computations and exercise control over the motion of animals. There are a number of pieces of this puzzle that would seem to be amenable to study using system-theoretic methods and, of course, this was part of Wiener's argument in advocating the study of cybernetics. Up until now success has been modest and one of the specific reasons lies in the fact that the available methods for treating nonlinear systems often fail for the type of pulse-like signals that play such a wide role in neurobiological systems. A Volterra series can represent the local behavior of an input-output system in a satisfactory way, but the domain of convergence of the series, thought of as a domain in the function space in which its inputs take on their values, is usually not large enough to include inputs that evoke behavior that is strongly nonlinear. This is not accidental; there are compelling reasons for it, as explained in [3]. Here we develop a more general scheme using expansions that allow more flexibility with respect to the domain of convergence. We use it to study a class of pulse driven systems and investigate the type of robustness associated with these highly nonlinear models.

If $f : \mathbb{C} \to \mathbb{C}$ is analytic at z_0 then we can expand f around that point and the expansion will converge within some open disk $\{z : |z - a| < r\}$ in the complex plane. In the study of maps $f : \mathbb{C}^n \to \mathbb{C}^n$ the disk is replaced by the cartesian product of such disks [4]. Suitable versions of the Taylor series representation is also available in some infinite dimensional settings, such as those described in [5]. These apply, for example, to analytic maps defined on Banach spaces and provide a suitable setting for the study of Volterra expansions.

If we are given a system of the form

$$\dot{x}(t) = f(x(t)) + G(x(t))u(t) \; ; \; x(0) = x_0$$

with f being a real-analytic $\mathbb{R}^n$-valued function and $G(x)$ a n by m matrix with real-analytic entries, then we can seek an expansion of x, thought of as a function on an interval $[0, a]$, in terms of u, thought of as a function on the same interval. In [6] we gave a general procedure for the construction of the kernels $\{W_i\}$ such that x is represented by

$$x(t) = W_0(t) + \int_0^t W_1(t, \sigma)u(\sigma)d\sigma + \int_0^t \int_0^{\sigma_1} W_2(t, \sigma_1, \sigma_2)u(\sigma_1) \otimes u(\sigma_2)d\sigma_2 d\sigma_1 + \dots$$

provided that when $u = 0$ there is no finite escape time and u is sufficiently small. As is clear, if f and G do not depend on time explicitly, and if $f(x_0) = 0$, then the kernels in the Volterra expansion are stationary in the sense that

$$W(t, \sigma_1, ..., \sigma_k) = W(0, \sigma_1 - t, ..., \sigma_k - t)$$

The Volterra series can be thought of as a Taylor series and is, therefore, determined uniquely by the system and its initial condition. In [3] we considered the problem of convergence on the interval $[0, \infty)$. We noted there that if all the eigenvalues of the Jacobian of f, evaluated at x_0 have negative real parts,

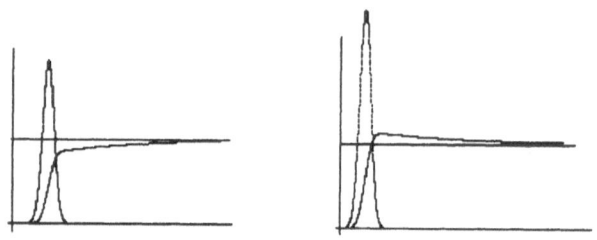

Figure 1: Pulses of different amplitudes and the responses of $\dot{x} = -\sin 2\pi x + u$.

then the kernels are formed from decaying exponentials. It was observed that if u approaches zero for large t, and if the Volterra series converges, then x necessarily approaches zero. One implication of this is that if we have a system with multiple equilibrium points, such as

$$\dot{x}(t) = -\sin 2\pi x(t) + u(t) \quad x(0) = 0$$

the domain of convergence of the Volterra series corresponding to an expansion around an asymptotically stable equilibrium solution can not be so large as to include both $u = 0$ and a compactly supported u that steers x from the initial equilibrium point occupied at $t = 0$ to another equilibrium point. Of course there exist a variety of "pulse-like" inputs that drive this particular system from the origin to $x = 1$. (See Figure 1.) If the Volterra series corresponding to an expansion about $x(t) = 0$ were to be evaluated for such a pulse it would necessarily diverge, for if it converged, $x(t)$ would approach zero, and not 1, as the kernels decayed to zero.

2 Closed Forms

In this paper we consider a modification of the Volterra series that allows us to capture, in the same domain of convergence, inputs that steer systems to different equilibrium points. Basic to our approach is the introduction of certain closed but not exact forms involving u and its derivatives, such as the derivative of $\tan^{-1}(\dot{u}/(u-a))$

$$\dot{\theta}(t) = \frac{\ddot{u}(t)(u(t) - a) - \dot{u}^2(t)}{\dot{u}^2(t) + (u(t) - a)^2}$$

We then seek expansions in a form that involves their integrals, e.g.

$$x(t) = w_0(t) + \frac{1}{2\pi} \int_0^t \dot{\theta}(t)dt + V(u, \dot{u})$$

with V being a Volterra expansion in terms of u and $\dot{u}$.

More precisely, we wish to consider iterated integrals involving $u_1, u_2, ..., u_m$ and the first k time derivatives of these functions. Unlike the iterated integrals that appear in the Volterra expansion, the type of term we introduce here does not depend explicitly on time. Let $u^{\{k\}}$ denote the $m(k+1)$ dimensional vector consisting of u and its first k time derivatives. An integral of the form

$$y_1(t) = \int_0^t \phi(u^{\{k\}}(\sigma))d\sigma$$

will be said to be a *closed first order integral term* provided that ϕ satisfies the (local) path-independent condition (see lemma 4 of [5])

$$\sum_{j=0}^k (-1)^j \frac{d^j}{dt^j} \frac{\partial^j \phi}{\partial u^j} \equiv 0$$

For example, both $\ddot{u}u$ and $u^{(4)}u^{(1)}$ are independent of path; $\dot{u}^2$ is not. A term of the form

$$y_2(t) = \int_0^t \phi(u^{\{k\}}(\sigma), y_1^{\{k\}}(\sigma), y_2^{\{k\}}(\sigma), ..., y_r^{\{k\}}(\sigma))d\sigma$$

with the y_j being closed first order integral terms and ϕ being a closed function of the u's and the y's will be called a *closed second order integral term*. Continuing in this way,

$$y_k(t) = \int_0^t \phi(u^{\{k\}}(\sigma), y_1^{\{k\}}(\sigma), y_2^{\{k\}}(\sigma), ..., y_r^{\{k\}}(\sigma))d\sigma$$

will be said to be a *closed k^{th} order integral term* if ϕ is closed and each of the y_i are closed integral terms of order $k-1$.

We characterize such terms as being "cohomological" for the following reason. A standard device for investigating the geometry of a differentiable manifold is to look for differentials such as

$$d\theta = \sum \alpha(m)dm_i$$

that satisfy the integrability condition

$$\frac{\partial \alpha_i}{\partial m_j} = \frac{\partial \alpha_j}{\partial m_i} .$$

and yet can not be expressed as the derivative of a smooth function define everywhere on the manifold. These closed but not exact differentials form a real vector space and constitute the first de Rham cohomology class of the manifold. If γ defines a closed path in the manifold the differential can be integrated along the path to get a number

$$p(\gamma) = \int_\gamma \sum \alpha(m)dm_i$$

Because of the condition on the partial derivatives, continuous deformations of the closed path will not change p. We may think of p as being a period of the differential associated with an equivalence class of closed paths. For our purposes it is convenient to divide $d\theta$ by dt and use forms involving u and its derivatives. Here, as in [7], we are interested in situations where derivatives of higher order occur. In this case the above exactness condition is replaced by the condition involving the higher order Euler-Lagrange operator as indicated above.

There are two new aspects associated with this type of expansion. Not only does the expansion involve both u and $\dot{u}$, but the expansion is only defined for those u's having the property that $u(t)$ and certain of its derivatives belong to a domain having particular topological features. In the situation just considered, we needed to ask that $\dot{u}^2(t) + (u(t) - a)^2$ never vanish. The line integral terms in these expansions capture effects that depend on the remote past. In the development given here, such forms are chosen to match certain structural properties of the differential equation. Having made a choice of these terms, the remainder of the expansion is uniquely determined. To use these techniques we must assume that u is differentiable and, even more, that u and its time derivative always belong to some set where the closed form is well defined. However, by restricting the input space and then accounting for the effect of such terms separately we can significantly influence the domain of convergence.

3   First Examples

We will develop expansions that involve not only the input, but also its time derivatives. This is not standard but it is essential in our context. In order to build some intuition about this we point out how it can be useful even in the case of linear systems.

Example 1: Consider the linear system

$$\dot{x}(t) = Ax + Bu(t)$$

Assume that A is invertible and that u is differentiable. Make the change of variables $z = x + A^{-1}bu$. Then z satisfies the equation

$$\dot{z}(t) = Az(t) + A^{-1}B\dot{u}(t)$$

If A has its eigenvalues in the right half-plane and if u is slowly varying then z is small. In any case

$$z(t) = e^{At}z(0) + \int_0^t e^{A(t-\sigma)}A^{-1}B\dot{u}(\sigma)d\sigma$$

and so x is given by

$$x(t) = -A^{-1}bu(t) + e^{At}x(0) + \int_0^t e^{A(t-\sigma)}A^{-1}b\dot{u}(\sigma)d\sigma$$

This shows that x can be expanded as a sum of a "slowly varying" term $-A^{-1}Bu(t)$ and a continuous functional of $\dot{u}$. Notice that instead of making the change of variable we could have, in this case, arrived at the same result using integration by parts.

Our second example deals with the highly nonlinear model mentioned above. It plays a role in our earlier papers [7,8] and, in the present context, allows us to introduce one of the important aspects of the representation we seek to establish.

Example 2: Consider

$$\dot{x}(t) = -\sin 2\pi x(t) + u(t)$$

The system is invariant with respect to the change of variables $x \mapsto x + k$ if k is an integer. There is a countable infinity set of asymptotically stable equilibrium points located at the integer multiples of 2π. Of course the behavior of this system is highly nonlinear but for certain types of inputs we can explain its behavior in qualitative terms. If $|x(0)|$ is less than $1/4$ and u is less than one then the system behaves more or less like the linear system $\dot{x} = -2\pi x + u$. On the other hand, if $|x(0)|$ is small and u is a function that is close to zero except for an occasional tall, narrow, positive pulse whose area is approximately 1 then x will stay near 0 until the first such pulse occurs, at which time it will move rapidly to a neighborhood of 1, stay near there until the next pulse occurs, at which time it will move to a neighborhood of 2, etc. It is important to point out that because of the asymptotic stability of the equilibria at integer values of x it is not necessary for the area under the pulses to be exactly 1. More specifically, it is quite possible for the difference between $x(t)$ and the integral

$$I(t) = \int_0^t u(\sigma)d\sigma$$

to become large. In order to get an expansion for x valid for such inputs it is necessary to incorporate the effect of a pulse in such a way as to reflect the "quantization" of x caused by its attraction to the periodically spaced asymptotically stable equilibria. Simple integration of u does not do this and we must replace the integral of u by a functional that counts pulses accurately.

In [7,8] we discussed the characterization of pulses in terms of the phase space representation of u. We noted that if u is pulse-like then in $(\dot{u}, u)$ space the trajectory of a pulse avoids some disk centered on the positive half of the u axis.(See figure 2.) With this in mind, we introduce cylindrical coordinates in $(\dot{u}, u)$-space, centered at $\dot{u}, u) = (0, a)$

$$u(t) = a + r(t)\cos\theta(t)$$

$$\dot{u}(t) = r(t)\sin\theta(t)$$

these equations imply that

$$r(t) = \sqrt{(u(t) - a)^2 + \dot{u}^2(t)}$$

and that

$$\theta(t) = \tan^{-1}\frac{\dot{u}(t)}{u(t) - a}$$

Thus

$$\dot\theta(t) = \frac{(\ddot u(t)(u(t) - a) - \dot u^2(t))}{(\dot u^2(t) + u(t) - a)^2}$$

The variable θ does not necessarily decrease by exactly 2π when u executes a pulse-like excursion. Its decrease will be approximately 2π but, and this is of crucial importance, after n pulses it will have decreased by an amount close to $2n\pi$ and the difference between its decrease and $2n\pi$ does not grow with n. This method of counting pulses is superior to integration of u for the purpose of determining their effect on the system being discussed. To convert these remarks into mathematics, we proceed by analogy with the approach used in example 1. Make a change of variable,

$$z(t) = x(t) + \frac{1}{2\pi} \int_0^t \dot\theta(t)dt$$

It is easy to see that z satisfies the differential equation

$$\dot z(t) = -\sin(2\pi z(t) - \theta(t)) + u(t) + \frac{1}{2\pi}\dot\theta(t)$$

Expanding the sine function we can rewrite this as

$$\dot z(t) = -\sin 2\pi z(t)\cos\theta(t) - \cos 2\pi z(t)\sin\theta(t) + u(t) + \frac{1}{2\pi}\dot\theta(t)$$

This system can be written as

$$\dot z(t) = -\sin 2\pi z(t) - \sin 2\pi z(t)u_1(t) - \cos 2\pi z(t)u_2(t) + u_3(t)$$

with

$$u_1(t) = (-1 + \cos\theta(t))$$
$$u_2(t) = \sin\theta(t)$$
$$u_3(t) = u(t) + \dot\theta(t)$$

Of course

$$x(t) = z(t) - \frac{1}{2\pi}\int_0^t \dot\theta(t)dt$$

and so

$$x(t) = -\frac{1}{2\pi}\int_0^t \dot\theta(t)dt + V(u, \dot u)$$

The Volterra series for z, as defined by the three input system, converges provided that the L_1 norms of the input terms are sufficiently small. Of course $\|u_1\|$ and $\|u_2\|$ are small if $u(t)$ is close to 0 most of the time. The situation with respect to u_3 is more complex. For some purposes, however, one can define a pulse train to be a function such that u_3 has a small L_1 norm. Figure 2 shows how u and $\dot\theta$ compare for a Gaussian pulse.

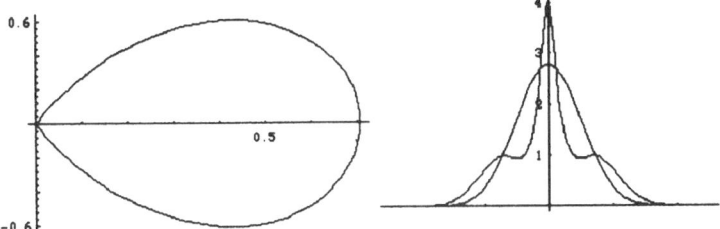

Figure 2: The phase space plot $\dot{u}$ versus u for u a Gaussian pulse and the corresponding plots of $u(t)$ and $\dot{\theta}(t)$.

4   Pulses, Periods and Associated Automata

Suppose $x(t)$ and $u(t)$ take on values in $\mathbb{R}^n$, and $\mathbb{R}^m$, respectively, and that we are given a system of the form

$$\dot{x}(t) = f(x(t)) + G(x(t))u(t)$$

Suppose, further, that f and G are real analytic and that f is periodic in the sense that $f(x + e_i) = f(x)$ for each standard basis element of $\mathbb{R}^n$. We let $\mathbb{Z}^n$ denote the integer lattice in $\mathbb{R}^n$ and let V^n denote the set of unit vectors in $\mathbb{R}^n$. Assume that $x = 0$ is an asymptotically stable equilibrium point for f, and hence that $f(e_i)$ is also an asymptotically stable equilibrium point for each $e_i \in V^n$. With respect to G, we assume that the vector fields $G(x)e_i$ are such that for $x_0 \in \mathbb{Z}^n$, the trajectory starting at x_0 at $t = 0$ passes through a point in $\mathbb{Z}^n$ at $t = 1$. Specifically, we assume that, the solution of the differential equation

$$\dot{x}(t) = G(x(t))e_i \ ; \ x(0) = p_j \in \mathbb{Z}^n$$

passes through $\hat{g}_i(p_j) \in \mathbb{Z}^n$ at $t = 1$. Systems of this form will be said to be *control periodic*.

We will now associate an automaton with such systems. The state space of this automaton is $\mathbb{Z}^n$; its input space V^m; and its evolution is given by

$$\hat{x}(k + 1) = \hat{G}(\hat{x}(k))\hat{u}(k)$$

with $\hat{G}(\hat{x})$ being a matrix whose columns are the vectors $\hat{g}_i(p_j)$ just introduced. We will call this the *associated automaton*.

We now turn to the problem of specifying what we mean by a pulse. Consider the $k + 1$-parameter family of scalar valued functions given by

$$u(\sigma, t_1, t_2, ..., t_k)(t) = \sum \frac{1}{\sqrt{2\pi\sigma}} e^{-(t-t_i)^2/2\sigma} \ ; \ |t_i - t_{i+1}| \geq \sigma$$

If $0 < \sigma \ll 1$ the graph of u consists of a set of tall narrow spikes, with the area under each spike being approximately 1. We will refer to a function of this form as being a *Gaussian pulse train of parameter σ*. If u is vector valued, we have an array of times $T = (t_{ij})$ with the ith component of u having a pulse centered at t_{ij}. We require $|t_{ij} - t_{kl}| \geq \sigma$ if $(i,j) \neq (k,l)$. We associate with a vector spike train a vector of differential forms. Let $a = 1/(2\sqrt{2\pi\sigma})$ and define

$$\dot{\theta}_i(t) = \frac{\ddot{u}_i(t)(u_i(t) - a) - \dot{u}_i^2(t)}{\dot{u}_i^2(t) + (u_i(t) - a)^2}$$

We can think of the vector $\dot{\theta}$ as a *winding rate* associated with u.

We now describe a method of topologizing a set of functions that contain spike trains. We are especially interested in how to do this effectively when σ is small. Consider the definition

$$||u||_{k\sigma} = \sup_t(|u^{(k)}(t)| + |u^{(k-1)}(t)| + ... + |u^{(0)}(t)| + \int_{t-\sigma}^t |u(\tau)|d\tau)$$

We will refer to this as the k, σ norm and define $P^m(\epsilon, k, \sigma)$ to be the set of all k-times differentiable functions that are within ϵ of some Gaussian pulse train whose pulses are of spread σ.

5 Another Example

While nicely illustrating the necessity of counting pulses in a topological way, Example 2 above is by no means fully representative of the behavior of the class of systems for which expansions in terms of closed integral forms is possible because it does not illustrate any aspects of the multiple input situation.

Example 3: Assume that $u \in P^2(\epsilon, k, \sigma)$ and consider the system

$$\dot{x}_1(t) = -\sin 2\pi x_1(t) + \cos(\pi x_2(t))u_1(t)$$

$$\dot{x}_2(t) = -\sin 2\pi x_2(t) + u_2(t)$$

The asymptotically stable equilibria are the lattice points $\mathbb{Z}^2 \subset \mathbb{R}^2$. The effect of an input in the form of a tall, narrow pulse of strength 1 applied to the second input channel is to advance x_2 by 1. The effect of an input in the form of a tall, narrow pulse of strength one, applied to the first input channel, is to increase $x_1(t)$ by 1 if $x_2(t)$ is an even integer and to decrease $x_1(t)$ by one if it is an odd integer.

Taking into account the nature of the input space, the behavior of this system can be characterized by saying that a u_1 pulse following an even number of u_2 pulses increases x_1 by 1 whereas a u_1 pulse following an odd number of u_2 pulses decreases x_1 by 1. We can express this in terms of an integral in the following way. Define $z(t) = (u_1(t) - a)\cos(\pi x_2(t))$ and consider the definition

$$\dot{\phi}(t) = \frac{\ddot{z}(t)z(t) - \dot{z}(t))^2}{\dot{z}(t) + z^2(t)}$$

Clearly ϕ counts the number of times the phase space trajectory of u_1 encircles $(a, 0)$ when x_2 is near an even integer minus the number of times it encircles

$(a, 0)$ when x_2 is near an odd integer. To obtain a closed integral form, however, we must approximate $x_2(t)$ by a closed integral form. The obvious choice is the winding number associated with u_2,

$$\hat{x}_2(t) = \int_0^t \frac{\ddot{u}_2(\sigma)(u_2(\sigma) - a) - \dot{u}_2(\sigma))^2}{\dot{u}_2(\sigma) + (u_2(\sigma) - a)^2} d\sigma$$

Now observe that x_1 is approximated by ϕ even if we replace x_2 by this integral and replace z by $\hat{z} = u_1 \cos \pi \hat{x}_2$. This then gives an integral expression that approximates $x_1(t)$ in terms of u_1, u_2 and certain derivatives,

$$x_1(t) = \int_0^t \phi_1(u^{\{2\}}(\sigma_1), \int_0^{\sigma_1} \phi_2(u^{\{2\}}(\sigma_2)) d\sigma_2) d\sigma_1$$

6 Functional Expansions

Putting together the ideas and definitions of the previous sections we can now state a theorem about the existence of convergent functional expansions for a class of pulse driven systems. Before doing so, however, we make a definition that will let us describe the class of systems for which our results hold.

We will say that a finite automaton whose input space is the lattice $\mathbb{Z}^m$ and whose state space is the lattice $\mathbb{Z}^n$ is *solvable* if there exists a finite set of vector-valued functions $\{\chi_j\}$, whose components are integer-valued, such that for all positive integers i

$$\hat{y}(i) = \chi_1(\sum_{i_1=0}^{i} \hat{u}(i_1), \chi_2(\sum_{i_2=0}^{i_1} \hat{u}(i_2), \ldots \chi_r(\sum_{i_r=0}^{i_{r-1}} \hat{u}(i_r))\ldots)))$$

Theorem. *Let*

$$\dot{x}(t) = f(x(t)) + G(x(t))u(t) \; ; \; x(0) = x_0$$

be a control periodic system. Suppose that the associated automaton

$$\hat{x}(k+1) = G(\hat{x}(k))\hat{u}(k)$$

is solvable. Then there exists an integer k, positive numbers ϵ_1, ϵ_2, and σ and an expansion

$$x(t) = x_0(t) + \Phi(u^{\{k\}}) + V(u^{\{k\}})$$

with Φ a closed integral term such that if $|x_0| \leq \epsilon_1$ then for each k-times differentiable function u in the Gaussian pulse space $P^m(\epsilon_2, k, \sigma)$ the series is convergent and represents x.

Proof: The first observation to be made is that if $x(0) \in \mathbb{Z}^n$ and if u consists of Dirac delta functions, then $x(t)$ jumps from one point in $\mathbb{Z}^n$ to another. Moreover, its trajectory is the same as the trajectory generated by the associated automaton in response to the input $u(0), u(1), \ldots u(k)$ with $u(i) = e_k$

if the ith pulse in the pulse train u is the result of a pulse on the kth component of u. Thus, in this case, we may replace the sums appearing in the definition of solvable by integrals, e.g.

$$\sum_{i_1=0}^{i} \hat{u}(i_1) = \int_0^t u(\sigma)d\sigma$$

Even better, we can replace the sums by closed integrals

$$\sum_{i_1=0}^{i} \hat{u}(i_1) = \int_0^t \frac{\dot{\theta}(\sigma)}{2\pi} d\sigma$$

The next step is to remark that we may pass from the χ's in the definition of solvable to the ϕ's in the definition of a closed integral by a suitable interpolation. The χ's are only defined on some integer lattice. There exists a smooth interpolation that interpolates between the lattice points letting us replace the χ's by smooth functions defined everywhere on $\mathbb{R}^m$ and agreeing with them on the lattice.

As explained previously, one can not use the integral of u if a robust expression is required. Instead we replace it by a suitable winding number. Reasoning as above, we make the change of variables

$$z(t) = x(t) - \chi_1(\int_0^t \frac{\dot{\theta}(\sigma_1)}{2\pi}, \chi_2(\int_0^{\sigma_1} \frac{\dot{\theta}(\sigma_2)}{2\pi}, \dots \chi_r(\int_0^{\sigma_{r-1}} \frac{\dot{\theta}(\sigma_r)}{2\pi} d\sigma_r)\dots d\sigma_2)d\sigma_1$$

and proceed to construct a Volterra series for z. For the reasons already explained, if $x(0)$ is near a lattice point and if $u \in P^m(\epsilon_2, k, \sigma)$ with ϵ_2 and σ sufficiently small, z will be close to zero and its Volterra series will converge. We omit further details.

Remark: The above theorem gives a satisfactory description of the pulse response an interesting class of first order systems. It is possible to carry out a similar analysis for a class of second order systems of the form

$$\ddot{x}(t) + M\dot{x}(t) + f(x(t)) = G(x(t))u(t)$$

except that in this case one must use inputs that are derivatives of pulses.

References

[1] George Zames, "Functional Analysis Applied to Nonlinear Feedback Systems", *IEEE Trans. on Circuit Theory*, Vol CT 10 (1963) pp. 392-404.

[2] M. A. Krasnosel'slii and A. V. Porkrovskii, *Systems with Hysteresis*, Springer-Verlag, Berlin, 1989.

[3] R. W. Brockett, "Convergence of Volterra Series on Infinite Intervals and Bilinear Approximations," in *Nonlinear Systems and Applications* (V. Lakshmikanthan, ed.). New York: Academic Press, 1977, pp. 39-46.

[4] Hassler Whitney, *Complex Analytic Varieties*, Addison-Wesley, Reading MA, 1972.

[5] Leopold Nachban, *Topology on Spaces of Holomorphic Mappings*, Springer-Verlag, Berlin, 1969.

[6] R. W. Brockett, "Volterra Series and Geometric Control Theory," *Automatica*, Vol. 12 (1976) pp. 167-176.

[7] R. W. Brockett, "Pulse Driven Dynamical Systems", in *Systems, Models and Feedback: Theory and Applications*, (Alberto Isidori and T. J. Tarn, Eds.), Birkhäuser, Boston, 1992, pp. 73–79.

[8] R. W. Brockett, "Dynamical Systems and Their Associated Automata", in *Systems and Networks: Mathematical Theory and Applications*, (Uwe Helmke, et al. Eds.) Akademie Verlag Berlin, 1994, (pp. 49-69).

AN APPROACH TO THE PROBLEMS OF COMPLEXITY AND HIERARCHY WITH AN APPLICATION TO A DETECTION PROBLEM

P.E. CAINES[1], T. MACKLING, C. MARTÍNEZ-MASCARÚA[2], AND Y.J. WEI
DEPARTMENT OF ELECTRICAL ENGINEERING AND
CENTRE FOR INTELLIGENT MACHINES
MCGILL UNIVERSITY

July 14, 1994

1. Introduction

A long standing problem within systems and control theory has been the definition and characterization of hierarchical systems. In systems and control theory, as in many areas of scientific and mathematical investigation, it is thought that various notions of hierarchical system may be formulated and that systems which, in some sense, possess hierarchical structure can exhibit properties of efficient information processing and control. Zames [12] characterized large systems as being those " whose salient feature is that they cannot be modeled with complete precision. The cost of such systems is a function of the complexity of the information processing (for identification, communications, numerical operations, etc.) needed to achieve a specified accuracy [of control]". Zames then states that "It is frequently possible to "organize" such systems by appropriate use of hierarchical partitioning and feedback to reduce complexity and indeed, often, to achieve accurate control with only a crude knowledge of the system". With this as one motivation, the article in question gave a formal definition of the metric complexity of causal linear systems and investigated its properties.

It is often speculated that an inherent property of cybernetic systems with limited information processing and control resources is that they must display hierarchical structures when designed (and, in particular, optimized) with respect to certain specified criteria. Furthermore, the almost universal perception of the importance and value of hierarchical structures has led to their use in the design of computational, industrial, transportation, economic and administrative systems.

Work supported by NSERC grant A1329 and PRECARN-NCE-IRIS-1 program B5.
[1] Canadian Institute for Advanced Research.
[2] Supported by grant 50596 from CONACyT, México.

The physical sciences provide explanations with hierarchical structures for many aspects of the natural universe; an obvious example is given by the relationship between models of the atomic, molecular, planetary and galactic systems. The role of hierarchical structures in theories of nature clearly carries the hypothesis that they form fundamental features of nature itself or, at least, that they are fundamental to the human perception of nature. In this context, it is perhaps worth mentioning two common requirements of hierarchical consistency between scientific theories: first, the usual condition that theories at adjacent levels of scale of description (granularity) of the universe must be consistent with respect to aggregation (or generalization) and decomposition (or analysis); and, second, the standard requirement of the logico-empirical scientific method that any theory which advances beyond or refines a standing theory must explain the phenomena successfully predicted by the superseded theory.

Given the significance of hierarchical systems outlined above, it is somewhat surprising that there is only a limited amount of formal theory investigating hierarchical systems, with some notable examples being provided by economics and optimization theory. In the decentralized optimization theory of Arrow and Hurwicz [1] and Hurwicz [5] one finds a theory of competitive market equilibrium in terms of the exchange of price information by the producing and consuming agents of the economy. Other work which relates the existence of hierarchical structures to the efficient processing of information for optimization is to be found in Marschak and Reichelstein [7], [8] (see the references therein for the work of the associated community of researchers). Consider a network of agents optimizing a cost function through partially shared action components. Marschak and Reichelstein state that "a necessary and sufficient condition for the existence of a dimensionally minimal hierarchical mechanism [i.e. an information and control structure that optimizes a static reward function and for which the message variable passing costs are minimal] is that there is a tree such that the subgraph containing only the agents concerned with any given action component is a subtree."

A principal feature of the economic studies is that they are primarily concerned with static optimization problems and the hierarchical structures which may appear in the organization of agents involved in the (possibly dynamic) optimization exercise. In the area of systems and control theory one is more concerned with the interaction between hierarchically organized controllers (with their associated information channels) and the resulting system dynamics of the overall controlled system.

Within the automata formulation of control theory for discrete event systems (DESs), the research of Zhong and Wonham [13] and Wong and Wonham [11] is devoted to the hierarchical structure of DESs; one of the contributions of [13] is the creation of aggregation methods which ensure that high level control events are well defined as functions (via "vocalized events") of the aggregated events of a given (low level) automaton. Furthermore, the issue of hierarchical control appears in the guise of subsystem co-ordination and observability in the work of Lin and Wonham [6] and Rudie and Wonham [9].

In this paper, we present and illustrate a formulation of the hierarchical decomposition of the information pattern and control structure of controlled dynamical systems in the case of discrete time finite machines. It rests upon the notion of dynamical consistency which is defined in Section 2.1 below. No optimality properties have, as yet, been established for the formulation given in this paper. However, it leads to a hierarchical decomposition of the trajectory control task which possesses certain efficiency properties in terms of (i) the transmission of observation and control signals and (ii) the computation of trajectories between any specified states. Furthermore, it fits common sense perceptions of the nature of hierarchical command structures. Intuitively, the control commands generated at any level of a hierarchical system constitute control problems; these are to be solved by a control agent of a system at a lower level in the hierarchy; moreover, by the very construction of the hierarchy, the higher level agent is assured that the control problem is indeed solvable by the lower level agent.

As we have demonstrated elsewhere [10], our formulation of hierarchical control fits the denotational properties of formal languages. This latter aspect permits the exploitation of the syntactic properties of COCOLOG (conditional observer and controller logic [3], [2]) controllers in the context of hierarchical systems. This corresponds to the idea that a symbol used by a high level agent represents a set of events, states, commands or control problems of a lower level agent. (The customary use of the real and complex number system, functions, operators, etc, in conventional control theory does not exhibit this denotational feature.) In addition, the logical language in which COCOLOG is expressed permits the use of existential and universal quantifiers (at the regulator level). Consequently, existential assertions, which appear to play an extremely useful role in hierarchical control, are purposefully included and exploited in the COCOLOG based formulation of hierarchical control.

2. Hierarchical Control Structures

Our approach to the definition of a hierarchical control structure for a finite state machine $\mathcal{M} = \langle X, U, \Phi \rangle$ (hereafter called the base machine) begins with a partition of its state space to obtain π; this partition defines the state space of a partition machine $\mathcal{M}^\pi = \langle X^\pi, U^\pi, \Phi^\pi \rangle$. Aggregation is accomplished from the fact that $|\pi| \leq |X|$. The dynamics of the partition machine are defined between $\overline{X}_i \in X^\pi$ and $\overline{X}_j \in X^\pi$ if, and only if, every state in X_i can be driven to some state in X_j without traversing other partition elements; in this case, the ordered pair $\langle X_i, X_j \rangle$ is said to be dynamically consistent. The existence of a dynamical relation between $\overline{X}_i$ and $\overline{X}_j$ implies the existence of a bundle of dynamical relations at the base machine control level between the states in X_i and those in X_j $(1 \leq i, j \leq |\pi|)$. This notion (presented formally in section 2.1) distinguishes our formulation to hierarchical control structures from other approaches to this problem. The consistency of two-level systems is also considered by Zhong and Wonham [13], where the ability to disable a low level event and its homomorphic image event (at the high level) is the consistency criterion employed.

Controllability issues for the partition machine $\mathcal{M}^\pi$ are then explored; if $\mathcal{M}^\pi$ is controllable, it will be called Between Block Controllable (BBC); if every partition element $X_i \in \pi$ is controllable in the classical sense, then $\mathcal{M}^\pi$ is said to be In-Block Controllable (IBC). If $\mathcal{M}^\pi$ is IBC, then $\mathcal{M}$ is controllable if, and only if, $\mathcal{M}^\pi$ is BBC. These results are the object of section 2.2.

Another relevant issue is the generation of partition machines, and the relationships among them. The IBC partitions of a given machine $\mathcal{M}$ form a non-distributive lattice, which we call the hierarchical lattice of IBC partitions, described in detail in section 2.3. This lattice is distinct from the lattice of partitions of a set X, $PAR(X)$, introduced bellow.

2.1. Dynamical Consistency.

Definition 1. *Given a set X, a **partition** π of X is a subset $\{X_1, \ldots, X_k\}$ of the power set $\mathcal{P}(X)$ such that $X_i \in \pi$ are non-empty pairwise disjoint, and $\cup_{i=1}^k X_i = X$. $X_i \in \pi$ are called (π-)blocks.* □

Definition 2. *Given a set X and two partitions of it, π_1 and π_2, $\pi_1 \leq \pi_2$ (π_1 is **stronger than** π_2) if every π_1-block is a subset of some π_2-block.* □

The least upper bound and the greatest lower bound of two partitions π_1, π_2 are defined, respectively, via *intersection* $\pi_1 \cap \pi_2$ and *chain union* $\pi_1 \cup^c \pi_2 = \{Z_1, Z_2, \ldots Z_r\}$; Z_i can be constructed by setting $Z_i = \cup_{n=1}^N Z_{i,n}$ ($N = |\pi_1| + |\pi_2|$) where $Z_{i,1} \triangleq X_i \in \pi_1$ and, given $Z_{i,n}$:

$$Z_{i,n+1} \triangleq \begin{cases} Z_{i,n} \cup \{X' \in \pi_1 : X' \cap Z_{i,n} \neq \emptyset\} & \text{if } n \text{ is even,} \\ Z_{i,n} \cup \{Y' \in \pi_2 : Y' \cap Z_{i,n} \neq \emptyset\} & \text{if } n \text{ is odd.} \end{cases}$$

Clearly $1 \leq |\pi_1 \cup^c \pi_2| \leq min(|\pi_1|, |\pi_2|)$.

The resulting lattice, $PAR(X)$, of all partitions of X is bounded above and below by the trivial partition $\pi_{tr} \triangleq \{X\}$ and the identity partition $\pi_{id} \triangleq \{\{x\} : x \in X\}$ respectively.

Let $\mathcal{M} = \langle X, U, \Phi \rangle$ be a finite machine, where X is the finite set of states, U the finite set of control actions and $\Phi : X \times U \to X$ the transition function – generally a partial function. Every partition π of X, has an associated partition machine $\mathcal{M}^\pi$ that will have X^π as its state space.

Given a partition π, and $X_i \in \pi$, $\overline{X_i}$ denotes the set X taken as an atomic entity (singleton). X^π denotes the set $\{\overline{X_i} : X_i \in \pi\}$. A^* is the set of all finite sequences of elements of a set A. For $\sigma, \sigma' \in A^*$, the notation $\sigma' \leq \sigma$ (respectively $\sigma' < \sigma$) means that σ' is the *initial segment* (*proper initial segment*) of σ. $|\sigma|$ denotes the length of σ.

Definition 3. *Given a machine $\mathcal{M} = \langle X, U, \Phi \rangle$, a sequence $\sigma = u_0 \ldots u_k \in U^*$ is called a **legal control sequence** (or simply, control sequence) of $\mathcal{M}$, if there is $x \in X$ such that $\Phi(x, u_0 \ldots u_l), 0 \leq l \leq k$ is defined. The state sequence generated by σ is called a **trajectory**. $Tra(x, y)$ will denote the set of all possible trajectories from x to y. $Con(\mathcal{M}) \subset U^*$ (respectively, $Tra(\mathcal{M}) \subset X^*$) will denote the set of all possible finite control sequences(respectively, trajectories) of $\mathcal{M}$.* □

Given a partition π of X, the elements of $Tra(\mathcal{M})$ fall into two categories: (1) those trajectories lying within a single block; (2) those trajectories lying in more than one block which can be decomposed into a sequence of sub-trajectories, say $t_1 t_2 \ldots t_m$, such that each segment t_i is either an internal or a direct trajectory as defined below.

Definition 4. *Given $\mathcal{M} = \langle X, U, \Phi \rangle$ and $\pi \in PAR(X)$, a sequence of states $t = x_1 \ldots x_l \in Tra(\mathcal{M})$ is called an **internal trajectory** with respect to π if $t \in X_i^*$ for some $X_i \in \pi$; t is called a **direct trajectory** from X_i to X_j $(X_i, X_j \in \pi, i \neq j)$ if $x_1 \ldots x_{l-1} \in X_i^*$ and $x_l \in X_j$.* $\square$

$Tra(\pi)_i^i$ (respectively, $Tra(\pi)_i^j$ for $i \neq j$) shall denote the set of all internal (respectively, direct) trajectories of $\mathcal{M}$ with respect to π.

A trajectory is specified entirely by its initial state and the control sequence applied to it; hence, there exists a mapping $\theta : X \times Con(\mathcal{M}) \to Tra(\mathcal{M})$:

$$\theta(x, \sigma) = x_0 \ldots x_p \in Tra(\mathcal{M}),$$

where $\sigma = u_0 \ldots u_{p-1} \in Con(\mathcal{M})$ and $x_0 = x$, $x_i = \Phi(x_{i-1}, u_{i-1})$, $1 \leq i \leq p$. θ is onto but not necessarily one-to-one.

Definition 5. *Given π, define the set of **block control events** U^π to be $\{\overline{U}_i^j;\ 1 \leq i, j \leq |\pi|\ \}$, where*

$$U_i^j \triangleq \theta^{-1}(Tra(\pi)_i^j) =$$
$$= \{(x, \sigma) : x \in X_i, \forall \sigma' < \sigma, (\Phi(x, \sigma') \in X_i \land \Phi(x, \sigma) \in X_j)\}$$

$\square$

The elements of the set U_i^i (respectively, U_i^j, with $i \neq j$) are called **internal** (respectively, **direct**) **control sequences** of X_i (respectively, *from X_i to X_j*). Obviously, the elements of U_i^j may vary in length. Let $P_1 : X \times Con(\mathcal{M}) \to X$ and $P_2 : X \times Con(\mathcal{M}) \to Con(\mathcal{M})$ denote projections, so that for a direct (or internal) control sequence $w \in U_i^j$, $P_1(w)$ is the initial state, and $P_2(w)$ is the control sequence corresponding to w.

Definition 6. *Given $\pi \in PAR(X)$ and $X_i, X_j \in \pi$, we say a **dynamically consistent condition** holds for the ordered block pair $\langle X_i, X_j \rangle$ if for each $x \in X_i$, there exists at least one direct trajectory from X_i to X_j which starts from x, i.e.*

$$\forall x \in X_i, \exists w \in U_i^j \in U^\pi (x = P_1(w)),\ or$$
$$X_i = \{x : x = P_1(w), w \in U_i^j\}.$$

*In this case we say $\langle X_i, X_j \rangle$ is a **dynamically consistent (DC) pair** and in the case $i = j$ we simply say X_i is **DC**.* $\square$

Definition 7. *Given* $\mathcal{M} = \langle X, U, \Phi \rangle$, $\pi \in PAR(X)$ *and* U^π, *the transition function* $\Phi^\pi : X^\pi \times U^\pi \rightarrow X^\pi$ *of the **dynamically consistent partition machine** $\mathcal{M}^\pi = \langle X^\pi, U^\pi, \Phi^\pi \rangle$ is* $\Phi^\pi(\overline{X}_i, \overline{U}_i^j) \triangleq \overline{X}_j$ *if and only if* $\langle X_i, X_j \rangle$ *is a DC pair.* $PAR(\mathcal{M})$ *shall denote the set of all dynamically consistent partition machines of* $\mathcal{M}$. □

2.2. Control Structures of a Partition Machine. Given $\mathcal{M} = \langle X, U, \Phi \rangle$, a DC transition function Φ^π is defined for every partition π of X. However, when the DC condition fails for many pairs of blocks, the resulting partition machine may have very limited dynamics. In addition, the controllability of $\mathcal{M}$ will not in general be reflected by $\mathcal{M}^\pi$. In fact, for some $\pi \in PAR(X)$, $\mathcal{M}^\pi$ may be highly uncontrollable (see below) while machine $\mathcal{M}$ is controllable. For an analysis of hierarchical machine behaviour, one should not only consider the high level dynamics defined between the elements of the partition machine but also the local dynamics within each submachine given by the partition elements $X_i \in \pi$. This leads us to consider the so-called in-block controllability, by which we mean submachine controllability.

Definition 8. $\mathcal{M} = \langle X, U, \Phi \rangle$ *is called **controllable** if* $|X| = 1$ *or if for any* $(x, y) \in X \times X$ *there exists a control sequence* $s \in U^*$ *which gives rise to a trajectory from* x *to* y, *that is to say*

$$\forall x, y \in X, (x \neq y \implies \exists s \in U^*, (\Phi(x, s) = y)).$$

□

For partition machines, we have

Definition 9. $\mathcal{M}^\pi = \langle X^\pi, U^\pi, \Phi^\pi \rangle$ *is called **in-block controllable** if every non-singleton submachine* $\mathcal{M}_i$ *is controllable, i.e.*

$$\forall \overline{X}_i \in X^\pi, \forall x, y \in X_i, (x \neq y \implies \exists w \in U_i^i, (\Phi(x, P_2(w)) = y)).$$

□

Definition 10. $\mathcal{M}^\pi$ *is called **between-block controllable** if*

$$\forall \overline{X}_i, \overline{X}_j \in X^\pi, (i \neq j \implies \exists S \in (U^\pi)^*, (\Phi^\pi(\overline{X}_i, S) = \overline{X}_j)).$$

In other words, $\mathcal{M}^\pi$ *is controllable as a finite machine.* □

$IBCP(\mathcal{M})$ (respectively, $BBCP(\mathcal{M})$) shall denote the set of all in-block controllable (respectively, between-block controllable) partition machines. By abuse of notation we shall often write $\pi \in IBCP(\mathcal{M})$ for $\mathcal{M}^\pi \in IBCP(\mathcal{M})$.

Theorem 1. [4] *Given* $\mathcal{M} = \langle X, U, \Phi \rangle$ *and* π, *if* $\mathcal{M}^\pi$ *is IBC, then* $\mathcal{M}$ *is controllable if, and only if,* $\mathcal{M}^\pi$ *is BBC.* □

Theorem 2. [4] $IBCP(\mathcal{M})$ *is closed under chain union.* □

2.3. The Hierarchical Lattice of IBCPs.

Definition 11. *Given* $\mathcal{M}^{\pi_1} = \langle X^{\pi_1}, U^{\pi_1}, \Phi^{\pi_1} \rangle$ *and* $\mathcal{M}^{\pi_2} = \langle X^{\pi_2}, U^{\pi_2}, \Phi^{\pi_2} \rangle \in$ $PAR(\mathcal{M})$, $\mathcal{M}^{\pi_1} \preceq \mathcal{M}^{\pi_2}$ *($\mathcal{M}^{\pi_2}$ is weaker than $\mathcal{M}^{\pi_1}$) if $\pi_1 \leq \pi_2$.* □

It is evident that chain union is an associative operation, hence Theorem 2 holds for the chain union of any finite number of partitions. By definition, $IBCP(\mathcal{M}) \neq \emptyset$ and $\mathcal{M}^{\pi_{id}}$ is the strongest element of $IBCP(\mathcal{M})$ with respect to $\preceq$. Furthermore, we can define the lower and upper bounds of $IBCP(\mathcal{M})$ respectively via:

$$inf_{IBCP(\mathcal{M})} \triangleq \mathcal{M}^{\pi_{id}}, \text{ and}$$

$$sup_{IBCP(\mathcal{M})} \triangleq \cup^c \{\mathcal{M}^\pi : \mathcal{M}^\pi \in IBCP(\mathcal{M})\}$$

If $\mathcal{M}$ is controllable, then $\mathcal{M}^{\pi_{tr}}$ is in-block controllable, and hence $sup_{IBCP(\mathcal{M})} = \mathcal{M}^{\pi_{tr}}$.

Unfortunately, intersection does not preserve in-block controllability.

Definition 12. *Given a finite machine $\mathcal{M}$ and two in-block controllable partitions π_1 and π_2, set*

$$\pi_1 \sqcap \pi_2 \triangleq \cup^c \{\pi'; \; \pi' \leq \pi_1, \pi' \leq \pi_2, \pi' \in IBCP(\mathcal{M})\}$$

□

Theorem 3. [4] $\langle IBCP(\mathcal{M}), \sqcap, \cup^c, \preceq \rangle$ *forms a lattice which is bounded above and below by $sup_{IBCP(\mathcal{M})}$ and $\mathcal{M}$ respectively. We call this lattice the **in-block controllable (partition) lattice**.* □

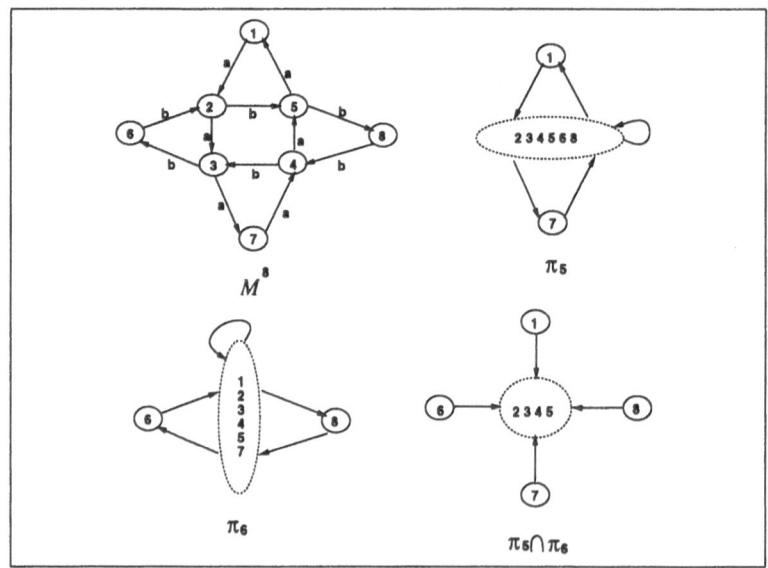

FIGURE 1. (a) $\mathcal{M}_8$ (b) $\mathcal{M}^{\pi_6}$ (c) $\mathcal{M}^{\pi_5}$ (d) $\mathcal{M}^{\pi_5 \sqcap \pi_6} \neq \mathcal{M}^{\pi_5 \cap \pi_6} = \mathcal{M}^{\pi_{iden}}$

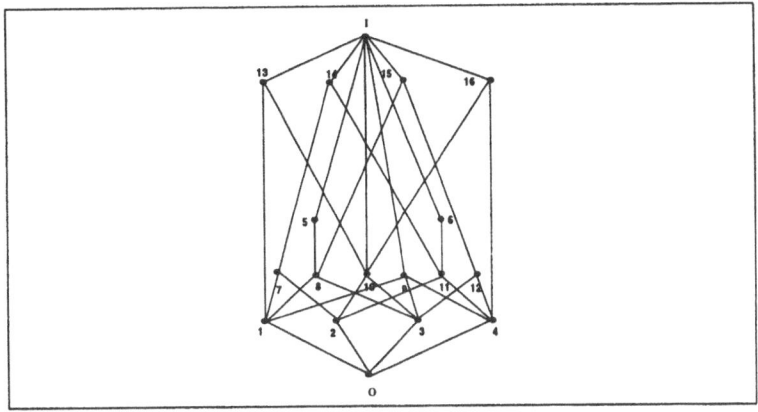

FIGURE 2. The $IBCP(\mathcal{M}_8)$

Example 1. *Figure 1 shows a base machine (called $\mathcal{M}_8$) and three of its partition machines. Machines $\mathcal{M}^{\pi_5}$ and $\mathcal{M}^{\pi_5}$ are elements of the $IBCP(\mathcal{M}_8)$, shown in Figure 2. The $IBCP(\mathcal{M}_8)$ contains 16 non-trivial elements. Most of them can be generated by the partitions π_i, $1 \le i \le 4$.*

2.4. Hierarchical Trajectory Control. Intuitively, control is performed in a hierarchical structure at all levels simultaneously; this is in the sense that a trajectory control problem for a machine $\mathcal{M}$ which is to be solved by a hierarchical controller is shared among the mutually ordered elements of a chain of regulators, where these regulators occupy positions down a chain through the $IBCP(\mathcal{M})$ between the top, $\mathcal{M}^{\pi_{tr}}$, and bottom, $\mathcal{M}^{\pi_{td}}$, elements. Similarly, the implementation of the feedback control of the system is distributed among the controllers along the same chain in the lattice.

Consider a problem where only two comparable partitions are employed in the hierarchical controller. The high level regulator generates a block trajectory from the block containing the current state to the block containing the target state. The low level regulator generates the trajectory from its current state to some entry state specified by the block trajectory. The generation of the block trajectory at the high level is independent of any detailed information concerning any individual block, while the generation of the trajectory by a low level regulator is independent of the chosen high level block trajectory (with the exception of the current command to exit the current block so as to enter a named adjacent block).

Given a controllable machine $\mathcal{M} = \langle X, U, \Phi \rangle$ in state $x \in X$, given a target state $y \in X$ and a partition π of X, the hierarchical control using partition machine $\mathcal{M}^\pi = \langle X^\pi, U^\pi, \Phi^\pi \rangle$ is performed as follows:

(1) Initial generalization step. This is performed finding i and j such that $x \in X_i$ and $y \in X_j$. Suppose, without loss of generality, that $i \neq j$.

(2) Computation of high level reach-ability. Find a sequence of high level control actions $U_i^m \ldots U_p^j$ that describes a high level trajectory starting at $\overline{X}_i$ and ending at $\overline{X}_j$ on machine $\mathcal{M}^\pi$.

(3) High level Control. Execution of the high level control trajectory as a sequence of sets of steps performed by the low level controller.

(4) Low level control. Each set of steps is as follows: suppose the current state is $x \in X_i$ and the high level control action to be performed is U_i^m. Then the set of corresponding low level commands is of the form $U_i^m(x) = u_1 \ldots u_k$, which are such that $\Phi(x, U_i^m(x)) \in X_m$.

(5) Perform the final leg of the trajectory. Once in $\overline{X}_j$, the high level controller orders the low level controller to perform the final leg of the trajectory control; this is done by delivering the problem of arriving at target state y from the current state. The solution to this problem is guaranteed by Theorem 1 and the fact that $\mathcal{M}$ is controllable.

3. Illustrative Application to a Detection Problem

This section is concerned with the application of the hierarchical control theory presented above to the control problem of playing a variant of the board game Clue. We believe that the structure of the state space (board position and information space structure) and game theoretic nature of this game present a sufficiently complex domain for the an illustration of our approach to hierarchical control.

Initially, we introduce the original rules of the game and then enumerate the simplifying modifications that we have introduced for this particular exercise. In section 3.3, we describe the state space complexity of 2-player Clue; the dynamics of the game are described in section 3.5 and we conclude with a sketch of an end game play (strategy) for one player.

3.1. Rules of 2-player Clue. Clue is a well-known board game; the game is played under the assumption that the master of an English country mansion (Mr.Boddy) has been murdered; the objective is to determine which of the house guests has committed the murder, with which weapon and in which room of the mansion. Figure 3 presents a schematic version of the board. Each of the dots outside of the rooms represents a position in the hallway of the house. Each dot at the entrance of a room represents a doorway. Each player has a token corresponding to one of the six house guests and may move its token after a throw of the dice. Once the token of a player is on a dot in a doorway, a player is considered to be inside a room, and all the doorways in the same room are identified so as to form a single position. The rooms in any pair of opposite corners are connected by a secret passage (indicated by the arrows in their corners).

Anyone of the six guests is a suspect; there are six possible weapons, and the murder could have taken place in any of the nine rooms of the house. Each of these elements is depicted on one card. The game is set up by separating the cards by category and then putting (without revealing them) one suspect, one weapon and one room card in an envelope (called the *Confidential* envelope);

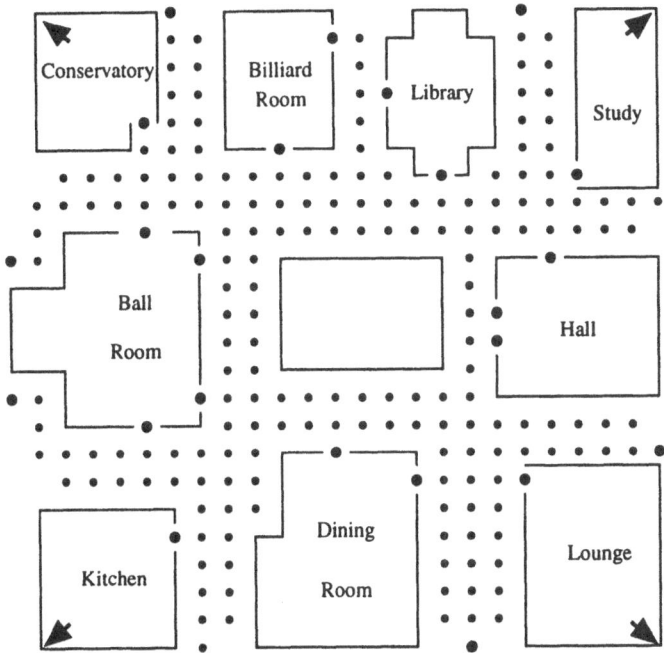

FIGURE 3. Diagram of the Clue game board

these three cards indicate the person who committed the murder, the weapon used and the room in which it occurred. Once this is done, all the cards are shuffled and dealt to the players. Each player chooses a token representing one of the suspects; this will determine the player's identity. Tokens are put in a set of fixed starting positions at the beginning of each game.

On each turn, each player tries to reach a room of the mansion. A player starts its turn by rolling a die or using a secret passage; if the die is rolled, the player's token can be moved the number of squares the corresponding number of rectangular steps, without entering at any time a square occupied by another token.

A room can be entered or left either by rolling the die or by using a secret passage. A room may not be re-entered on a single turn; if a door is blocked by an opponent, it cannot be used to enter or leave the room.

As soon as a player enters a room, it can declare a suspicion; this is the mechanism through which the players try to determine the contents of the *Confidential* envelope. The suspect's and weapon's tokens are moved into the room at the occurrence of a suggestion; then the suggestion is made that the crime was committed in that room, by that suspect and with that weapon.

Once the suggestion is made, the opponents are obligated –in turn– to prove it false: the first player to the left with a card denoting an element involved in the suggestion must show it to the player who made the suggestion without revealing it to anybody else.

A player can also make an accusation when in a room. Only one accusation per game can be made by each player. The accusation is again a statement involving a person, a weapon and a room; the room in this case need not be the same in which the player's token is. The player looks at the cards in the *Confidential* envelope to verify its accusation. If the accusation is right, it is declared the winner; otherwise it cannot play anymore and remains in the game only to refute suggestions.

3.2. Modifications to obtain the rules of 2-player Clue.

The rules of the modified version of Clue called *2-player Clue* which will be considered for the application of the hierarchical control theory are:

(1) Only two players are involved. This facilitates the analysis of the state space, in particular concerning board positions.

(2) The players are not suspects. This is in order to avoid the motions on the board caused by a player α making a suggestion that involves player β, and hence moving β's token in a manner not in general planned by β.

(3) There is no die to roll. This removes the stochastic element from the game, and each player is allowed to choose a number (from 1 to 6) of positions to move. Note that the game is still non-deterministic, since an opponents' moves are not necessarily predictable.

(4) Cards of the three different types are shuffled independently, and then handed to players in order; that is, a player is always guaranteed to receive at least one suspect card, two weapon cards and always four room cards.

(5) A player can choose to wait inside a room even when he can leave it. This will translate to passing a turn to the other player.

3.3. State space of 2-player Clue.

Let $P = \{p_1, \ldots, p_4\}$ denote the set of four suspects (recall that the two players are beyond suspicion), let $W = \{w_1, \ldots, w_6\}$ denote the set of six weapons and let $R = \{r_1, \ldots, r_9\}$ denote the nine rooms. $Cards = P \cup W \cup R$ denotes the set of all cards.

Let $B = \{1, \ldots, 203, I, \ldots, IX\}$ represent the board; the arabic numerals represent the 203 positions outside of the rooms, and the roman numbers represent the positions within rooms. $B \otimes B$ shall represent the subset of the free product $B \times B$ for which the set of pairs for which two players occupy the same position outside a room are omitted.

The state of a game of 2-player Clue at play k can be described by:

- *Board position.* This component describes the board position of the tokens of the two players. It can be represented by an ordered pair $Pos_k \triangleq \langle Pos_k^\alpha, Pos_k^\beta \rangle \in B \otimes B$.

- *Initial information.* This component describes the initial information of the players given by the cards they have been dealt. It can be denoted by an ordered pair of mutually disjoint sets $Inf_0 \triangleq \langle Inf_0^\alpha, Inf_0^\beta \rangle$, where Inf_0^i ($i \in \{\alpha, \beta\}$) is a set $\{c_1^i, \ldots, c_8^i\}$ and $c_j^i \in Cards$. c_j^i respect modification (4) above.

- *Current information.* This component describes the current information tion each player has about the innocence, lack of knowledge or guilt of a person, weapon and room (allowing ourselves a certain abuse of language). It can be denoted by an ordered pair $Inf_k \triangleq \langle Inf_k^\alpha, Inf_k^\beta \rangle$, where Inf_k^i ($i \in \{\alpha, \beta\}$) is an ordered triad $\langle P_k^i, W_k^i, R_k^i \rangle$, P_k^i is an ordered triad $\langle p_1, p_2, p_3 \rangle_k^i$, $p_j \in \{0, ?, 1\}$, W_k^i is an ordered 4-tuple $\langle w_1, \ldots, w_4 \rangle_k^i$, $w_j \in \{0, ?, 1\}$, and R_k^i is an ordered 5-tuple $\langle r_1, \ldots, r_5 \rangle_k^i$, $r_j \in \{0, ?, 1\}$. $\{0, ?, 1\}$ denote the knowledge of innocence, ignorance or knowledge of guilt. Let $\langle p_1, p_2, p_3 \rangle_k^i$ be as above, $j, l, m \in \{1, 2, 3\}$ ($j \neq l \neq m \neq j$); then $p_j = 1$ iff $p_l = 0$ and $p_m = 0$. Similar rules apply to weapons and rooms.

The cards in the *Confidential* envelope are the complement of the cards handed to both players: $Conf = Cards \backslash (Inf_0^\alpha \cup Inf_0^\beta) = \{conf_p, conf_w, conf_r\}$ ($conf_p \in P$, $conf_w \in W$ and $conf_r \in R$).

The state of 2-player Clue at play k is given by

$$x_k \triangleq \langle Pos_k, Inf_0, Inf_k \rangle$$

The α player's projection of x_k, henceforth called the state of player α at play k is

$$x_k^\alpha \triangleq \langle Pos_k, Inf_0^\alpha, Inf_k^\alpha \rangle$$

similarly, the state of player β at play k is

$$x_k^\beta \triangleq \langle Pos_k, Inf_0^\beta, Inf_k^\beta \rangle$$

Mappings $id_p^i : \{1, 2, 3\} \rightarrow (P \backslash Inf_0^i) \cup \{null\}$, $id_w^i : \{1, \ldots, 4\} \rightarrow (W \backslash Inf_0^i) \cup \{null\}$ and $id_r^i : \{1, \ldots, 5\} \rightarrow (R \backslash Inf_0^i)$, for $i \in \{\alpha, \beta\}$, relate the current information vectors to the people, weapons and rooms that they refer to; these mappings are defined at the beginning of the game, and remain constant for all k; $id_p^\alpha(p_j) = null$ indicates, for example, that the element is not in use during the game because the corresponding player received two people cards. The same applies to id_p^β, id_w^α and id_w^β.

Suppose that player β received three weapon cards. Let *Candlestick, Knife, Rope* $\notin Inf_0^\beta$. Then initially $W_1^\beta = \langle ?, ?, ?, 0 \rangle$ and $\langle id_w^\beta(1), id_w^\beta(2), id_w^\beta(3), id_w^\beta(4) \rangle = \langle Knife, Rope, Candlestick, null \rangle$. The ways in which this initial information may evolve throughout the game are shown in Figure 4. Each arrow denotes acquisition of information, and the relation expressed is transitive.

Once the cards have been dealt, the total number of states of the 2-player Clue game is

$$\underbrace{\left[\binom{203}{2} + 2(203 \cdot 9) + 9^2 \right]}_{\text{Board positions}} \cdot \underbrace{\left[\frac{2^{(3-1)} \cdot 2^{(4-1)} \cdot 2^{(5-1)}}{2} \right]^2}_{\text{Information}} \approx 1.6 \text{ billion}$$

This calculation accounts for 3 people, 4 weapons and 5 rooms remaining suspect less either a weapon or a person, depending on which kind of card was

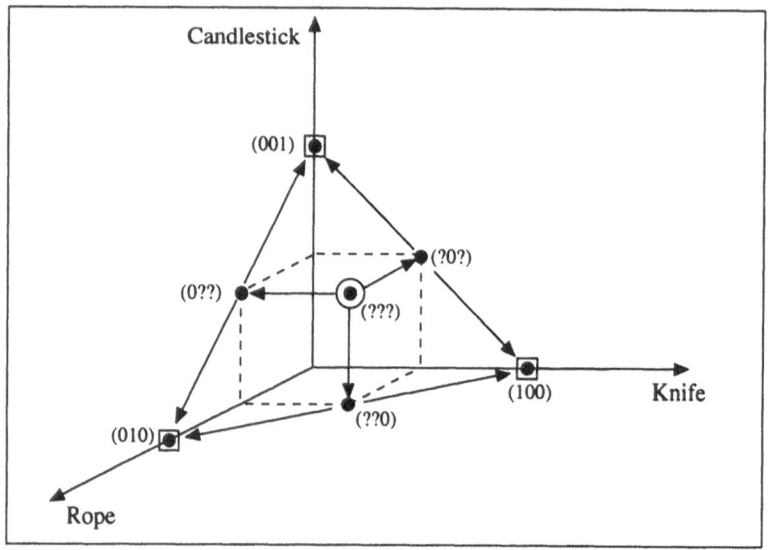

FIGURE 4. Evolution of the state space

received (see modification (4) above). Similarly, the number of states that one player can visit once cards are dealt is

$$\left[\binom{203}{2} + 2(203 \cdot 9) + 9^2\right] \cdot \left[\frac{2^{(3-1)} \cdot 2^{(4-1)} \cdot 2^{(5-1)}}{2}\right] \approx 6.8 \text{ million}$$

3.4. Control problem. Let the initial state for players α and β be given by

$$x_1^\alpha \triangleq \langle Pos_1, Inf_0^\alpha, Inf_1^\alpha \rangle$$
$$x_1^\beta \triangleq \langle Pos_1, Inf_0^\beta, Inf_1^\beta \rangle$$

and let $Conf = \{conf_p, conf_w, conf_r\}$ be the cards in the confidential envelope.

The control problem for player α is to steer the system to arrive at a state whose projection for player α is $x_k^\alpha \triangleq \langle Pos_k, Inf_0^\alpha, Inf_k^\alpha \rangle$ such that $Inf_k^\alpha = \langle P_k^\alpha, W_k^\alpha, R_k^\alpha \rangle$ and

$P_k^\alpha = \langle s_1, s_2, s_3 \rangle_k^\alpha$ and $s_j = 1$ for some $j \in \{1, 2, 3\}$ and $id_p^\alpha(j) = conf_p$;

$W_k^\alpha = \langle s_1, \ldots, s_4 \rangle_k^\alpha$ and $s_j = 1$ for some $j \in \{1, \ldots, 4\}$ and $id_w^\alpha(j) = conf_w$, and

$R_k^\alpha = \langle s_1, \ldots, s_5 \rangle_k^\alpha$ and $s_j = 1$ for some $j \in \{1, \ldots, 5\}$ and $id_r^\alpha(j) = conf_r$.

The condition expressed above is necessary and sufficient for player α to win, which is its control objective; a similar control problem faces player β.

The dynamics of the game will be expressed in the following section. The hierarchical approach will be used to reduce the number of states considered at a high control level; this situation can be compared with that of the Chief Inspector directing the Police Agent through the telephone to make the inquiries, while global strategy issues are handled by the Inspector.

3.5. Dynamics of 2-player Clue. The state transition function is defined as an alternating sequence of actions of players α and β. Thus, a play is two consecutive transitions, one performed by each player (a play is a pair of ploys).

There are two state transition functions, $\Phi^\alpha: X \times U^\alpha \rightarrow X$ and $\Phi^\beta: X \times U^\beta \rightarrow X$, which will alternate to form the total state transition function $\Phi: X \times U^\alpha \times U^\beta \rightarrow X$. That is to say, let $x_k \in X$, $u_k^\alpha \in U^\alpha$ and $u_k^\beta \in U^\beta$, then,

$$x_{k+1} \triangleq \Phi(x_k; u_k^\alpha, u_k^\beta) \triangleq \Phi^\beta(\Phi^\alpha(x_k, u_k^\alpha), u_k^\beta)$$

The input space U is defined as follows:

$$U \triangleq U^\alpha \cup U^\beta$$
$$U^\alpha, U^\beta \triangleq \{move, \langle p, w, r \rangle_1, \ldots, \langle p, w, r \rangle_n, wait\}$$

Where:
  move stands for up to six rectangular moves, and
  $\langle p, w, r \rangle_i$ represents a question ($i \in \{1, \ldots, n\}, n = 4 \cdot 6 \cdot 9 = 216$).

We now consider the definition of the base machine $\mathcal{M} = \langle X, U, \Phi \rangle$ done, and move on to find a partition that will enable hierarchical control.

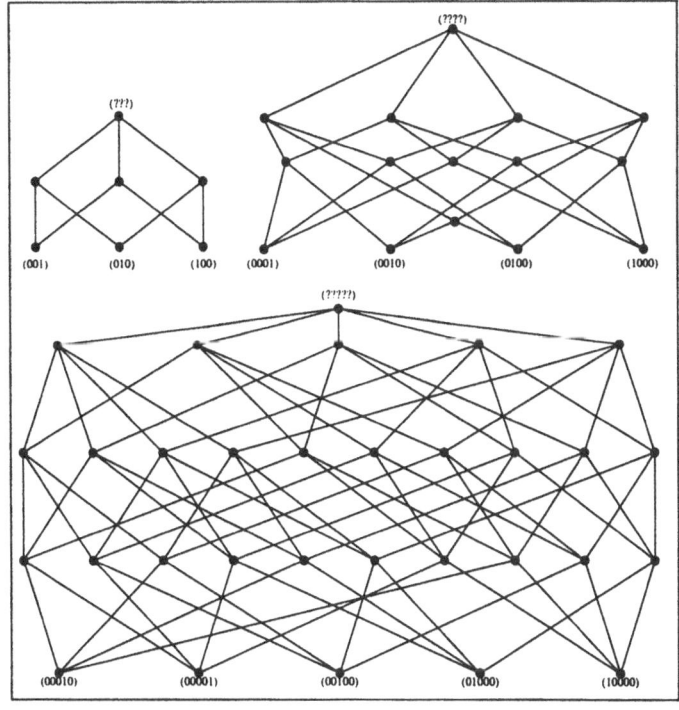

FIGURE 5. Evolution of the information part of the state space for one player

Figure 5 shows three partial orders depicting the evolution of the information part of the state space; these partial orders correspond to the cases where there are three, four and five elements of suspicion, and traversing them from the top (complete ignorance) element to any bottom (complete information) element describes the acquisition of information about the suspects.

3.6. A partition machine for 2-player Clue. In order to obtain a partition machine $\mathcal{M}^{\pi}$ consistent with the hierarchical control theory we have presented, we will obtain a partition of the state space and the DC dynamics between its elements.

This partition will be done in two steps; first, work will be done on the projection of state space on board positions (the so-called Board Space); then, work will be done on the projection of state space on the knowledge about the suspect elements (persons, weapons and rooms), the so-called Information Space. Figure 6 depicts such projection onto Information and Board Space for player α.

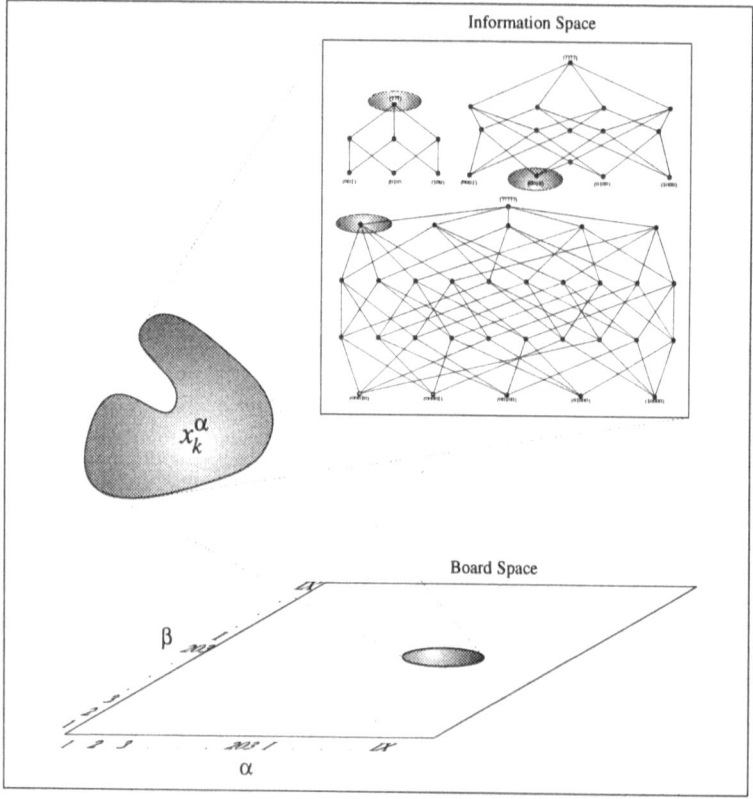

FIGURE 6. Projection of the state of player α at time k onto its Information and Board Spaces

The relative independence of these two dimensions facilitates the partition to be performed on the projections of the state for player α.

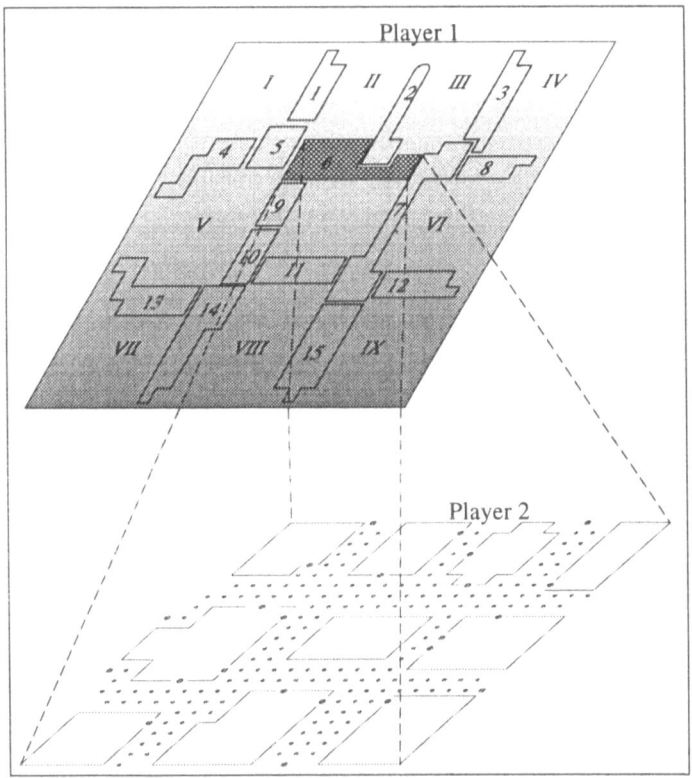

FIGURE 7. High level board product space

3.6.1. *The Partition of Board Space.* The projection of the state space onto board positions is described by the ordered pair $\langle Pos_k^\alpha, Pos_k^\beta \rangle \in B \otimes B$.

Figure 7 depicts the concept of high level board product space. In it, the position of player 1 on the board is indicated only by the partition of the board he is at. The position on the board will then be expressed by $B^\pi \otimes B$ ($B^\pi \triangleq \{1, \ldots, 15, I, \ldots, IX\}$).

The criteria used to partition this projection of the state space are:

- *Block controllability.* The opponent steers the state on Board Space. Hence, we want to partition the Board Space in such a way that such motion (regarded as noise) will not steer the state out of a board partition element.
- *Identification of the rooms.* The partition machine must discriminate among rooms and among different doors of the same room. This allows some efficiency aspects to be handled at the partition machine control level.
- *DC Controllability of the partition machine.* The partition machine must be BBC.

The partition of board space is depicted in Figure 8. Block 2 is not com-

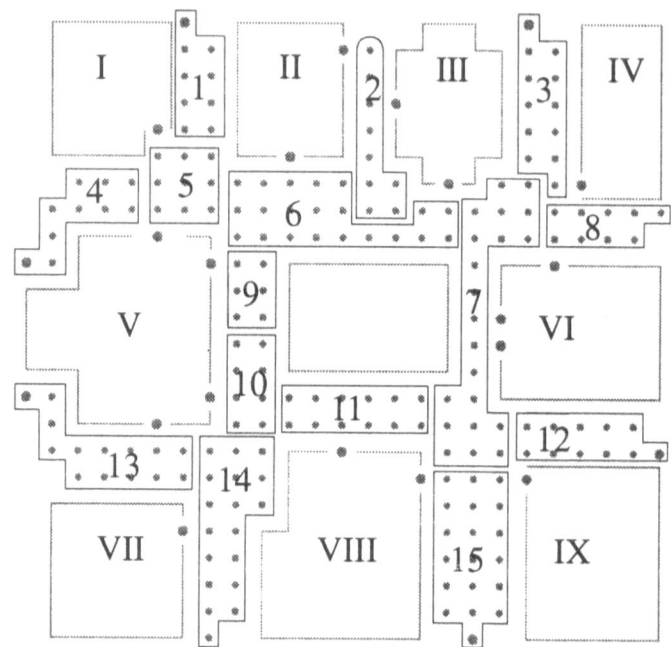

FIGURE 8. Partition of Board Space

pletely controllable. Hence, the partition obtained is not an IBCP. However, the DC partition machine is BBC, as shown in Figure 9.

3.6.2. *The Partition of Information Space.* Information space does not present the same problems as board space, since there is no influence of the opponent's moves on player α's position on its projection onto information space.

However, the higher control needs specific information about the information space for the rooms; this is in order to steer the board position in order to achieve the correct acquisition of information. This allows only to partition the components of suspects and weapons. This partition is shown in figure 10. For suspects and weapons, I and K denote sets of states of ignorance and knowledge respectively. For rooms, $\tilde{I}$ and $\tilde{K}$ represent analogous sets of states; these last symbols are useful only to schematize the dynamics of the DC partition machine on this projection of the state of player α (shown in Figure 11).

3.7. **A Climactic Scenario.** Suppose the state of player α has the projections onto board space and information space shown in Figure 12. Thus, the high level controller "knows" that the weapon and room have been established, and furthermore, it has access to the specific room in which the murder took place (let us suppose that it was the study room, board partition element 7). The high level controller also "knows" that the suspect has not been discovered; a reasonable control action is to ask the low level control to drive the position component to room 7, and try to gather information about the person

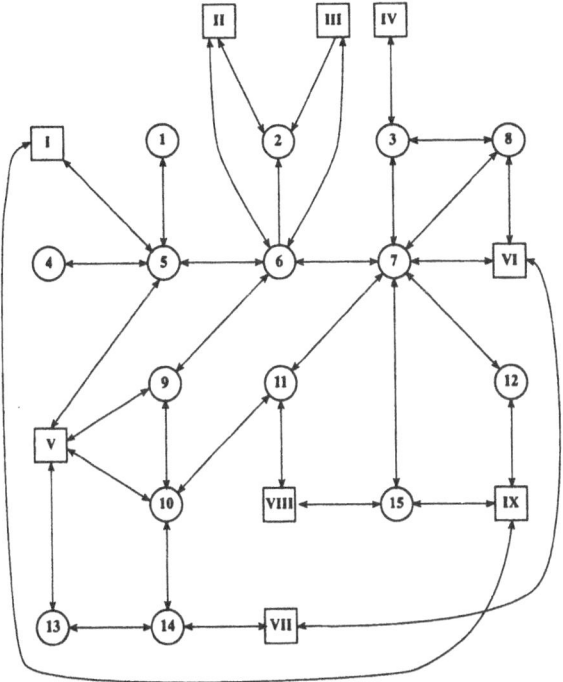

FIGURE 9. DC board partition machine dynamics

by making a suggestion. Once this is done, the person who committed the murder is found, and the accusation can be made, leading player α to win the game, and thus satisfy the control objective.

REFERENCES

1. J. J. Arrow and L. Hurwicz. Decentralization and computation in resource allocation. In R.W. Pfouts, editor, *Essays in Economics and Econometrics*. University of North Carolina Press, 1960.
2. P.E. Caines and S. Wang. Cocolog: A conditional observer and controller logic for finite machines. In *Proc. of 29th Conference on Decision and Control*, pages 2845–2850, Honolulu, HA, December 1990.
3. P.E. Caines and S. Wang. Cocolog: A conditional observer and controller logic for finite machines. To appear in *SIAM J. Cont. and Opt.*, 1994.
4. P.E. Caines and Y.J. Wei. The hierarchical lattices of a finite machine. Submitted to *Systems and Control Letters*, 1994.
5. L. Hurwicz. On informationally decentralized systems. In B. McGuire and R.Radner, editors, *Decision and Organization*. North Holland, 1972.
6. F. Lin and W.M. Wonham. Decentralized control and coordination of discrete-event systems with partial observation. *IEEE Trans. on Automatic Control*, 35, December 1990.
7. T. Marschak and S. Reichelstein. Communication requirements for indvidual agents in networks and hierarchies. Technical report, Haas School of Business, U. C. Berkeley, December 1993.
8. T. Marschak and S. Reichelstein. Network mechanisms, informational efficiency and hierarchies. Technical Report December, Haas School of Business, U. C. Berkeley, 1993.

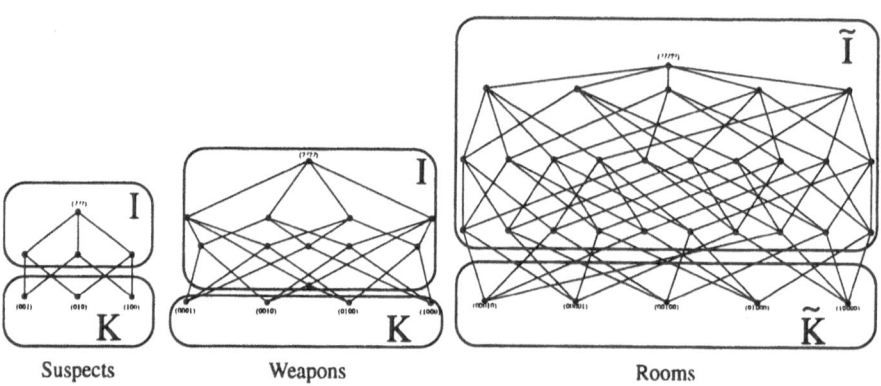

FIGURE 10. Partition of information space

9. K. Rudie and W.M. Wonham. Think globally, act locally: Decentralized supervisory control. *IEEE Trans. on Automatic Control*, 37(11):1692–1708, November 1992.
10. Y-J. Wei and P.E. Caines. Hierarchical cocolog for finite machines. In A. Bensoussan and J. Lions, editors, *Proceedings of the 11th INRIA International Conference on the Analysis and Optimization of Systems*, volume 199 of *Lecture Notes in Control and Information Sciences*, pages 29–38, Sophia Antipolis, France, June 1994. INRIA, Springer Veralag.
11. K.C. Wong and W.M. Wonham. Hierarchical and modular control of descrete-event systems. In *Proc. of Thirtieth Annual Allerton Conference on Communication, Control and Computing*, pages 614–623, September 1992.
12. G. Zames. On the metric complexity of causal linear systems: ϵ- entropy and ϵ- dimension for continuous time. *IEEE Transactions on Automatic Control*, AC-24(2):222–230, April 1979.
13. H. Zhong and W.M. Wonham. On the consistency of hierarchical supervision in discrete-event systems. *IEEE Trans. on Automatic Control*, 35(10):1125–1134, November 1990.

DEPARTMENT OF ELECTRICAL ENGINEERING, MCGILL UNIVERSITY, 3480 UNIVERSITY ST, MONTRÉAL, P.Q., H3A 2A7, CANADA
E-mail address: peterc@cim.mcgill.ca

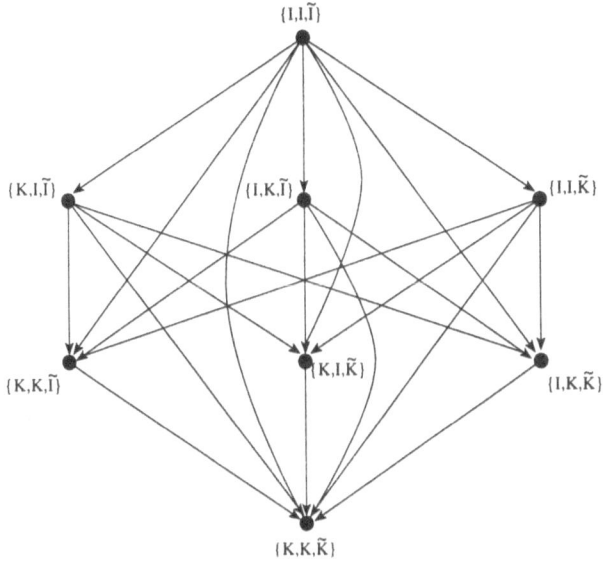

FIGURE 11. DC dynamics for the partition of Information Space

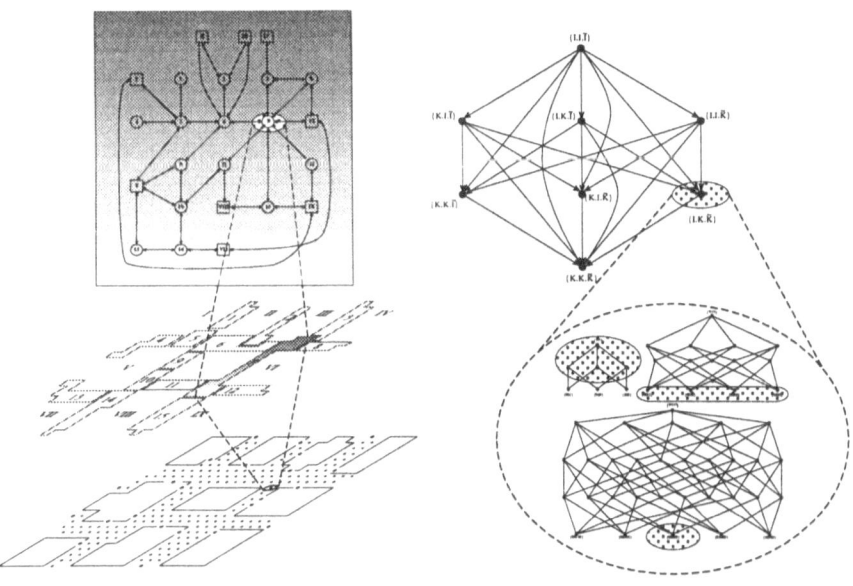

FIGURE 12. Position at the beginning of the end

A Unified Framework for Identification and Control

Munther A. Dahleh and Mitchell M. Livstone
Electrical Engineering and Computer Science
Massachusetts Institute of Technology
Cambridge, MA 02139
USA

Abstract

In this paper we examine some of the recent complexity results in worst case deterministic, or control–oriented system identification. We use these as motivation for introducing a unified approach for iterative system identification and control. The approach is an iterative procedure for refining the uncertainty set via robust control based model invalidation and can be viewed as a systematic way of efficiently searching for a controller delivering a certain desired level of performance to the plant. As a result, either the performance goal will be met or the entire uncertainty set will be invalidated in accordance with our modeling and control method biases. We will comment on the computations involved in such a procedure and provide some results for a particular model structure.

1 Introduction

Recently, there has been much research activity in the area of worst–case, or control–oriented system identification. The motivation can be attributed to new advances in robust control theory which did not interface well with existing theory of classical system identification. Thus, the main focus of current research has been the design of algorithms that yield nominal models along with measures of uncertainty which are well suited for robust control design [5, 4, 16, 10, 7]. Unfortunately, these worst–case algorithms tend to provide error bounds which are very conservative in practice [6] and are therefore of limited utility.

This provides motivation for the new approach which was first introduced in [2]. This approach combines system identification and robust control resulting in an efficient search for a controller which delivers the desired level of performance to the plant. The identification is actually accomplished through invalidating sets of models based on the performance of the actual closed loop system. In this framework we view the model as an auxiliary tool on which we will base our control design. From this point of view, the quality of the model is judged based on the performance that its controller can deliver to the actual plant, rather than how well it can predict the open loop behavior of the plant [15].

Some iterative identification/control schemes have recently been examined by a number of researchers [17, 8, 14]. The main shortcoming of these is the lack of a well formulated goal for the *actual* closed loop system and it is not clear what the proposed schemes actually achieve and in some cases try to achieve.

In the following sections we state some of the recent results on the time complexity of worst–case system identification algorithms and use these to motivate the new iterative approach. The new approach is then discussed at a general level. This is followed by a description of a particular model structure which is currently being studied. We then show in detail the computations associated with this special model.

2 Worst–Case System Identification

Worst–case identification in the presence of bounded noise has been studied recently in the $\mathcal{H}_\infty$ setting [5, 4] as well as the l_1 setting [16, 10]. In the $\mathcal{H}_\infty$ setting, one starts with a set of noisy samples of the frequency response of an LTI system assumed to lie in some compact subset of $\mathcal{H}_\infty$ (e.g., having all poles inside a disk of radius $\rho < 1$). The algorithm fits an LTI system of order n to this data in such a way that certain consistency conditions are satisfied. In particular, one wants the worst–case error to tend to zero as the cardinality of the data as well as the order of the identified model tend to infinity in a certain way.

In the l_1 setting one starts with time domain input/output data (length n) assumed to have been generated by the experiment $y = h * u + d$, w where h is assumed to be in some compact subset of l_1 and d is magnitude bounded by some $\epsilon > 0$. The algorithm typically delivers a finite impulse response model of order N. Furthermore, asymptotic results are provided in [16] where it is shown that the worst–case error is bounded below by 2ϵ and there exist special inputs which can asymptotically drive the error within a factor of two of the lower bound. The following section provides some of the recent complexity results for worst–case identification in l_1, all of which show that even suboptimal worst–case estimates require prohibitively long experiments.

2.1 Asymptotic Results on Error Bounds

The first complexity results were given in [16] and [10] and show that identification of an order N FIR system within the optimal error requires experiments of length $\mathcal{O}(2^{N-1})$. This motivated Dahleh, et.al. [3] as well as Poolla [12] to independently consider the time complexity of algorithms achieving a worst–case error within a factor $K > 1$ of the optimal identification error.

The result in [3] shows that the minimum experiment length required to achieve a worst–case error of $2K\epsilon$ for an FIR system of order N is bounded below by

$$2^{Nf(1/K)-1} - N + 2\lceil N/K \rceil - 1$$

where $f(\alpha) = 1 + \left(\frac{1-\alpha}{2}\right)\log\left(\frac{1-\alpha}{2}\right) + \left(\frac{1+\alpha}{2}\right)\log\left(\frac{1+\alpha}{2}\right)$, and the work in [12] provides similarly disappointing results.

Recently, Chen [1] has studied time complexity of identification with less conservative noise descriptions [11]. The results here are equally disappointing and it seems that time complexity for worst–case identification is exponential for any noise description that is not purely stochastic in nature.

These time complexity results show that experiments must be prohibitively long and that in practice the error bounds furnished by the theory are too conservative for robust control design. This is the motivation behind the approach which will be described in the following section.

3   A Unified Approach to Identification and Control

The goal of control–oriented identification is a controller which achieves a certain level of performance with the plant. When this performance cannot be achieved for the large model set provided by the identification algorithms, identification alone is not enough to achieve our goals. The new iterative approach for identification and control was introduced by Dahleh and Doyle [2] at a very general level. The main idea is to partition the model set and then efficiently invalidate subsets in the partition while searching for a subset which yields a controller that achieves the desired performance for the actual process. A subset is invalidated if there exists a controller which achieves a certain level of robust performance for this set but fails to achieve the same level of performance for the actual process. The key issues are choosing the model structure/parameterization and deciding how to partition the model set.

In general, the selection of the model parameterization is a process that requires engineering insight as well as careful consideration of available robust control techniques. Although the concepts in this paper apply to general time invariant models and partitions, the following model structure and partitioning will be used to develop the ideas.

$$\mathcal{M} = \{G(\theta) + \Delta G \mid \theta \in \Theta \subset R^n \ , \ \Delta G \in \Delta\}$$

where $G(\theta)$ is a finite dimensional LTI model and Δ is some subset of $\mathcal{H}_\infty$. This is a model which can be used in Set Membership Identification (SMID), for example. The partitioning will be performed with respect to Θ while ΔG is assumed to represent the inherent nonparametric uncertainty. Thus, we will refer to Θ as the model set and supress the ΔG part which is fixed for each parameter value in Θ.

3.1 Iterative Procedure

Before stating the iterative procedure, a few comments relating to some of the steps in the procedure should to be discussed. An important requirement is that performance must be testable in finite time to invalidate a subset based on a failed performance test. This necessitates the use of finite time/finite signal types of performance criteria. Another consideration is the minimum size of the partition, which depends on the size of the nonparametric uncertainty and desired performance level. This is true because as the parameter set decreases in size, the nonparametric uncertainty becomes the dominant factor in limiting robust performance.

A parameter set Θ is falsified in the following way. If we find a controller that meets the desired performance (robustly) for the mode set $\{G(\Theta) + \Delta G\}$ but fails the performance test when connected to the plant, we invalidate the set Θ. The result in [9] shows that one is justified in falsifying models in this way since the input/output data collected during this failed test would also invalidate the set $\{G(\Theta) + \Delta G\}$ (i.e., no system in this set could produce such input/output data). Of course, we can also use system identification to falsify other parameter values based on the data collected *during* the failed performance test. This process continues until either the performance test is passed for some controller, or the entire model set is invalidated. Invalidation of the entire model set means this model parameterization is not a good description of the process in view of the performance objectives and must be changed.

We begin with the following three assumptions.

Assumption 1 We are given some *a priori* information and a performance objective that is testable in finite time.

Assumption 2 There exists a robust control technique which implies the performance objective above (possibly conservative).

Assumption 3 We accept this robust control technique in the sense that if it cannot come up with a controller satisfying a given performance objective, we assume that no controller can satisfy it.

The iterative procedure consists of the following steps.

1. Label the initial model set Θ_0 and set $k = 0$.

2. Can the desired performance be achieved for Θ_k by some K_k? If yes, go to (4).

3. Refine Θ_k in the following way (to achieve better performance):

   (a) Find j such that the performance is most sensitive with respect to the jth parameter, θ_j.

   (b) Split Θ_k along the jth dimension, resulting in the two sets X_0 and X_1, with $\Theta_k = X_0 \cup X_1$.

   (c) (Skip if $k = 0$) If X_0 is smaller than the smallest allowable partition size we invalidate Θ_k by decrementing k by 1, and go to (2).

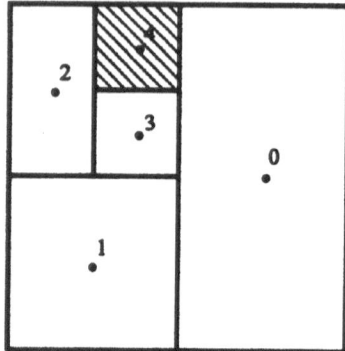

Figure 1: 2D Iteration Example

(d) Find $q \in \{0, 1\}$ such that the best performance which can be achieved for X_q is better than the one for X_{1-q}. Let K_{k+1} be the controller which delivers this performance to X_q.

(e) Set $\Theta_k = X_{1-q}$, $\Theta_{k+1} = X_q$, increment k by 1, and go to (2).

4. Connect K_k to the plant and test for performance

5. If the performance is satisfied, stop.

6. If $k > 0$ invalidate Θ_k by decrementing k by 1 and go to (2). Otherwise, choose a new model parameterization and go to (1).

This procedure has several nice properties. By choosing the smallest allowable partition size to be nonzero, we are guaranteed termination in finite time. Every time a set is split, the memory requirement is only increased by one unit (containing the center and side lengths information, for example) so there is no geometric or exponential explosion. The search is optimistic, always seeking the best set in the partition. At first thought it seems that this may potentially exhibit very bad worst–case behavior. For if the only controller which achieves the performance for the actual process is one that is designed for a "bad" set, the "good" sets will have to be invalidated first. However, the "good" sets will be invalidated quickly because they will typically be large and will not need to be split as many times as the "bad" sets. The following figure illustrates how the iterations might proceed in the case when Θ has dimension 2. In this example, the shaded box 4 is invalidated, the counter is decremented from 4 to 3, and the procedure resumes by focusing on box 3.

3.2 Computations

The computations involved in the above iterative procedure depend on the robust control design methodology which of course depends on the model structure. The computationally difficult steps are steps 2, 3d, and possibly 3a. Note that in steps 2 and 3d, we are trying to synthesize controllers meeting either the desired or the best possible performance levels, with step 3d having to solve two such problems. Step 3a which computes the sensitivity of performance with respect to each parameter can also be very difficult depending on the robust control design. If this step is too difficult to compute, one can select the longest dimension of Θ_k and split the set along this direction. In the following sections, we will consider these computations for a special model currently being studied.

4   Computations for a Special Model

In this section we develop a detailed implementation of the proposed scheme, based on a stable SISO, fixed–pole model with parametric as well as non-parametric uncertainty. The corresponding robust control design is a rank one mixed–μ synthesis which has been shown to be equivalent to a quasi-convex optimization problem [13]. At this time, this is the most sophisticated model (with a minimum l_2 tracking error performance) which has the rank one structure and thus guarantees an arbitrarily–close–to–optimal robust controller solution in finite time.

4.1   The Mixed Uncertainty Model

The model we use is a special case of the perturbed coprime factor model used in [13] with parametric and nonparametric perturbations in the numerator. We assume that we have a rough idea of the pole locations of the process due to either known physical properties or simple input–output experiments which can be performed prior to this procedure. Let the plant model be given by

$$G_{\delta,\Delta} = \frac{\sum_{k=0}^{n-1}(\theta_k^{(c)} + \eta_k \delta_k)z^k + W\Delta}{A(z)}$$

where $A(z)$ is a fixed stable polynomial of order n or greater, $|\delta_k| \leq 1$ and $\eta_k \geq 0$ for $0 \leq k \leq n-1$, $\|\Delta\|_{H_\infty} < 1$, and W is a unit in RH_∞. The weighting function, W, represents the size of the unmodeled dynamics across frequency, while the η_k's give the size of the parameter variation. We also define the nominal system as $G_{0,0} \equiv G$. Typically, W will be small at low frequencies and large at high frequencies. This allows one to express most of the low frequency uncertainty in the parameter variation and allow the unmodeled dynamics to dominate the uncertainty at high frequencies.

The above model can be expressed in terms of the uncertain coprime model:

$$G_{\delta,\Delta} = \frac{N + \delta^T N_\delta + \Delta N_\Delta}{M + \delta^T M_\delta + \Delta M_\Delta}$$

with $\delta \in R^n$, $\|\delta\|_\infty < 1$, $\Delta \in RH_\infty$, $\|\Delta\|_\infty < 1$, $N_\delta \in RH_\infty^n$, $N_\Delta \in RH_\infty$, $M_\delta = 0$, and $M_\Delta = 0$. Furthermore, since the process is assumed to be stable, we can choose $M = 1$ and $N = G$, the nominal system. Note that the sizes

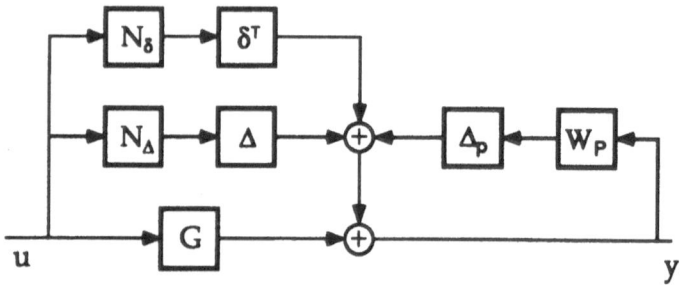

Figure 2: Rank One Structure of the Model

of the n uncertain parameters are captured by the η's in N_δ, and according to the above definitions

$$N_\delta = \frac{[\eta_0 \ \eta_1 z \cdots \eta_{n-1} z^{n-1}]^T}{A(z)}$$

The performance objective is to minimize a frequency weighted l_2 tracking error for some set of command inputs. However, for robust control design we will use the following weighted induced l_2 norm criterion: $\|W_p S\|_\infty < \gamma^{-1}$, where $S(z)$ is the sensitivity transfer function from the commanded input to the tracking error. We next transform this robust performance problem into a robust stability problem in the standard way. The new robust stability problem is:

Stabilize G_a for all $\|\Delta\|_\infty < 1$ and $\|\Delta_p\|_\infty < \gamma$.

Here, $G_a \equiv G_{\delta,\Delta}(1 - \Delta_p W_p)^{-1}$ and can also be expressed in terms of the uncertain coprime model:

$$G_a = \frac{N + \delta^T N_\delta + \Delta N_\Delta + \Delta_p N_{\Delta_p}}{M + \delta^T M_\delta + \Delta M_\Delta + \Delta_p M_{\Delta_p}}$$

with $N_{\Delta_p} = 0$, $M_{\Delta_p} = W_p$, and the rest of the quantities defined as above. Figure 2 shows the model and makes the rank one structure apparent.

We initially assume that the behavior of the process can be adequately explained by the model $y = G_{\delta,\Delta} u + d$, where d is some noise sequence satisfying $\|d\|_\infty \leq \epsilon$ for some $\epsilon > 0$, and begin the iterative procedure.

4.2 Rank One Synthesis

In this section we briefly review the result of Rantzer and Megretski [13] and show how it specializes to our fixed-pole model. The rank one result uses a separating hyperplane argument to come up with a convex parameterization of all robustly stabilizing controllers for rank one uncertainty lying in a convex set. In our special case, after appropriate selection of the Bezout factors, the result can be stated in the following simplified form.

The model G_a is robustly stable for all $\|\delta\| < 1$, $\|\Delta\|_\infty < 1$, and $\|\Delta_p\|_\infty < \gamma$ if, and only if

$$\inf_{\substack{\alpha \text{ positive real} \\ \beta \in RH_\infty}} \phi(\alpha, \beta) < \gamma \quad , \text{ where}$$

$$\phi(\alpha, \beta) = \sup_{\omega \in [0, 2\pi]} \frac{|W_p([\alpha + \beta]G + \alpha)|_{(e^{i\omega})}}{\Re\{\alpha(e^{i\omega})\} - \|\Re\{N_\delta[\alpha + \beta]\}_{(e^{i\omega})}\|_d - |N_\Delta(\alpha + \beta)|_{(e^{i\omega})}}$$

and $\|\cdot\|_d$ is dual of the norm used to measure δ. The controller is then given by $K = (Q+1)/(G+GQ+1)$, where $Q = \beta/\alpha$. This optimization is quasi–convex in α and β and a solution arbitrarily close to the optimal can be computed in finite time.

4.3 Computations for the Special Model

The computations associated with this special model are then as follows. The robust control design is accomplished by solving the convex optimization problem above. This is the most costly computation in the procedure. According to Assumption 2, the minimum value attained by the optimization which is a (weighted) l_2 gain must imply the finite time performance objective.

In this case it is also straightforward to solve for the parameter which has the most impact on performance. After solving the quasi–convex problem above, we have a feasible pair $(\alpha(z), \beta(z))$ as well as the worst frequency, ω_0 which maximizes the right hand side of the above equation. Using this and recalling the form of N_δ, one can show that the parameter which has the greatest impact on performance is given by $\theta_{k_{max}}$, where

$$k_{max} = \arg \max_{k \in [0, n-1]} \left| \Re \left\{ e^{i\omega_0 k} \left(\frac{\beta(e^{i\omega_0}) + \alpha(e^{i\omega_0})}{A(e^{i\omega_0})} \right) \right\} \right| \tag{1}$$

This is the parameter which is split in step 3a of the procedure.

The last issue is that of deciding on the smallest allowable partition size. Choosing the limiting partition size is important for making sure that the procedure terminates in a reasonable amount of time. The following result provides some insight into how this value should be picked. The result really provides a worst–case (with respect to the process) lower bound on the size of the parametric uncertainty, even in the absence of noise. We ask the following question:

What is the largest $\epsilon_0 > 0$ such that

$$\left\{ \frac{\sum_{k=0}^{n-1} (\theta_k + \delta_k) z^k}{A(z)} \mid \|\delta\|_\infty \leq \epsilon_0 \right\} \subseteq \left\{ \frac{\sum_{k=0}^{n-1} \theta_k z^k}{A(z)} + W(z)\Delta(z) \mid \|\Delta\|_\infty \leq 1 \right\}$$

This says that if the actual process P is equal to $\sum \theta_k z^k / A(z)$, then the consistent set will always contain the ϵ_0 ball around θ for any input. Another way of thinking about this is that for the nominal system, $\sum \theta_k z^k / A(z)$, all parameters in the ϵ_0 ball around vector θ are indistinguishable because of the unmodeled dynamics. This suggests that it does not make much sense to partition the parameter space much finer than ϵ_0 because the achievable robust performance is not likely to improve. The following result provides a lower bound for ϵ_0.

Theorem 4.1 *Given the problem stated above and assuming that W is minimum phase,*

$$\epsilon_0 \geq \frac{1}{n\rho} \quad \text{where}$$

$$\rho = \|\frac{1}{A(z)W(z)}\|_\infty$$

Proof: The unmodeled dynamics can make the parameter set $\{theta + \delta \mid \|\delta\| \leq \epsilon_0\}$ indistinguishable if and only if, for every parameter in this set there is some $\|\Delta\|_\infty \leq 1$ which satisfies

$$\Delta(z) = \frac{\sum_{k=0}^{n-1} \delta_k z^k}{A(z)W(z)}$$

If we next define

$$f(\gamma) = \sup_{\|\delta\|_\infty \leq \gamma} \|\frac{\sum_{k=0}^{n-1} \delta_k z^k}{A(z)W(z)}\|_\infty$$

this means that $\epsilon_0 = \sup\{\gamma : f(\gamma) \leq 1\}$. Furthermore, it is easy to see that f is a nondecreasing function. One can also show that $f(\gamma) \leq n\gamma\rho$, where $\rho = \|1/AW\|_\infty$. Using these two facts, we have

$$\epsilon_0 = \sup\{\gamma : f(\gamma) \leq 1\} \geq \sup\{\gamma : n\gamma\rho \leq 1\} = \frac{1}{n\rho}$$

5 Conclusion

The recent formulation of worst–case system identification leads to algorithms that typically provide error bounds which are too conservative for use with robust control design. We have reviewed some of the recent time complexity results for worst–case identification which show that useful error bounds around estimates require experiments which are prohibitively long. This motivates the new iterative approach to identification and control. The iterative procedure uses robust control based model invalidation to essentially perform an efficient search for a controller which delivers the desired performance to the plant. We have shown in detail the computations associated with the iterative procedure for a special model structure. The generality of the model structure is limited by the available robust control techniques and so this procedure will become more sophisticated as robust control theory develops.

6 Acknowledgments

This work was supported in part by the Airforce Office of Scientific Research under Grant AFOSR-91-0368, by the National Science FOundation under Grant NSF 9157306-ECS, and by C.S. Draper Laboratory under Grant DL-H-467128 and Draper IR& D Project No. 438.

References

[1] Kuang-Hang Chen. System identification with corellated noise descriptions. Master's thesis, Massachusetts Institute of Technology, Cambridge, MA, May 1994.

[2] M.A. Dahleh and J. Doyle. "From Data to Control". In *Proc. Workshop on Modeling of Uncertainty in Control Systems*. Springer–Verlag, 1992.

[3] M.A. Dahleh, T.V. Theodosopoulos, and J.N. Tsitsiklis. "The sample complexity of worst–case identification of FIR linear systems". *Systems Control Lett.*, 20(3):157–166, March 1993.

[4] G. Gu and P. Khargonekar. "Linear and Nonlinear Algorithms for Identification in H^∞ With Error Bounds". *IEEE Transactions on Automatic Control*, 34(8):831–847, July 1992.

[5] A.J. Helmicki, K. Jacobson, and C. Nett. "Control Oriented System Identification in H_∞". *IEEE Transactions on Automatic Control*, 36(10):1163–1176, October 1991.

[6] P. Khargonekar. "System Identification in Frequency Domain: Theory and Examples". *Proceedings Conf. Feedback Control, Nonlinear Systems, and Complexity*, May 1994.

[7] R. Kosut, M. Lau, and S. Boyd. "Set-Membership Identification of Systems with Parametric and Nonparametric Uncertainty". *IEEE Trans. on Auto. Control*, 37(7):929–941, July 1992.

[8] W. Lee, B. Anderson, R. Kosut, and I. Mareels. "On Adaptive Robust Control and Control-Relevent System Identification". *Proc. 1992 American Control Conference, Chicago, IL*, pages 2834–2841, June 1992.

[9] M.M. Livstone, M.A. Dahleh, and J.A. Farrell. "A Framework for Control Based Model Invalidation". To Appear in the 1994 American Control Conf., Baltimore, MD.

[10] P.M. Mäkilä . "Robust Identification and Galois Sequences". Technical Report Rep. 91-1, Åbo Akademi (Swedish University of Åbo), Åbo, Finland, January 1991.

[11] Fernando Paganini. "Set Descriptions of White Noise and Worst Case Induced Norms". Technical report, California Institute of Technology), Pasadena, CA, February 1993.

[12] K. Poolla and A. Tikku. "On the Time Complexity of Worst–Case System Identification". *IEEE Transactions on Automatic Control*, 39(5):944–950, May 1994.

[13] A. Rantzer and A. Megretski. "A Convex Parameterization of Robustly Stabilizing Controllers". Technical report, The Royal Institute of Technology, Stockholm, Sweden, 1993.

[14] R. Schrama and P. Van den Hof. "An Iterative Scheme for Identification and Control Design Based on Coprime Factorizations". *Proc. 1992 American Control Conference, Chicago, IL*, pages 2842–2846, June 1992.

[15] Ruud Schrama. *"Approximate Identification and Control Design"*. PhD thesis, Delft University of Technology, Delft, The Netherlands, March 1992.

[16] D. Tse, M.A. Dahleh, and J. Tsisiklis. "Optimal Asymptotic Identification Under Bounded Disturbances". *Proc. 1992 American Control Conference, Chicago, IL*, pages 679–685, July 1992.

[17] Z. Zang, R. Bitmead, and M. Gevers. "H_2 Iterative Model Refinement and Control Robustness Enhancement". *Proc. 1991 Conference on Decision and Control, Brighton, England*, pages 279–284, December 1991.

Intelligent Control: Some Preliminary Results

Edward J. Davison* and Michael Chang
Systems Control Group
Department of Electrical & Computer Engineering
University of Toronto
Toronto, Ontario
Canada M5S 1A4

Abstract

In today's industrial world, it is commonplace to use control in the design and operation of complex systems. However, severe limitations in potential performance are often present in such systems when large unexpected structural changes occur in the system, e.g. 'conventional' control schemes generally do not have the ability to control systems which are subject to unplanned extreme changes; we will call controllers which have this ability as being 'intelligent'.

This paper introduces a controller, based on the results of [8], which has intelligent-like features. The controller is nonlinear and contains a switching device which applies a sequence of LTI controllers to the system, and which has the property that after a finite time (i.e. 'learning' time), switching ceases, resulting in the system being controlled by a LTI controller. The paper describes studies on implementing this class of controllers on 'MARTS', a highly interacting multivariable experimental hydraulic system; in particular, the paper shows how the new proposed 'intelligent' controller can successfully control MARTS, unlike conventional controllers, when a catastrophic change occurs in the system.

1   Introduction

Although there has been considerable interest in so-called 'intelligent' control research for some time now, e.g. see [1]-[5], there is no precise definition yet as to what exactly constitutes intelligent control. In this paper, the following working definition will therefore be made:

Definition 1 *An intelligent controller for a plant is a controller which successfully carries out its mandate for the plant's nominal operating conditions,*

---

*This work has been supported by the Natural Sciences and Engineering Research Council of Canada under Grant No. A4396.

as well as for any unexpected *(i.e.* unplanned*) events which may occur.*

This definition is 'loose' in the sense that the wording is not well defined; however, the flavour of the definition is clear. For example, if one *designs* a controller so that the controller successfully controls the nominal plant, as well as when the plant is in some failure mode '*x*', say, then the controller has the property that it displays certain integrity properties (which is very desirable), but the controller is not 'intelligent'; the controller is 'intelligent' only if it successfully controls the plant, either in nominal mode or failure mode '*x*', in spite of the fact that failure mode '*x*' *was not anticipated* in the design of the controller.

In the conventional design of controllers for multivariable systems, the general approach often adopted is to find a model for the plant (which is often very difficult to do), and thence design a controller based on this model. This approach will not be adopted here, however, since we would like the controller to be able to control the plant when subject to *unanticipated* (i.e. unknown) structural changes in the plant.

Thus in this paper, it will be assumed that the plant can be described by a finite-dimensional LTI system and is open loop asymptotically stable, but that a model of the plant is *unknown*. Such an approach has been used in [6], where certain steady-state experiments were assumed to be allowed to be performed on the plant, in conjunction with certain 'on-line tuning' methods, and the method proposed has been successfully applied in industry with excellent results, e.g. see [7]. Recently, various extensions to [6] have been obtained, in which as little as possible *a priori* plant information is assumed, e.g. see [8]-[13]. The motivation of this interest stems from the fact that it is generally difficult and often impossible to obtain an accurate model representation of an actual industrial plant. It is also to be noted that although standard adaptive control techniques reduce the amount of information required from that used in conventional controller methods, detailed plant information (e.g. a knowledge of the upper bound on the order of the plant, the relative degree of the plant, etc.) is generally still needed. Note too that the controllers proposed in [8]-[13] are nonlinear controllers, in which a sequence of LTI controllers are switched to the plant using a 'tuning procedure', until an appropriate LTI controller is found, at which point switching stops. (See also [14], [15] for other types of switching controllers.)

The proposed intelligent controller to be studied in this paper is directly based on the results of [8]-[13]; in particular, the controller is an extension of the integral controller described in [8], and a detailed description of the controller is given in [9], [20]. This paper describes studies on implementing this controller on a highly interacting experimental multivariable system called 'MARTS'; in particular, the paper shows how the proposed intelligent controller can successfully control MARTS when a catastrophic change occurs in the system.

2 Notation

Let $\mathbf{R}$, $\mathbf{R}^+$, and $\mathbf{N}$ denote respectively the set of real, positive real, and natural numbers, $\mathbf{R}^n$ the n-dimensional real vector space, and $\mathbf{R}^{m \times n}$ the set of $m \times n$ real matrices. For any $x, y \in \mathbf{N}$,

$$x \bmod y := x - \text{floor}\left(\frac{x}{y}\right) y$$

where floor$(\cdot)$ rounds the expression $(\cdot)$ down to the nearest integer.

With $x \in \mathbf{R}^n$, denote its ∞-norm to be

$$\|x\|_\infty := \max_{1 \le i \le n} |x_i|.$$

For any arbitrary $A \in \mathbf{R}^{n \times n}$, let $\lambda(A)$ denote the eigenvalues of A, and let $\mathbf{Re}(\lambda)$ and $\mathbf{Im}(\lambda)$ be the real and imaginary parts of λ respectively. A is said to be *stable* if $\mathbf{Re}(\lambda(A)) < 0$ and *unstable* otherwise.

For the more general case when $A \in \mathbf{R}^{m \times n}$, A^T denotes its matrix transpose, rank(A) its rank, and, if A has full row rank, $A^\dagger := A^T(AA^T)^{-1}$ denotes its pseudo-inverse. In addition, the induced ∞-norm of A is defined by

$$\|A\|_\infty := \max_{1 \le i \le m} \sum_{j=1}^n |a_{ij}|.$$

3 Experimental Apparatus

The MARTS (Multivariable Apparatus for Real Time Control Studies) facility used for all experiments consists of an interconnection of industrial commercial actuators and sensors which monitor and control a nonlinear hydraulic system as described in figure 1 and table 1. In figure 1, the by-pass, drainage, and interconnecting valves are all adjustable manually to enable the selection of desired equilibrium column heights and to control the degree of interaction existing between both columns. Actuator valves for both columns also enable one to individually apply positive constant system disturbances w_1 and w_2. A Texas Instruments (TM 990/101/MA-1) real-time digital computer using the PDOS operating system with a 12 bit A/D, D/A board and a variable sampling time $\ge$ 10 msec. (not shown) is also used to control the system. For a detailed description of the apparatus, the reader is referred to [16].

The control objective using MARTS is to regulate the level of both column heights (if possible) for all initial conditions and disturbances applied; we note, however, that the physical dimensions of this system are relatively large (e.g. the height of the column $\approx$ 1.2 meters), the plant is only marginally open loop stable, and that for a relatively small interconnection valve opening (e.g. $\theta = 30°$), the plant is a *highly* interacting 2 input/2 output system.

Part	Description
Drainage valve	1/2 inch globe valve
By-pass valve	1/2 inch globe valve
Interconnection valve	1/2 inch ball valve
Pump	Iwaki centrifugal pump (1/12 HP)
Level sensor	Taylor 3400T Series pressure transmitter
Control valve	Foxboro V4A 1/2 inch linear valve
Control valve transducer	Foxboro E69P pneumatic valve positioner

Table 1: Summary of the major components of MARTS.

3.1 Linearized Model of MARTS

To demonstrate the general difficulty in obtaining an accurate model representation of an unknown system, consider the interconnection structure of MARTS shown in figure 2.

Let $i \in \{1, 2\}$, and define

$$
\begin{aligned}
u_i &:= \text{input to control valve } i \\
y_i &:= \text{measured output (height) of column } i \\
A_j &:= \text{cross sectional area, } j \in \{1, 2, 3\} \\
g &:= \text{acceleration due to gravity} \\
h_i &:= \text{liquid height in column } i \\
Q_{in_i}(u_i) &:= \text{input flow to column } i \\
Q_i &:= \text{output flow from column } i \\
\theta &:= \text{interconnection valve angle (degrees); } 0° \leq \theta \leq 90° \\
C_d &:= \text{coefficient of discharge} \\
C_d(\theta) &:= \text{coefficient of discharge with respect to } \theta \\
C_d(0) &:= 0 \text{ (ie. } \theta = 0° \Leftrightarrow \text{interconnection valve is shut).}
\end{aligned}
$$

On ignoring the time delay occurring between the input signal $u_i(t)$ and the output flow rate $Q_i(t)$, on assuming that any actuator valve non-linearities have been eliminated (by using nonlinear compensation gains), and defining $h_i := h_{i_s} + \delta h_i$ and $u_i := u_{i_s} + \delta u_i$, where h_{i_s} and u_{i_s} are respectively the steady state height of and input to column i and control valve i, the following linearized model can be obtained for the case when $h_{1_s} > h_{2_s}$ (see [9]):

$$
\begin{bmatrix} \delta \dot{h}_1 \\ \delta \dot{h}_2 \end{bmatrix} \approx \underbrace{\begin{bmatrix} -\beta_1 - \gamma(\theta) & \gamma(\theta) \\ \gamma(\theta) & -\beta_2 - \gamma(\theta) \end{bmatrix}}_{=: A} \begin{bmatrix} \delta h_1 \\ \delta h_2 \end{bmatrix} + \underbrace{\begin{bmatrix} 1 & 0 \\ 0 & 1 \end{bmatrix}}_{=: B} \begin{bmatrix} \delta u_1 \\ \delta u_2 \end{bmatrix} \quad (1)
$$

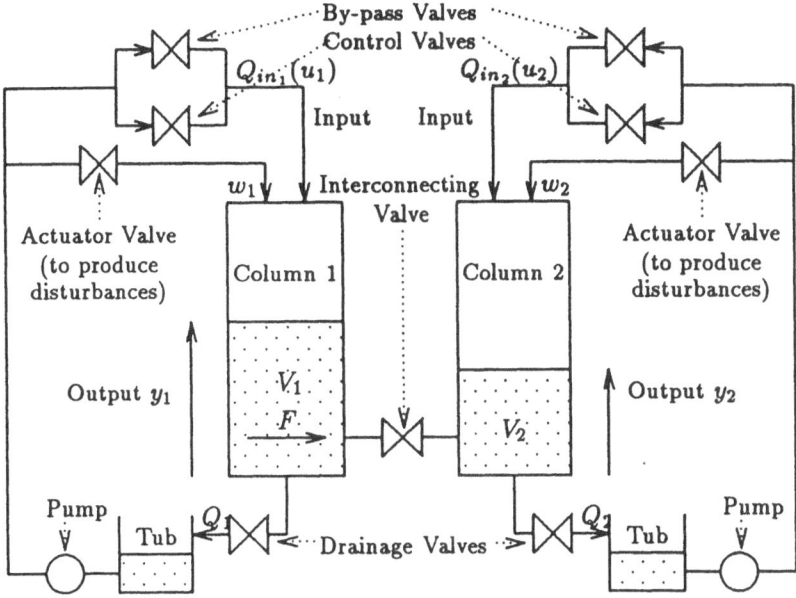

Figure 1: Schematic diagram of the MARTS setup (not to scale).

$$\begin{bmatrix} \delta y_1 \\ \delta y_2 \end{bmatrix} = \underbrace{\begin{bmatrix} 1 & 0 \\ 0 & 1 \end{bmatrix}}_{=:\, C} \begin{bmatrix} \delta h_1 \\ \delta h_2 \end{bmatrix}$$

where

$$\beta_i := \frac{C_d A_2}{A_1} \sqrt{\frac{g}{2h_{is}}}, \quad \gamma(\theta) := \frac{C_d(\theta) A_3}{A_1} \sqrt{\frac{g}{2(h_{1s} - h_{2s})}}.$$

Here, δy_i is the perturbed liquid level of column i ($i \in \{1, 2\}$), and δu_i is the perturbed input liquid flow from control valve i entering column i.

A similar derivation for the case when $h_{1s} < h_{2s}$ yields the same basic format for equation (1), with

$$\gamma(\theta) := \frac{C_d(\theta) A_3}{A_1} \sqrt{\frac{g}{2(h_{2s} - h_{1s})}}$$

being the only slight modification needed. In addition, for the case when $h_{1s} \cong h_{2s}$ and $\theta \neq 0°$, one can also show [20] that the system behaves as a single column apparatus obeying the equations

$$\dot{\mathcal{X}} = -\beta \mathcal{X} + \delta u_1 + \delta u_2$$
$$\delta y_i = \frac{1}{2} \mathcal{X}$$

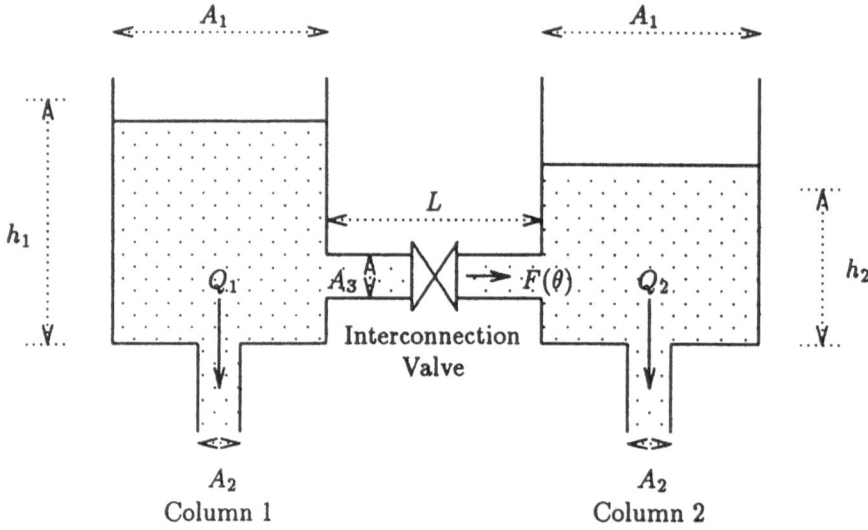

Figure 2: Cross sectional view of the interconnected columns in MARTS (not to scale).

where $\mathcal{X} := \delta h_1 + \delta h_2$, and the (realistic) assumption is made that $\beta := \beta_1 \cong \beta_2$. Hence, equation (1) may be seen to be equally valid under this particular condition.

For our experimental setup with the interconnection valve angle θ set at $30°$, $(\beta_1, \beta_2, \gamma(\theta))$ were experimentally found to be $(0.0237, 0.0260, 0.0135)$, which implies that the MARTS system is approximately described by:

$$\begin{bmatrix} \delta \dot{h}_1 \\ \delta \dot{h}_2 \end{bmatrix} = \begin{bmatrix} -0.0372 & 0.0135 \\ 0.0135 & -0.0395 \end{bmatrix} \begin{bmatrix} \delta h_1 \\ \delta h_2 \end{bmatrix} + \begin{bmatrix} 1 & 0 \\ 0 & 1 \end{bmatrix} \begin{bmatrix} \delta u_1 \\ \delta u_2 \end{bmatrix} \quad (2)$$

$$\begin{bmatrix} \delta y_1 \\ \delta y_2 \end{bmatrix} = \begin{bmatrix} 1 & 0 \\ 0 & 1 \end{bmatrix} \begin{bmatrix} \delta h_1 \\ \delta h_2 \end{bmatrix}.$$

In this case, one can verify that $(-0.0248, -0.0519)$ are the asymptotically stable eigenvalues of A.

Alternatively, using singular value analysis and the model identification and reduction algorithm given in [17], the following discrete time model, with a sampling time period of $T = 2$ seconds, was experimentally obtained:

$$\begin{aligned} x(k+1) &= \mathcal{F}x(k) + Gu(k) \\ y(k) &= Hx(k) \end{aligned} \quad (3)$$

with $[\ \mathcal{F} \mid G \mid H\ ]$ given by

$$\begin{bmatrix} 0.9783 & -0.0032 & 0.1674 & 0.8759 & 0.5623 & -1.0041 \\ -0.0150 & 0.9636 & -0.4568 & 0.3873 & 1.6699 & 0.6109 \end{bmatrix},$$

which yields $(0.9811, 0.9609)$ for the eigenvalues of $\mathcal{F}$. Further information concerning the derivation and determination of these models and parameters is given in [20].

Remark 3.1 The above procedures show the difficulty of obtaining "good and accurate" mathematical models for a system; although both models (C, A, B) (2) and $(H, \mathcal{F}, G)$ (3) obtained *approximately* describe the behaviour of the system, the models in fact are not consistent with each other (eg. with $T = 2$, $\lambda(e^{TA}) = (0.901, 0.952) \notin \lambda(\mathcal{F})$). $\diamond$

It is to be noted and emphasized that the models obtained above will not actually be used in the proposed 'intelligent' controller design for the plant.

4   Problem Statement/Main Assumptions

Assume that the system to be controlled behaves as a finite-dimensional LTI model described by the equations

$$
\begin{aligned}
\dot{x} &= Ax + Bu + Ew \\
y &= Cx + Fw \\
e &:= y_{ref} - y
\end{aligned}
\tag{4}
$$

where $x \in \mathbf{R}^n$ is the state, $u \in \mathbf{R}^m$ is the control input, $y \in \mathbf{R}^r$ is the output to be regulated, $w \in \mathbf{R}^q$ is the disturbance, and $e \in \mathbf{R}^r$ is the difference between the specified reference input y_{ref} and the output y, and assume that y_{ref} and w are piecewise constant signals. In this case, it is desired to solve the *robust servomechanism problem* (RSP) [18] for equation (2). Assume also that A is stable (not essential), that n, A, B, C, E, and F are unknown, and that there exists a solution to the robust servomechanism problem for equation (4) [18], i.e. that

$$
\text{rank} \begin{pmatrix} A & B \\ C & 0 \end{pmatrix} = n + r;
\tag{5}
$$

let $T := -CA^{-1}B$.

Note that due to the physical nature of the system and the industrial sensor and actuators used, time lag and other nonlinear effects have inherently also been ignored.

5   Conventional Controller Design Results

For comparative purposes, in this section, we present experimental results obtained using the high performance controller design method given in [18], [19], assuming that a mathematical model of the system is available. More specifically, with the resultant closed loop system obtained by augmenting the servocompensator [18] written as

$$
\begin{bmatrix} \dot{x} \\ \dot{\eta} \end{bmatrix} = \begin{bmatrix} A & 0 \\ -C & 0 \end{bmatrix} \begin{bmatrix} x \\ \eta \end{bmatrix} + \begin{bmatrix} B \\ 0 \end{bmatrix} u + \begin{bmatrix} 0 \\ I \end{bmatrix} y_{ref} + \begin{bmatrix} E \\ -F \end{bmatrix} w
\tag{6}
$$
$$
e = y_{ref} - y
$$

and on using the performance index [18]

$$J_\epsilon := \int_0^\infty (e^T e + \epsilon \dot{u}^T \dot{u}) dt \tag{7}$$

with $\epsilon > 0$, then the optimal controller which minimizes this performance index is given by

$$\dot{u} = \begin{bmatrix} K_0^\epsilon & K_1^\epsilon \end{bmatrix} \begin{bmatrix} \dot{x} \\ e \end{bmatrix}.$$

Using the MARTS model given in equation (2) (for $\theta = 30°$), with $\epsilon = 1$, the following controller gains are obtained:

$$(K_0^1 \mid K_1^1) = \begin{pmatrix} -1.3776 & -0.0131 & 1.0000 & 0.0000 \\ -0.0131 & -1.3753 & 0.0000 & 1.0000 \end{pmatrix},$$

which implies that the following controller is now obtained [18] for the MARTS system:

$$u = -K_0^1 e + K_1^1 \int_0^t e(\tau) d\tau. \tag{8}$$

5.1 Reset Windup Control Concerns

For SISO systems, a well known effect which may occur when integral control is used is so-called "control reset-windup", which results when a control signal obtained by using integral control reaches saturation; in this case, once the control signal is no longer saturated, large transients in the system response may occur if no means for 'stopping' the integral action (during saturation) have been provided. The same effect may also occur in multivariable systems. To prevent such unnecessary transient deviation, all controllers in this paper were therefore provided with a controller reset windup mechanism [20]. For example, figures 3 and 4 give a comparison of the difference of response obtained by using such a reset windup controller, and with no such reset windup mechanism, for the case of digitally implementing integral controller 1 of [8, pg. 513] using

$$f(k) = k^k, \quad g(k) = \frac{10}{2^k}, \quad \beta = 2, \quad K = \begin{bmatrix} 0.0510 & -0.0198 \\ -0.0061 & 0.0239 \end{bmatrix} \tag{9}$$

with a sampling period of $T = 2$ seconds on the MARTS system with $\theta = 30°$ for the case of $h_1(0) \cong 3000$, $h_2(0) \cong 1800$, and $y_{ref}(t) = (3000, 2500)$ applied at $t = 0$. It is seen that the controller reset windup mechanism is most effective at preventing undesirable transient behaviour.

5.2 Experimental Results Obtained

When controller (8) is digitally implemented on the MARTS system using a sampling period of $T = 0.4$ seconds, the experimental results of figure 5 are

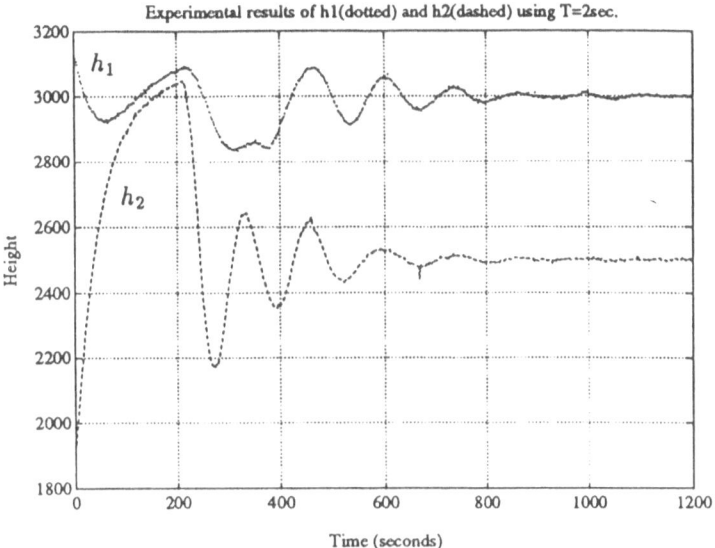

Figure 3: Plot of h_1 (dotted) and h_2 (dashed) using integral controller (9), $T = 2$sec., $\theta = 30°$, and no reset windup mechanism.

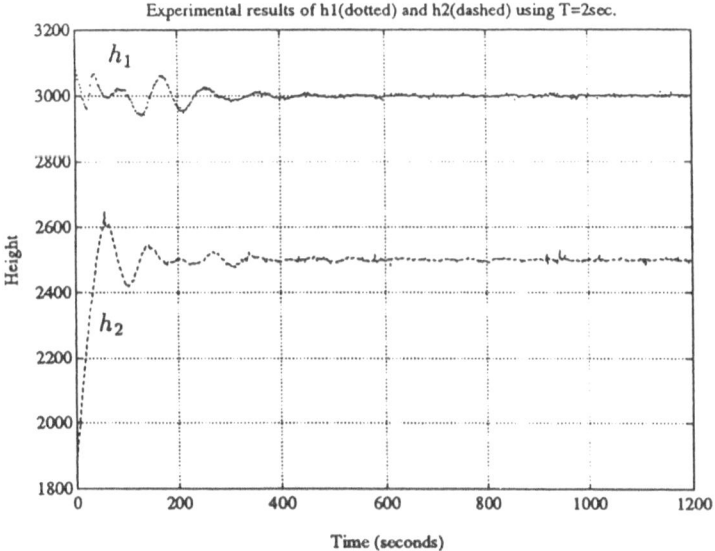

Figure 4: Plot of h_1 (dotted) and h_2 (dashed) using integral controller (9), $T = 2$sec., $\theta = 30°$, and a reset windup mechanism applied.

obtained for the case when $\theta = 30°$ and $\epsilon = 1$ with:

$$(y^1_{ref}(t), y^2_{ref}(t)) := \left\{ \begin{array}{ll} (3000, 2500), & 0 \leq t < 500 \\ (3500, 2000), & 500 \leq t < 1000 \\ (3000, 2500), & 1000 \leq t < 1500 \\ (3500, 2000), & 1500 \leq t < 2000 \text{ seconds} \end{array} \right\}$$

as the reference input signal. It is seen that the controller produces excellent control. This figure also includes the results of implementing controller (8), designed using $\epsilon = 1000$, on the MARTS system with a sampling period of $T = 0.4$ seconds; it is observed that the two responses are almost identical, which shows that the controller's performance is insensitive to the choice of ϵ for $0 < \epsilon < 1000$.

When the same controller (8) is digitally implemented on the MARTS system with $T = 0.4$ seconds for the case when $\theta = 40°$ and $\epsilon = 1$ with:

$$(y^1_{ref}(t), y^2_{ref}(t)) := \left\{ \begin{array}{ll} (3000, 2500), & 0 \leq t < 500 \\ (3500, 2500), & 500 \leq t < 1000 \\ (2500, 2500), & 1000 \leq t < 1500 \\ (3500, 2000), & 1500 \leq t < 2000 \text{ seconds} \end{array} \right\}$$

the results of figure 6 are obtained, which show that the controller design is robust, i.e. due to prolonged saturation effects resulting from $\theta = 40°$, the given reference signal $y_{ref}(t)$ is 'unattainable' for $500 \leq t < 1000$ and $1500 \leq t < 2000$ seconds; in spite of this, however, the system's response behaves in a desirable qualitative manner.

5.3 Experimental Results Obtained using a 'Conventional' Controller when an 'Unexpected' Event Occurs

In order to study the effect when 'unexpected' events occur using the conventional controller (8), the following gross change in the MARTS configuration was made at $t = 1000$ seconds: with the controller (8) implemented on the MARTS system (which is designed for the case when $\theta = 30°$ and $\epsilon = 1$), and with $T = 0.4$ seconds and $\theta = 30°$, the plant's configuration was suddenly changed at $t = 1000$ seconds by artificially reversing the output leads, with the following (physical) reference input applied:

$$(y^1_{ref}(t), y^2_{ref}(t)) := \left\{ \begin{array}{ll} (3000, 2500), & 0 \leq t < 500 \\ (3500, 2000), & 500 \leq t < 1000 \\ (3000, 2500), & 1000 \leq t < 1500 \\ (3500, 2000), & 1500 \leq t < 2000 \text{ seconds} \end{array} \right\}.$$

In this case, the response of figure 7 was experimentally obtained, which shows that the controller (8) *fails* to bring about tracking/regulation for such a severe configuration change. Figure 8 shows that a similar failure occurs on using controller (8) with $T = 0.4$ seconds and $\theta = 0°$ on the MARTS system. Such a failure of the controller (8) to bring about tracking/regulation is not unexpected, since a *drastic* configuration change has occurred at $t = 1000$ seconds.

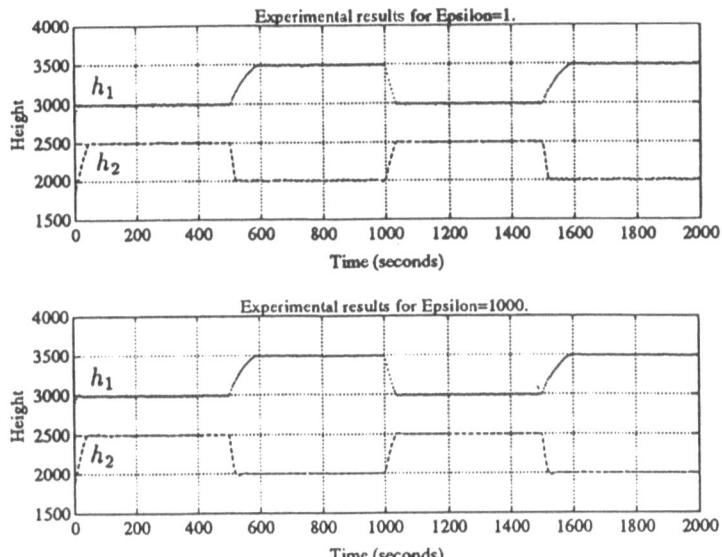

Figure 5: Experimental proportional-integral results of h_1 (dotted) and h_2 (dashed) using $T = 0.4$ seconds and $\theta = 30°$ with conventional controller (8) applied.

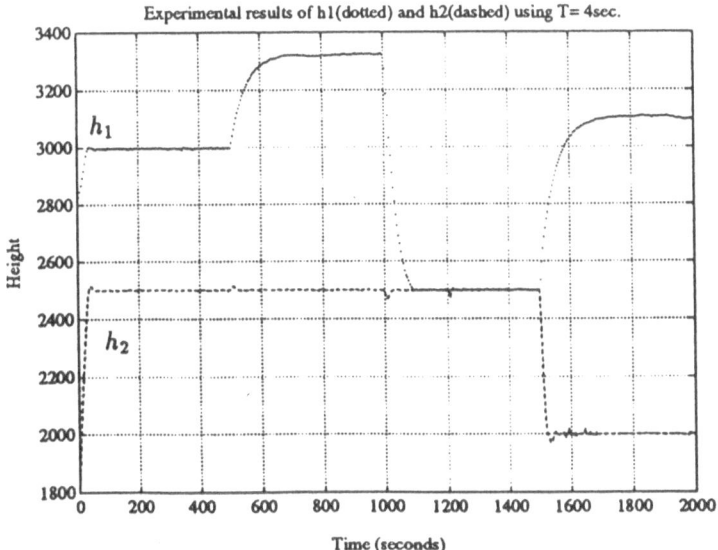

Figure 6: Experimental proportional-integral results of h_1 (dotted) and h_2 (dashed) using $T = 0.4$ seconds, $\epsilon = 1$, and $\theta = 40°$ with conventional controller (8) applied.

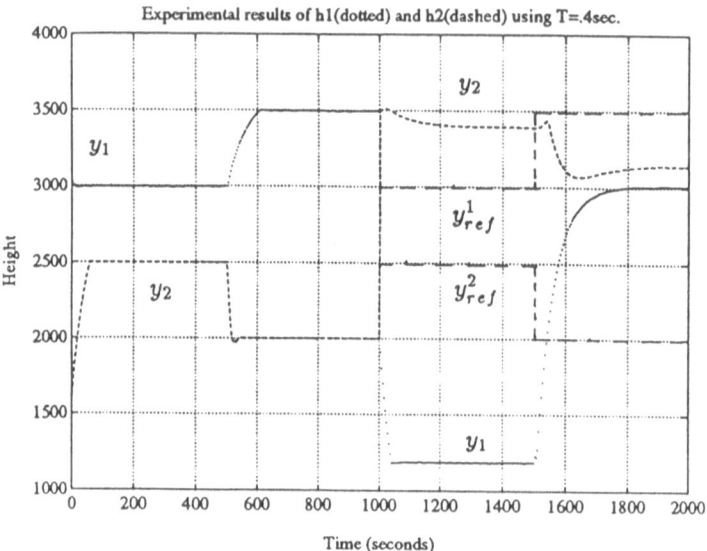

Figure 7: Experimental proportional-integral results for y_1 (dotted) and y_2 (dashed) with $T = 0.4$ seconds, $\theta = 30°$ and the outputs reversed at $t = 1000$ seconds, showing failure of conventional controller (8).

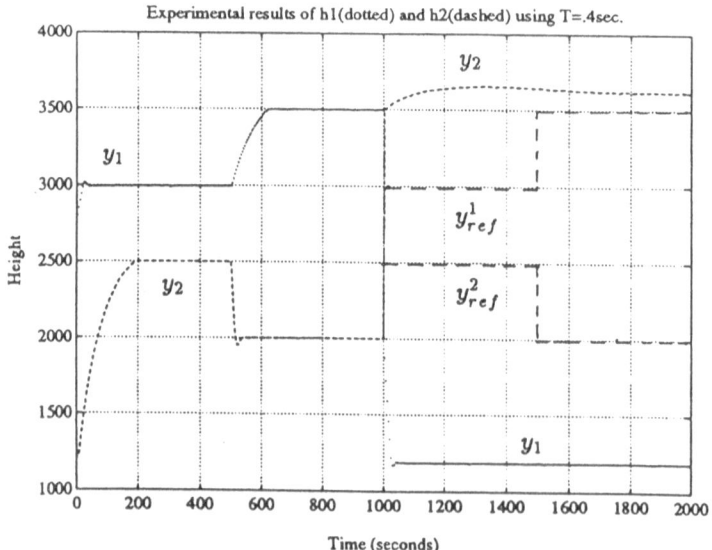

Figure 8: Experimental proportional-integral results for y_1 (dotted) and y_2 (dashed) with $T = 0.4$ seconds, $\theta = 0°$ and the outputs reversed at $t = 1000$ seconds, showing failure of conventional controller (8).

6 An Approach to 'Intelligent Control' [9]

The following approach to 'intelligent control' is proposed [9]. The approach generalizes the integral controller type of results given in [8] to include proportional type terms, and consists of applying a sequence of LTI controllers to the plant via a 'switching controller' (e.g. see [8]-[9], [11]-[13]); in this case, switching of the controller eventually ceases to occur, resulting in an appropriate LTI controller which is applied to the plant. It will experimentally be shown that such a controller displays 'intelligent-like' features when applied to the MARTS plant. The following preliminary result is initially required:

Lemma 6.1 ([22]) Given $r \in \mathbb{N}$, then there exist an integer $p \in \mathbb{N}$ and a set of matrices $W := \{W_1, W_2, \ldots, W_p\}$ such that for every $T \in \mathbb{R}^{r \times r}$ of full rank, there exists an $i \in \{1, 2, \ldots, p\}$ so that $-TW_i$ is stable.

In particular, for $r = 2$, define

$$W := \left\{ \begin{array}{ccc} \underbrace{\begin{bmatrix} 1 & 0 \\ 0 & 1 \end{bmatrix}}_{W_1}, & \underbrace{\frac{1}{2}\begin{bmatrix} -1 & -\sqrt{3} \\ \sqrt{3} & -1 \end{bmatrix}}_{W_2}, & \underbrace{\frac{1}{2}\begin{bmatrix} -1 & \sqrt{3} \\ -\sqrt{3} & -1 \end{bmatrix}}_{W_3}, \\ \underbrace{\begin{bmatrix} 1 & 0 \\ 0 & -1 \end{bmatrix}}_{W_4}, & \underbrace{\frac{1}{2}\begin{bmatrix} -1 & -\sqrt{3} \\ -\sqrt{3} & 1 \end{bmatrix}}_{W_5}, & \underbrace{\frac{1}{2}\begin{bmatrix} -1 & \sqrt{3} \\ \sqrt{3} & 1 \end{bmatrix}}_{W_6} \end{array} \right\}.$$

The following definitions are also required:

Definition 2 A function $g : \mathbb{N} \to \mathbb{R}^+$ is a *tuning function* ($g \in$ TF) if $\lim_{k \to \infty} g(k) = 0$. If $g \in$ TF and there exist constants $\epsilon_0 > 0$ and $\tau > 1$ so that $g(k) = \frac{\epsilon_0}{\tau^k}$ for $k \in \mathbb{N}$, then define $g \in$ TF' to be a *modified tuning function*.

Definition 3 A function $f : \mathbb{N} \to \mathbb{R}^+$ is a *bounding function* ($f \in$ BF) if it is strictly increasing and if, for every $c \in \mathbb{R}^+$ and $n \in \mathbb{N}$, there exists a $k > n$ such that $\dfrac{f(k+1)}{f(k)} > c$. If $f \in$ BF, and if $\lim_{k \to \infty} \dfrac{f(k+1)}{f(k)} \to \infty$, then define $f \in$ BF' to be a *modified bounding function*.

Definition 4 A function $f : \mathbb{N} \to \mathbb{R}^+$ is said to be a *strong bounding function* ($f \in$ SBF) if $f \in$ BF' and if, for all finite $\alpha > 1$, for all finite c_0, c_1, c_2, c_3 in $\mathbb{R}^+$, and for all finite $n \in \mathbb{N}$,

$$\frac{f(i)}{c_0 \alpha^{i-n} + c_1 \sum_{j=1}^{i-n} \alpha^j + c_2 \sum_{j=1}^{i-n-1} \alpha^{j+1} f(i-j-1) + c_3 f(i-1)} \to \infty$$

as $i \to \infty$.

6.1 Proposed 'Intelligent' Controller

It is desired now to synthesize a controller to solve the robust servomechanism problem (RSP) with constant disturbance/reference input signals for a plant assuming that:

(a) The plant can be described by the system model (4).
(b) The plant is open loop asymptotically stable.
(c) A solution exists to the RSP, i.e. condition (5) holds.
(d) The plant model (4) is *unknown* to the control designer for the plant. $\qquad\qquad$ (10)

Consider now the following controller:

$$\eta(t) \;=\; \int_0^t \epsilon(\tau)e(\tau)d\tau + \eta_0 \qquad\qquad (11)$$
$$u(t) \;=\; K(t)(\eta(t) + \rho\epsilon(t)e(t))$$

where

$$\epsilon(t) := g(k), \quad t \in (t_k, t_{k+1}], \quad k \in \{1,2,3,\ldots\}$$

$t_1 := 0$, and where, for each $k \ge 2$ such that $t_{k-1} \ne \infty$, t_k is defined by

$$t_k := \left\{ \begin{array}{ll} \text{minimum value of } t \text{ such that} \quad \text{if this minimum exists} \\ \quad \text{i)} \quad t > t_{k-1}, \text{ and} \\ \quad \text{ii)} \quad \|\eta(t)\|_\infty = f(k-1) \\ \infty \qquad\qquad\qquad\qquad\qquad \text{otherwise} \end{array} \right\}$$

with $\rho \ge 0$,

$$f \in \left\{ \begin{array}{ll} \text{BF}', & \rho = 0 \\ \text{SBF}, & \rho > 0 \end{array} \right\},$$

$g \in \text{TF}'$, and where, for the sake of simplicity of presentation, it is assumed that $m = r = 2$; in this case,

$$K(t) = W_i, \quad i \in \{1,2,\ldots,6\}, \quad i = ((k-1) \bmod 6) + 1, \quad t \in (t_k, t_{k+1}]$$

for $k \in \{1,2,3,\ldots\}$, where W_i, $i \in \{1,2,\ldots,6\}$, are defined in lemma 6.1.

Remark 6.1 The 'cyclic' switching action of controller (11) *which we will use* can be summarized by the following table.

k	1	2	3	4	5	6	7
t	$(t_1, t_2]$	$(t_2, t_3]$	$(t_3, t_4]$	$(t_4, t_5]$	$(t_5, t_6]$	$(t_6, t_7]$	$(t_7, t_8]$
K	W_1	W_2	W_3	W_4	W_5	W_6	W_1
f	$f(1)$	$f(1)$	$f(1)$	$f(1)$	$f(1)$	$f(1)$	$f(2)$
g	$g(1)$	$g(1)$	$g(1)$	$g(1)$	$g(1)$	$g(1)$	$g(2)$
k	8	9	10	11	12	13	$\cdots$
t	$(t_8, t_9]$	$(t_9, t_{10}]$	$(t_{10}, t_{11}]$	$(t_{11}, t_{12}]$	$(t_{12}, t_{13}]$	$(t_{13}, t_{14}]$	$\cdots$
K	W_2	W_3	W_4	W_5	W_6	W_1	$\cdots$
f	$f(2)$	$f(2)$	$f(2)$	$f(2)$	$f(2)$	$f(3)$	$\cdots$
g	$g(2)$	$g(2)$	$g(2)$	$g(2)$	$g(2)$	$g(3)$	$\cdots$

In addition, after each switch at time t_k, $\eta(t_k^+)$ will be reset to be zero. ◇

The following result is now obtained:

Theorem 6.1 ([9]) Consider a plant described by (4) subject to assumptions (10) with $\|\eta(0)\|_\infty < f(1)$, and let controller (11) be applied to the plant; then

i) there exist a finite time $t_{ss} \geq 0$, a constant $\epsilon_{ss} > 0$ and a matrix K_{ss} such that $\epsilon(t) = \epsilon_{ss}$ and $K(t) = K_{ss}$ $\forall t \geq t_{ss}$;

ii) the controller state $\eta(t)$ and plant state $x(t)$ are bounded $\forall t \geq 0$; and

iii) if y_{ref} and w are constant signals, then for almost all y_{ref} and w, $e(t) \to 0$ as $t \to \infty$.

6.2 Simulation Results Obtained

Consider the following single input/single output LTI plant taken from [21]:

$$Y(s) = \frac{2}{s+1} \cdot \frac{229}{s^2 + 30s + 229} U(s) \tag{12}$$

where the measured plant output and reference input are given respectively as

$$y_m(t) := y(t) + 0.5\sin(8t)$$
$$y_{ref}(t) := 2.$$

Upon applying controller (11) to this system with $\eta(0) := 0$, $x(0) := 0$, $(W_1, W_2) := (1, -1)$, and

$$(\rho, g(k)) := \left(7, \frac{10}{3^k}\right)$$

$$f(k) := \left\{ \begin{array}{ll} 4k, & 1 \leq k \leq 10 \\ 20(k-10)^2 \exp((k-10)^3), & k > 10 \end{array} \right\}$$

the output response shown in figure 9 is obtained (using a similar cyclic switching action as given in remark 6.1). As one can see, the controller performs *quite well* considering the relatively little *a priori* plant information required and the presence of an output sinusoidal disturbance term. The final value of $K(t)$ is 1, and the final value of $\epsilon(t)$ is $\frac{10}{27}$, making the final LTI closed loop system asymptotically stable.

6.3 Experimental Results Obtained

When controller (11) is digitally implemented on the MARTS system with a sampling period of $T = 0.4$ seconds,

$$(\rho, g(k)) := \left(10, \frac{10}{3^k}\right)$$

$$f(k) := \left\{ \begin{array}{ll} 10k, & 1 \leq k \leq 10 \\ 100(k-10)^2 \exp((k-10)^3), & k > 10 \end{array} \right\}, \tag{13}$$

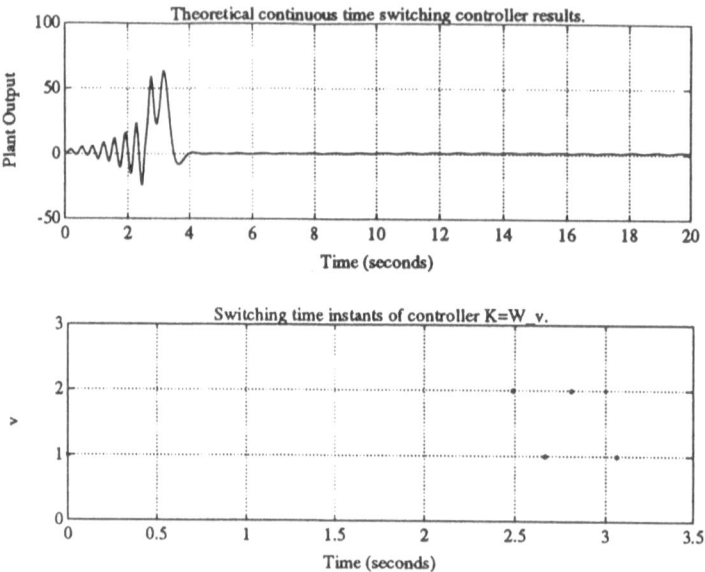

Figure 9: Simulated output results and switching time instants obtained using 'intelligent' controller (11).

the experimental results of figure 10 are obtained for the case when $\theta = 30°$ with:

$$(y^1_{ref}(t), y^2_{ref}(t)) := \left\{ \begin{array}{ll} (3500, 2000), & 0 \leq t < 1000 \\ (3000, 2500), & 1000 \leq t < 2000 \\ (3500, 2000), & 2000 \leq t < 3000 \text{ seconds} \end{array} \right\}$$

as the reference input. It is seen that the controller produces excellent control, i.e. after a 'learning' period of approximately 136 seconds, the controller has 'tuned' itself, and has effectively become a LTI controller for $t \geq 136$ seconds.

6.4 Experimental Results Obtained using Proposed Controller (11) when an 'Unexpected' Event Occurs

As was done for the case of the conventional controller (8), in order to study the effect when 'unexpected' events occur using the proposed controller (11), the following change in the MARTS configuration was made at $t = 1500$ seconds: with the controller (11) and the parameter values given in equation (13) implemented on the MARTS system, and with $T = 0.4$ seconds and $\theta = 30°$, the plant's configuration was suddenly changed at $t = 1500$ seconds by artificially reversing the output leads at $t = 1500$ seconds with the following (physical) reference input applied:

$$(y^1_{ref}(t), y^2_{ref}(t)) := (3000, 2500), \quad 0 \leq t < \infty.$$

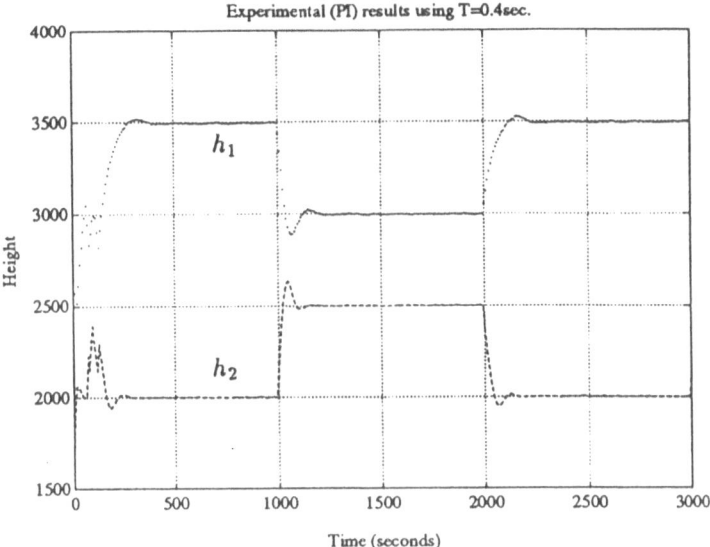

Figure 10: Experimental proportional-integral self-tuning results for h_1 (dotted) and h_2 (dashed) using proposed 'intelligent' controller (11) on the nominal plant with $\theta = 30°$ and $T = 0.4$ seconds.

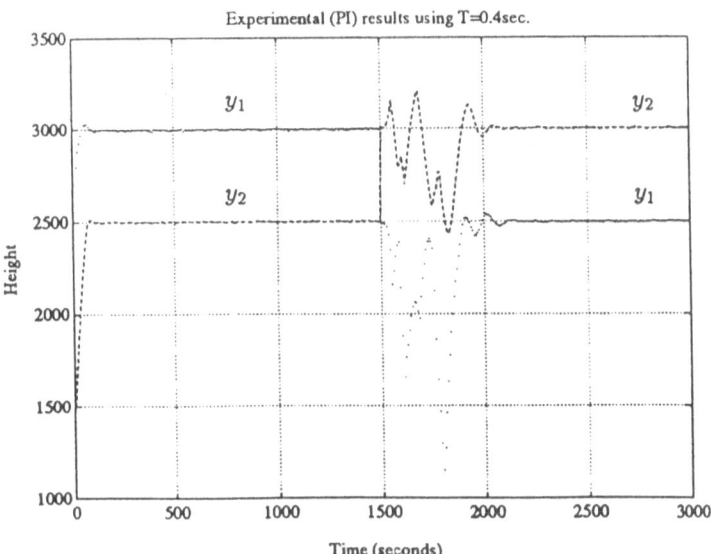

Figure 11: Experimental proportional-integral self-tuning results for y_1 (dotted) and y_2 (dashed) with $\theta = 30°$, $T = 0.4$ seconds and the outputs reversed at $t = 1500$ seconds, showing successful operation of proposed 'intelligent' controller (11).

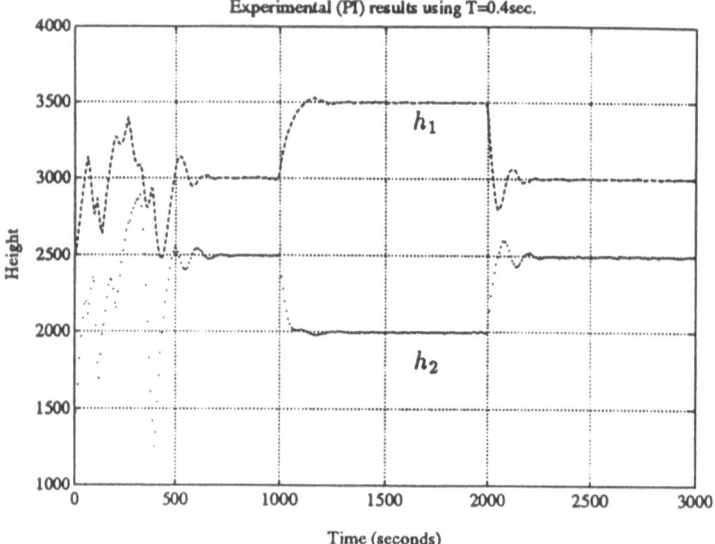

Figure 12: Experimental proportional-integral self-tuning results for h_1 (dashed) and h_2 (dotted) with $\theta = 30°$, $T = 0.4$ seconds and the outputs reversed at $t = 0$ seconds, showing successful operation of proposed 'intelligent' controller (11).

In this case, the response of figure 11 was experimentally obtained, which shows that the proposed controller (11) *is quite successful* in bringing about tracking/regulation for such a severe configuration change, i.e. at $t = 1500$ seconds, the controller goes through an additional 'learning' period for approximately 368 seconds, and at $t = 1868$ seconds, the switching controller has stopped switching, resulting in an appropriate LTI controller being applied to the system.

Figure 12 gives a response of the plant for the case when the output leads are reversed at $t = 0$ seconds, with the parameter values given in (13) implemented and the following (physical) reference input applied:

$$(y_{ref}^1(t), y_{ref}^2(t)) := \left\{ \begin{array}{ll} (3000, 2500), & 0 \leq t < 1000 \\ (3500, 2000), & 1000 \leq t < 2000 \\ (3000, 2500), & 2000 \leq t < 3000 \text{ seconds} \end{array} \right\}.$$

Again, it can be seen that the proposed controller (11) has been *quite successful* in adjusting to the severe configuration change applied to the plant.

7 Conclusions

Due to increased demands on control system specifications – in particular, on the demands that a control system should be able to carry out its specified mandate *in spite of gross changes which may occur in the plant's configuration* – conventional control design approaches need to be enhanced. A class of controllers (called 'intelligent controllers') is proposed in this paper to deal with this type of problem. These controllers have the significant feature of not requiring that a mathematical model of the plant be known, which implies that if a 'drastic change' in the plant's configuration occurs, say, then the controller can 're-adjust' itself to deal with the new modified plant.

This paper, in particular, evaluates the performance of such an 'intelligent controller' by applying the controller to a highly interacting experimental multivariable plant consisting of industrial sensors/actuators, and shows that the proposed controller can successfully 'adapt', unlike 'conventional controllers', to drastic changes in the plant's configuration. The results of these experiments therefore give encouragement as to the development of a full theory of 'intelligent control'.

References

[1] L. Boullart, A. Krijgsman, and R. A. Vingerhoeds, editors, *Application of Artificial Intelligence in Process Control*, Pergamon Press, Oxford, 1992, Lecture notes Erasmus intensive course.

[2] P. Antsaklis, "Task Force on Intelligent Control Report", *IEEE Control Systems Magazine*, vol. 14, pp. 4–5, 58–66, June 1994.

[3] K. Furata, editor, *1st Workshop on Intelligent Control*, Tokyo, May 1993. Tokyo Institute of Technology.

[4] *8'th IEEE Symposium on Intelligent Control (ISIC '93)*, Chicago, August 1993.

[5] *Proceedings of the 1994 IEEE/RSJ/GI International Conference on Intelligent Robots and Systems*, Munich, September 1994.

[6] E. J. Davison, "Multivariable Tuning Regulators: The Feedforward and Robust Control of a General Servomechanism Problem", *IEEE Transactions on Automatic Control*, vol. 21, pp. 35–47, February 1976.

[7] G. W. M. Coppus, S. L. Shah, and R. K. Wood, "Robust multivariable control of a binary distillation column", *IEE Proceedings*, vol. 130, pt. D, pp. 201–208, September 1983.

[8] D. E. Miller and E. J. Davison, "The Self-Tuning Robust Servomechanism Problem", *IEEE Transactions on Automatic Control*, vol. 34, pp. 511–523, May 1989.

[9] M. Chang and E. J. Davison, "Control of Unknown Systems using Switching Controllers: an Experimental Study", *in Proceedings of the 1994 American Control Conference*, pp. 2984–2989, 1994.

[10] D. E. Miller and E. J. Davison, "An Adaptive Controller Which Provides Lyapunov Stability", *IEEE Transactions on Automatic Control*, vol. 34, pp. 599–609, June 1989.

[11] D. E. Miller and E. J. Davison, "Adaptive Control of a Family of Plants", in D. Hinrichsen and B. Mårtensson, editors, *Control of Uncertain Systems: Proceedings of an International Workshop, Bremen, West Germany, June 1989*, pp. 197–219. Birkhäuser Press, Boston, 1990.

[12] D. E. Miller and E. J. Davison, "An Adaptive Controller Which Provides an Arbitrarily Good Transient and Steady-State Response", *IEEE Transactions on Automatic Control*, vol. 36, pp. 68–81, January 1991.

[13] D. E. Miller and E. J. Davison, "An Adaptive Tracking Problem with a Control Input Constraint", *Automatica*, vol. 29, pp. 877–887, July 1993.

[14] Minyue Fu and B. Ross Barmish, "Adaptive Stabilization of Linear Systems Via Switching Control", *IEEE Transactions on Automatic Control*, vol. 31, pp. 1097–1103, December 1986.

[15] A. S. Morse, "Supervisory Control of Families of Linear Set-Point Controllers", *in Proceedings of the 32'nd IEEE Conference on Decision and Control*, 1993.

[16] E. J. Davison, "Description of Multivariable Apparatus for Real Time Control Studies (MARTS)", Systems Control Report 8514, University of Toronto, Department of Electrical Engineering, November 1985.

[17] S. Kung, "A new identification and model reduction algorithm via singular value decompositions", *in Twelfth Asilomar Conference on Circuits, Systems & Computers*, pp. 705–714, 1978.

[18] E. J. Davison and B. Scherzinger, "Perfect Control of the Robust Servomechanism Problem", *IEEE Transactions on Automatic Control*, vol. 32, pp. 689–702, August 1987.

[19] E. J. Davison and I. Ferguson, "The design of controllers for the multivariable robust servomechanism problem using parameter optimization methods", *IEEE Transactions on Automatic Control*, vol. 26, pp. 93–110, February 1981.

[20] Michael Chang, "Adaptive Control Applied to Unmodelled Multivariable Systems", Master's thesis, University of Toronto, Department of Electrical and Computer Engineering, April 1993.

[21] C. E. Rohrs, L. Valavani, M. Athans, and G. Stein, "Robustness of Continuous-Time Adaptive Control Algorithms in the Presence of Unmodeled Dynamics", *IEEE Transactions on Automatic Control*, vol. 30, pp. 881–889, September 1985.

[22] B. Mårtensson, *Adaptive Stabilization*, PhD thesis, Department of Automatic Control, Lund Institute of Technology, Lund, Sweden, 1986.

Metric Uncertainty and Nonlinear Feedback Stabilization

Tryphon T. Georgiou
Dept. of Electrical Engineering
University of Minnesota
Minneapolis, MN 55455, U.S.A.

Malcolm C. Smith
Department of Engineering
University of Cambridge
Cambridge, CB2 1PZ, U.K.

Dedicated to George Zames on the occasion of his 60th birthday

Abstract. In this paper we define a generalization of the gap metric for nonlinear systems. We prove two results which specialize to known results for linear systems. First, we show that a feedback system remains stable if the distance between the plant and its perturbation is less than the inverse of the norm of a certain parallel projection operator. Second, we show that every "continuously robust" metric on nonlinear systems is equivalent to the metric introduced in this paper.

For anyone working on the foundations of control theory it is a privilege to pay tribute to George Zames, who has contributed such fertile and fundamental ideas to the field. Over the past 35 years his work has had the most profound influence on the development of control science as well as being an inspiration to countless researchers. To honour his work we seek, in this paper, to synthesize two of his major ideas: the small gain theorem[1] in nonlinear feedback and gap metric uncertainty[2]. To date, these two important tools have experienced largely separate evolutions.

The setting for an input-output/operator view of feedback control is provided by the block diagram of Fig. 1. The plant $\mathbf{P}$ and compensator $\mathbf{C}$ are operators, possibly nonlinear, unbounded, with $\mathbf{P} : \mathcal{D}_\mathbf{P} \to \mathcal{Y}$ (where $\mathcal{D}_\mathbf{P} \subseteq \mathcal{U}$ is the domain of $\mathbf{P}$), and $\mathbf{C} : \mathcal{D}_\mathbf{C} \to \mathcal{U}$ (where $\mathcal{D}_\mathbf{C} \subseteq \mathcal{Y}$ is the domain of $\mathbf{C}$). The signals u_i ($i \in \{0, 1, 2\}$) belong to a Banach space $\mathcal{U}$ and the signals y_i ($i \in \{0, 1, 2\}$) belong to a Banach space $\mathcal{Y}$. Typical choices for $\mathcal{U}$ and $\mathcal{Y}$ are the $\mathcal{L}_p$ ($p \in \{1, 2, \infty\}$) spaces of vector-valued functions on the continuous or discrete semi-infinite time axis.

The problem of feedback stabilization is to ensure that, for any $(u_0, y_0) \in \mathcal{U} \times \mathcal{Y}$, there exist unique signals $u_1, u_2 \in \mathcal{U}$ and $y_1, y_2 \in \mathcal{Y}$ such that the following feedback equations hold:

$$u_0 = u_1 + u_2,$$
$$y_0 = y_1 + y_2,$$
$$y_1 = \mathbf{P}u_1,$$
$$u_2 = \mathbf{C}y_2.$$

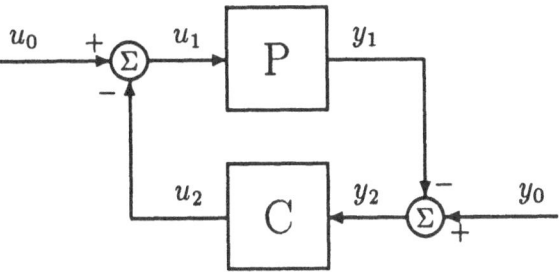

Figure 1: Standard feedback configuration.

An additional requirement is that the operator

$$\mathbf{H}_{P,C} : \mathcal{W} \to \mathcal{W} \times \mathcal{W},$$

where $\mathcal{W} := \mathcal{U} \times \mathcal{Y}$, defined via the feedback equations by

$$\mathbf{H}_{P,C} : \begin{pmatrix} u_0 \\ y_0 \end{pmatrix} \to \left(\begin{pmatrix} u_1 \\ y_1 \end{pmatrix}, \begin{pmatrix} u_2 \\ y_2 \end{pmatrix} \right),$$

is bounded. In this paper we will use the incremental gain $\|\cdot\|_\Delta$ for operators $\mathbf{A} : \mathcal{X} \to \mathcal{Z}$,

$$\|\mathbf{A}\|_\Delta := \sup_{x_1, x_2 \in \mathcal{X},\ x_1 \neq x_2} \frac{\|\mathbf{A}x_1 - \mathbf{A}x_2\|}{\|x_1 - x_2\|},$$

where $\|\cdot\|$ denotes the norm in the Banach spaces $\mathcal{X}, \mathcal{Z}$. Thus we say that the feedback configuration of Fig. 1, denoted by $[\mathbf{P}, \mathbf{C}]$, is *stable* if $\mathbf{H}_{P,C}$ is well-defined and $\|\mathbf{H}_{P,C}\|_\Delta < \infty$. Henceforth, without loss of generality, we will assume that the zero elements in $\mathcal{U}$ and $\mathcal{Y}$ are included in the domains of the plant and compensator respectively, and that $\mathbf{P}0 = 0$, $\mathbf{C}0 = 0$.[3] Note that, under these conditions, uniqueness of solution of the feedback equations implies $\mathbf{H}_{P,C}\,0 = 0$.

Consider the *graph* of $\mathbf{P}$

$$\mathcal{G}_P := \left\{ \begin{pmatrix} u \\ \mathbf{P}u \end{pmatrix} : u \in \mathcal{U},\ \mathbf{P}u \in \mathcal{Y} \right\} \subset \mathcal{W}$$

and the *inverse graph* of $\mathbf{C}$

$$\mathcal{G}_C' := \left\{ \begin{pmatrix} \mathbf{C}y \\ y \end{pmatrix} : y \in \mathcal{Y},\ \mathbf{C}y \in \mathcal{U} \right\} \subset \mathcal{W}.$$

Let $\mathcal{M} := \mathcal{G}_\mathbf{P}$, $\mathcal{N} := \mathcal{G}'_\mathbf{C}$, and define the operator

$$\Sigma_{\mathcal{M},\mathcal{N}} : \mathcal{M} \times \mathcal{N} \to \mathcal{W} : (m, n) \to m + n.$$

The stability of the feedback system in Fig. 1 is equivalent to $\Sigma_{\mathcal{M},\mathcal{N}}$ having an inverse defined on the whole of $\mathcal{W}$ which is bounded. In fact, if $\Sigma_{\mathcal{M},\mathcal{N}}$ has a bounded inverse, the inverse is precisely $\mathbf{H}_{\mathbf{P},\mathbf{C}}$.[4] It can be shown that a necessary condition for $[\mathbf{P}, \mathbf{C}]$ to be stable is that $\mathcal{M}$, $\mathcal{N}$ are closed subsets of $\mathcal{W}$.[5]

Let $\mathcal{X}$, $\mathcal{Y}$ be closed subsets of a Banach space $\mathcal{W}$. We define[6]

$$\vec{\delta}(\mathcal{X}, \mathcal{Y}) \quad := \quad \begin{cases} \inf \{\|\mathbf{\Phi} - \mathbf{I}\|_\Delta : \mathbf{\Phi} \text{ is invertible on } \mathcal{W}, \\ \qquad\qquad \text{with } \mathbf{\Phi}0 = 0, \text{ and } \mathbf{\Phi}\mathcal{X} = \mathcal{Y}\}, \\ \\ \infty \text{ if no such operator } \mathbf{\Phi} \text{ exists}, \end{cases}$$

$$\delta(\mathcal{X}, \mathcal{Y}) \quad := \quad \max \left\{ \vec{\delta}(\mathcal{X}, \mathcal{Y}), \vec{\delta}(\mathcal{Y}, \mathcal{X}) \right\},$$

and

$$d(\mathcal{X}, \mathcal{Y}) \quad := \quad \log(1 + \delta(\mathcal{X}, \mathcal{Y})).$$

Then $d(\cdot, \cdot)$ is a metric on the closed subsets of $\mathcal{W}$.[7] In general $\delta(\cdot, \cdot)$ does not satisfy the triangular inequality. Yet both $d(\cdot, \cdot)$ and $\delta(\cdot, \cdot)$ define the same topology. Moreover, $\delta(\cdot, \cdot)$ is a natural generalization of the gap (see Note 7). Hence, below, we will express our conclusions in terms of $\delta(\cdot, \cdot)$.

We now present a basic lemma which will be used in the proof of our two main results to follow. The lemma can also be used to prove the small gain theorem (see Note 1).

Lemma. Let $\mathbf{A} : \mathcal{X} \to \mathcal{X}$ where $\mathcal{X}$ is a Banach space and suppose $\|\mathbf{A}\|_\Delta \leq \zeta < 1$. Then

$$(\mathbf{I} + \mathbf{A})x = y \tag{1}$$

has a unique solution for all $y \in \mathcal{X}$. Furthermore,

$$\|(\mathbf{I} + \mathbf{A})^{-1} - \mathbf{I}\|_\Delta \leq \frac{\zeta}{1 - \zeta}.$$

Proof. The equation (1) can be rearranged as

$$x = y - \mathbf{A}x =: \mathbf{B}x.$$

Since

$$\frac{\|\mathbf{B}x_1 - \mathbf{B}x_2\|}{\|x_1 - x_2\|} = \frac{\|\mathbf{A}x_1 - \mathbf{A}x_2\|}{\|x_1 - x_2\|} \leq \zeta,$$

the contraction mapping theorem[8] applies to show that $\mathbf{B}$ has a unique fixed point. Since y is arbitrary this proves the first part of the lemma. Next note that

$$\|(\mathbf{I} + \mathbf{A})x_1 - (\mathbf{I} + \mathbf{A})x_2\| \geq \|x_1 - x_2\| - \|\mathbf{A}x_1 - \mathbf{A}x_2\|$$
$$\geq (\frac{1}{\zeta} - 1)\|\mathbf{A}x_1 - \mathbf{A}x_2\|$$
$$= (\frac{1}{\zeta} - 1)\|(\mathbf{I} - (\mathbf{I} + \mathbf{A}))x_1 - (\mathbf{I} - (\mathbf{I} + \mathbf{A}))x_2\|$$

which implies that

$$\frac{\|((\mathbf{I} + \mathbf{A})^{-1} - \mathbf{I})y_1 - ((\mathbf{I} + \mathbf{A})^{-1} - \mathbf{I})y_2\|}{\|y_1 - y_2\|} \leq \frac{\zeta}{1 - \zeta}. \qquad \square$$

Let $\Pi_i : \mathcal{W} \times \mathcal{W} \to \mathcal{W}$ denote the natural projection onto the ith component $(i = 1, 2)$ of $\mathcal{W} \times \mathcal{W}$. If $[\mathbf{P}, \mathbf{C}]$ is stable define $\Pi_{\mathcal{M}//\mathcal{N}} := \Pi_1 \mathbf{H}_{\mathbf{P},\mathbf{C}}$. Then

$$\Pi_{\mathcal{M}//\mathcal{N}} : \mathcal{W} \to \mathcal{M}$$

is a nonlinear idempotent (parallel projection) operator onto the graph of $\mathbf{P}$.[9] The theorem below gives a basic robustness result for uncertainties measured by $\delta(\cdot, \cdot)$ which generalizes a standard result from the linear case.[10]

Theorem 1. Let $[\mathbf{P}, \mathbf{C}]$ be stable. If

$$\delta(\mathcal{G}_{\mathbf{P}}, \mathcal{G}_{\mathbf{P}_1}) < \|\Pi_{\mathcal{M}//\mathcal{N}}\|_\Delta^{-1},$$

then $[\mathbf{P}_1, \mathbf{C}]$ is stable.

Proof. Since $\delta(\mathcal{G}_{\mathbf{P}}, \mathcal{G}_{\mathbf{P}_1}) < \infty$ there exists a bounded invertible transformation $\Phi : \mathcal{W} \to \mathcal{W}$ such that

$$\|\Phi - \mathbf{I}\|_\Delta < \|\Pi_{\mathcal{M}//\mathcal{N}}\|_\Delta^{-1}, \qquad (2)$$

and $\Phi\mathcal{M} = \mathcal{M}_1$ (where $\mathcal{M}_1 = \mathcal{G}_{\mathbf{P}_1}$). Define

$$\hat{\Phi} : \mathcal{M} \times \mathcal{N} \to \mathcal{M}_1 \times \mathcal{N}$$

by $\hat{\Phi}(m, n) = (\Phi m, n)$ and note that $\hat{\Phi}$ is invertible. Consider

$$\Sigma_{\mathcal{M}_1, \mathcal{N}} \hat{\Phi} \Sigma_{\mathcal{M}, \mathcal{N}}^{-1} : \mathcal{W} \to \mathcal{W}.$$

It follows that

$$\Sigma_{\mathcal{M}_1, \mathcal{N}} \hat{\Phi} \Sigma_{\mathcal{M}, \mathcal{N}}^{-1} = (\Sigma_{\mathcal{M}, \mathcal{N}} + \Sigma_{\mathcal{M}_1, \mathcal{N}} \hat{\Phi} - \Sigma_{\mathcal{M}, \mathcal{N}}) \Sigma_{\mathcal{M}, \mathcal{N}}^{-1}$$
$$= \mathbf{I} + (\Sigma_{\mathcal{M}_1, \mathcal{N}} \hat{\Phi} - \Sigma_{\mathcal{M}, \mathcal{N}}) \Sigma_{\mathcal{M}, \mathcal{N}}^{-1}$$
$$= \mathbf{I} + (\Phi - \mathbf{I}) \Pi_1 \Sigma_{\mathcal{M}, \mathcal{N}}^{-1}$$
$$= \mathbf{I} + (\Phi - \mathbf{I}) \Pi_{\mathcal{M}//\mathcal{N}}. \qquad (3)$$

The third step above follows from

$$(\Sigma_{\mathcal{M}_1,\mathcal{N}}\hat{\Phi} - \Sigma_{\mathcal{M},\mathcal{N}})(m,n) = \Sigma_{\mathcal{M}_1,\mathcal{N}}(\Phi m, n) - \Sigma_{\mathcal{M},\mathcal{N}}(m,n)$$
$$= \Phi m - m$$
$$= (\Phi - I)m.$$

From (2) we conclude that

$$\|(\Phi - I)\Pi_{\mathcal{M}//\mathcal{N}}\|_\Delta < 1.$$

Hence, from (3) and the lemma, we conclude that $I + (\Phi - I)\Pi_{\mathcal{M}//\mathcal{N}}$ is invertible. Therefore, $\Sigma_{\mathcal{M}_1,\mathcal{N}}\hat{\Phi}\Sigma_{\mathcal{M},\mathcal{N}}^{-1}$ is invertible, and so is $\Sigma_{\mathcal{M}_1,\mathcal{N}}$. Consequently, $[\mathbf{P}_1, \mathbf{C}]$ is stable. □

The statement of Theorem 1 can be strengthened by replacing $\delta(\cdot, \cdot)$ by $\vec{\delta}(\cdot, \cdot)$, with the same proof being valid. A similar result to Theorem 1 can be proved for uncertainty in $\mathbf{C}$, in which case the upper bound is replaced by $\|\Pi_{\mathcal{N}//\mathcal{M}}\|_\Delta^{-1}$.[11] The result can be further extended to give guarantees on combined plant-controller uncertainty.[12]

Our second main result is a generalization to the nonlinear case of [21, Theorem 1] which asserts that open-loop uncertainties which correspond to small closed-loop errors are precisely those that are small in the gap metric. The result shows, in the language of [21], that every "continuously robust" metric on nonlinear systems is equivalent to $\delta(\cdot, \cdot)$.[13]

Theorem 2. Let $[\mathbf{P}, \mathbf{C}]$ be stable and consider a sequence of plants $\mathbf{P}_i$ for $i = 1, 2, \ldots$. Then the following statements are equivalent:

(a) $[\mathbf{P}_i, \mathbf{C}]$ is stable for sufficiently large i and $\|H_{\mathbf{P},\mathbf{C}} - H_{\mathbf{P}_i,\mathbf{C}}\|_\Delta \to 0$,

(b) $\delta(\mathcal{G}_{\mathbf{P}}, \mathcal{G}_{\mathbf{P}_i}) \to 0$.

Proof. $(a) \Rightarrow (b)$. Let $\mathcal{M}_i := \mathcal{G}_{\mathbf{P}_i}$ and define

$$\Phi_i := I_{\mathcal{W}} - \Pi_1(\Sigma_{\mathcal{M},\mathcal{N}}^{-1} - \Sigma_{\mathcal{M}_i,\mathcal{N}}^{-1}).$$

Since $\|H_{\mathbf{P},\mathbf{C}} - H_{\mathbf{P}_i,\mathbf{C}}\|_\Delta \to 0$, it follows that $\|\Pi_1(\Sigma_{\mathcal{M},\mathcal{N}}^{-1} - \Sigma_{\mathcal{M}_i,\mathcal{N}}^{-1})\|_\Delta \to 0$. Using the lemma we conclude that there exists an integer i_0 so that Φ_i is invertible for all $i \geq i_0$. Since the graph of any system we consider contains the origin, it follows that $\Phi_i 0 = 0$ for all i. In order to complete the proof we show that, for $i \geq i_0$, $\Phi_i \mathcal{M} = \mathcal{M}_i$. Let $m \in \mathcal{M}$. Then

$$\Phi_i m = m - \Pi_1\Sigma_{\mathcal{M},\mathcal{N}}^{-1}m + \Pi_1\Sigma_{\mathcal{M}_i,\mathcal{N}}^{-1}m$$
$$= \Pi_1\Sigma_{\mathcal{M}_i,\mathcal{N}}^{-1}m \in \mathcal{M}_i.$$

Let now $m_i \in \mathcal{M}_i$ and $x = \Phi_i^{-1}m_i$. Writing

$$m_i = x - \Pi_1\Sigma_{\mathcal{M},\mathcal{N}}^{-1}x + \Pi_1\Sigma_{\mathcal{M}_i,\mathcal{N}}^{-1}x$$

where $(x - \Pi_1 \Sigma_{\mathcal{M},\mathcal{N}}^{-1} x) \in \mathcal{N}$ and $\Pi_1 \Sigma_{\mathcal{M}_i,\mathcal{N}}^{-1} x \in \mathcal{M}_i$ we see that $\Pi_1 \Sigma_{\mathcal{M}_i,\mathcal{N}}^{-1} x = m_i$ and $x - \Pi_1 \Sigma_{\mathcal{M},\mathcal{N}}^{-1} x = 0$. Therefore $x \in \mathcal{M}$, which means that $\Phi_i \mathcal{M} = \mathcal{M}_i$. At the same time $\|\Phi_i - \mathbf{I}\|_\Delta \to 0$. Similarly, if

$$\Psi_i := \mathbf{I}_W + \Pi_1(\Sigma_{\mathcal{M},\mathcal{N}}^{-1} - \Sigma_{\mathcal{M}_i,\mathcal{N}}^{-1}),$$

it follows that there exists an integer i_0 so that Ψ_i is invertible for all $i \geq i_0$, $\Psi_i \mathcal{M}_i = \mathcal{M}$, and $\|\Psi_i - \mathbf{I}\|_\Delta \to 0$. (An alternative approach would be to use $\Psi_i = \Phi_i^{-1}$ instead. It can be shown using the lemma that

$$\|\Psi_i - \mathbf{I}\|_\Delta \leq \frac{\|\Phi_i - \mathbf{I}\|_\Delta}{1 - \|\Phi_i - \mathbf{I}\|_\Delta}.$$

Hence, $\Psi_i \mathcal{M}_i = \mathcal{M}$ and $\|\Psi_i - \mathbf{I}\|_\Delta \to 0$ as well.) Therefore, $\delta(\mathcal{G}_P, \mathcal{G}_{P_i}) \to 0$.

$(b) \Rightarrow (a)$. Suppose $\delta(\mathcal{G}_P, \mathcal{G}_{P_i}) \to 0$. Then there exists Φ_i invertible so that $\Phi_i \mathcal{M} = \mathcal{M}_i$ and $\|\Phi_i - \mathbf{I}\|_\Delta \to 0$. Using the notation of the proof of Theorem 1 we see, from (3) and the lemma, that $\Sigma_{\mathcal{M}_i,\mathcal{N}} \hat{\Phi}_i \Sigma_{\mathcal{M},\mathcal{N}}^{-1}$ is invertible for sufficiently large i and

$$\|\Sigma_{\mathcal{M}_i,\mathcal{N}} \hat{\Phi}_i \Sigma_{\mathcal{M},\mathcal{N}}^{-1} - \mathbf{I}\|_\Delta \to 0$$

as $i \to \infty$. Hence $\Sigma_{\mathcal{M}_i,\mathcal{N}}$ is invertible for sufficiently large i and $\|\Sigma_{\mathcal{M}_i,\mathcal{N}}^{-1}\|_\Delta$ is uniformly bounded. Now observe that

$$\hat{\Phi}_i \Sigma_{\mathcal{M},\mathcal{N}}^{-1} - \Sigma_{\mathcal{M}_i,\mathcal{N}}^{-1} = \Sigma_{\mathcal{M}_i,\mathcal{N}}^{-1} \Sigma_{\mathcal{M}_i,\mathcal{N}} \left(\hat{\Phi}_i \Sigma_{\mathcal{M},\mathcal{N}}^{-1} - \Sigma_{\mathcal{M}_i,\mathcal{N}}^{-1} \right)$$

$$= \Sigma_{\mathcal{M}_i,\mathcal{N}}^{-1} \left(\Sigma_{\mathcal{M}_i,\mathcal{N}} \hat{\Phi}_i \Sigma_{\mathcal{M},\mathcal{N}}^{-1} - \mathbf{I} \right)$$

since $\Sigma_{\mathcal{M}_i,\mathcal{N}}$ is linear. Thus

$$\|\hat{\Phi}_i \Sigma_{\mathcal{M},\mathcal{N}}^{-1} - \Sigma_{\mathcal{M}_i,\mathcal{N}}^{-1}\|_\Delta \to 0$$

as $i \to \infty$. From the identity

$$\Sigma_{\mathcal{M},\mathcal{N}}^{-1} - \Sigma_{\mathcal{M}_i,\mathcal{N}}^{-1} = \left(\mathbf{I} - \hat{\Phi}_i\right) \Sigma_{\mathcal{M},\mathcal{N}}^{-1} + \hat{\Phi}_i \Sigma_{\mathcal{M},\mathcal{N}}^{-1} - \Sigma_{\mathcal{M}_i,\mathcal{N}}^{-1}$$

we see that $\|\Sigma_{\mathcal{M},\mathcal{N}}^{-1} - \Sigma_{\mathcal{M}_i,\mathcal{N}}^{-1}\|_\Delta \to 0$. $\qquad \square$

Notes

1. The small gain theorem was introduced as an analysis tool for studying the stability of interconnected dynamical systems (cf. [19], [12], [2]). The

theorem applies to the feedback configuration of Fig. 1 under the assumption that $\mathbf{P}$ and $\mathbf{C}$ are stable. The precise statement is: if $\|\mathbf{P}\|_\Delta \cdot \|\mathbf{C}\|_\Delta < 1$ then the feedback system of Fig. 1. is stable. This result, together with bounds for $\|\mathbf{H}_{\mathbf{P},\mathbf{C}}\|_\Delta$ (cf. [19], [2]) can be derived from the basic lemma stated in the text.

2. The gap metric was introduced into the control literature by G. Zames and A.K. El-Sakkary [21] as an analysis tool for studying perturbations of dynamical systems, and for assessing the robustness of stability of feedback interconnections. The metric defines a distance between (possibly unstable) systems by measuring the aperture between their respective graphs. The gap metric was originally used in functional analysis by B. Sz.-Nagy and by M.G. Krein and M.A. Krasnosel'skii for studying perturbations of Hilbert space operators (see [9] and the references therein).

3. If $\mathbf{P}$ and $\mathbf{C}$ do not satisfy these assumptions, biases can be introduced as suggested in [19]. Specifically, take any $b_1 \in \mathcal{D}_\mathbf{P}$ and define the operator $\tilde{\mathbf{P}}$ with domain $\mathcal{D}_{\tilde{\mathbf{P}}} = \mathcal{D}_\mathbf{P} - b_1$ by: $\tilde{\mathbf{P}}u = \mathbf{P}(u + b_1) - \mathbf{P}b_1$. We can define an operator $\tilde{\mathbf{C}}$ similarly. It can now be seen that the incremental gains of $\mathbf{H}_{\mathbf{P},\mathbf{C}}$ and $\mathbf{H}_{\tilde{\mathbf{P}},\tilde{\mathbf{C}}}$ are the same.

4. If $[\mathbf{P}, \mathbf{C}]$ is stable then $\mathbf{H}_{\mathbf{P},\mathbf{C}}\mathcal{W} \subseteq \mathcal{M} \times \mathcal{N}$. Furthermore, given any pair

$$\left(\begin{pmatrix} u_1 \\ y_1 \end{pmatrix}, \begin{pmatrix} u_2 \\ y_2 \end{pmatrix} \right) \in \mathcal{M} \times \mathcal{N},$$

it is easy to see that for $u_0 = u_1 + u_2$ and $y_0 = y_1 + y_2$ the feedback equations are satisfied. Hence

$$\left(\begin{pmatrix} u_1 \\ y_1 \end{pmatrix}, \begin{pmatrix} u_2 \\ y_2 \end{pmatrix} \right) = \mathbf{H}_{\mathbf{P},\mathbf{C}} \begin{pmatrix} u_1 + u_2 \\ y_1 + y_2 \end{pmatrix}.$$

Therefore, $\mathbf{H}_{\mathbf{P},\mathbf{C}}$ sets up a bijective correspondence between $\mathcal{W}$ and $\mathcal{M} \times \mathcal{N}$ such that

$$\Sigma_{\mathcal{M},\mathcal{N}} \circ \mathbf{H}_{\mathbf{P},\mathbf{C}} = \mathbf{I}_\mathcal{W},$$
$$\mathbf{H}_{\mathbf{P},\mathbf{C}} \circ \Sigma_{\mathcal{M},\mathcal{N}} = \mathbf{I}_{\mathcal{M} \times \mathcal{N}}.$$

The two graphs $\mathcal{M}, \mathcal{N}$ are said to *coordinatize* $\mathcal{W}$ (see [4], [6]). This property provides a characterization of stability of a feedback system. Other contributions on the significance of graph representations and on the complementarity of graphs as a condition for nonlinear feedback stability include [20], [11], [8], [16], [14].

5. The proof given in [5] applies directly to the nonlinear case. Specifically, take a Cauchy sequence in $\mathcal{M}$ and set $\begin{pmatrix} u_0 \\ y_0 \end{pmatrix}$ equal to an element of this sequence. Observe that $u_1 = u_0$ and $y_1 = y_0$ is the solution of the feedback equations. Now take limits and note that $\mathcal{M}$ must contain the limit point since $\mathcal{W}$ is complete and $\mathbf{H}_{\mathbf{P},\mathbf{C}}$ is continuous.

6. The definition of $\bar{\delta}(\cdot,\cdot)$ can be relaxed slightly without affecting the metric property of $d(\cdot,\cdot)$, or the subsequent theorems, in that Φ need not be defined on the whole of $\mathcal{W}$. It is sufficient to define: $\bar{\delta}(\mathcal{X},\mathcal{Y}) :=$ inf $\{\||(\Phi - \mathbf{I})|_{\mathcal{X}}\|_\Delta : \Phi$ is an invertible map from $\mathcal{X}$ onto $\mathcal{Y}$ with $\Phi 0 = 0\}$, and ∞ if no such operator Φ exists.

7. This metric is a slight modification of a metric introduced in [10, page 562] for subspaces of a Banach space. The definition of the metric in [10] differs from the one in this paper in that ∞ is replaced by 1. The value ∞ is necessary in our case since we seek to measure distances between closed subsets of a Banach space which may not necessarily be subspaces. When the underlying space is a Hilbert space, $\delta(\cdot,\cdot)$ coincides with the usual gap metric [1, Corollary 3.3]. We now show that $d(\cdot,\cdot)$ is a metric. Clearly, $d(\mathcal{X},\mathcal{Y}) \geq 0$. If $d(\mathcal{X},\mathcal{Y}) = 0$ then $\delta(\mathcal{X},\mathcal{Y}) = 0$. It follows that, given any $\epsilon > 0$, there exists an invertible mapping Φ_ϵ such that, $\Phi_\epsilon 0 = 0$, $\Phi_\epsilon \mathcal{X} = \mathcal{Y}$, and $\|\Phi_\epsilon - \mathbf{I}\|_\Delta < \epsilon < 1$. For any given $y \in \mathcal{Y}$ there exists a family $x_\epsilon \in \mathcal{X}$ such that $\Phi_\epsilon x_\epsilon = y$. Moreover,

$$\begin{aligned}\|y - x_\epsilon\| &= \|\Phi_\epsilon x_\epsilon - x_\epsilon\| = \|(\Phi_\epsilon - \mathbf{I})x_\epsilon\| \\ &< \epsilon\|x_\epsilon\| < \frac{\epsilon}{1 - \epsilon}\|y\|,\end{aligned}$$

where the first inequality follows from the fact that the incremental gain bounds the norm. Hence we can find a sequence $x_\epsilon \in \mathcal{X}$ such that $x_\epsilon \to y$. Since $\mathcal{X}$ is closed then $y \in \mathcal{X}$ which means that $\mathcal{X} \subseteq \mathcal{Y}$. The reverse inclusion is proved similarly, so $\mathcal{X} = \mathcal{Y}$. Next we note that $d(\mathcal{X},\mathcal{Y}) = d(\mathcal{Y},\mathcal{X})$. Finally we show that the triangular inequality holds:

$$d(\mathcal{X},\mathcal{Z}) \leq d(\mathcal{X},\mathcal{Y}) + d(\mathcal{Y},\mathcal{Z})$$

where $\mathcal{X}$, $\mathcal{Y}$, $\mathcal{Z}$ are all closed subsets of $\mathcal{W}$. If either of $d(\mathcal{X},\mathcal{Y})$, $d(\mathcal{Y},\mathcal{Z})$ is ∞, then the statement is obvious. So assume that both $d(\mathcal{X},\mathcal{Y})$, $d(\mathcal{Y},\mathcal{Z})$ are less than ∞ and, consequently, that there exist invertible mappings Φ_i $(i = 1, 2)$ for which $\Phi_1 \mathcal{X} = \mathcal{Y}$ and $\Phi_2 \mathcal{Y} = \mathcal{Z}$. Then $\Phi = \Phi_2\Phi_1$ is invertible and maps $\mathcal{X}$ onto $\mathcal{Z}$. From the identity

$$\Phi_2\Phi_1 - \mathbf{I} = (\Phi_2 - \mathbf{I})\Phi_1 + \Phi_1 - \mathbf{I}$$

we obtain

$$\begin{aligned}\|\Phi_2\Phi_1 - \mathbf{I}\|_\Delta &\leq \|\Phi_2 - \mathbf{I}\|_\Delta\|\Phi_1\|_\Delta + \|\Phi_1 - \mathbf{I}\|_\Delta \\ &\leq \|\Phi_2 - \mathbf{I}\|_\Delta(\|\Phi_1 - \mathbf{I}\|_\Delta + 1) + \|\Phi_1 - \mathbf{I}\|_\Delta.\end{aligned}$$

This shows that

$$\log(1 + \|\Phi_2\Phi_1 - \mathbf{I}\|_\Delta) \leq \log(1 + \|\Phi_1 - \mathbf{I}\|_\Delta) + \log(1 + \|\Phi_2 - \mathbf{I}\|_\Delta)$$

and completes the proof.

We remark that if the requirement $\Phi 0 = 0$ is removed in the definition of $\vec{\delta}(\cdot, \cdot)$ then d becomes a *pseudo-metric*, because in this case $d(\mathcal{X}, \mathcal{X} + w) = 0$ for any $w \in \mathcal{W}$. That is, the δ-distance between graphs of systems differing only by a constant bias term, is zero. However, Theorem 1 is still valid in this case.

8. The Contraction Mapping Theorem of S. Banach has the following statement. Let $\mathcal{X}$ be a complete metric space with metric $d(\cdot, \cdot)$ and let $f : \mathcal{X} \to \mathcal{X}$ be a contraction mapping, i.e., $d(f(x_1), f(x_2)) \leq \zeta d(x_1, x_2)$ for some constant $\zeta < 1$ and all $x_1, x_2 \in \mathcal{X}$. Then there is a unique fixed point, i.e., there exists a unique $x \in \mathcal{X}$ such that $f(x) = x$ (see [15, p. 130]).

9. In [4] the notion of a (nonlinear) parallel projection operator was defined and studied. It was shown that there is an intimate connection between the existence of such operators and the stability of feedback systems.

10. For $\mathbf{P}$, $\mathbf{P}_1$ and $\mathbf{C}$ linear, Theorem 1 is the sufficiency part of [7, Theorem 5] and [5, Theorem 3] which asserts that the inverse of the norm of the parallel projection is the stability margin for gap metric uncertainty. Thus, the bound of Theorem 1 is tight for linear systems. The proofs of necessity given in [7], [5] do not generalize in an obvious way to the nonlinear case, and it is likely that special assumptions on $\mathbf{P}$ and $\mathbf{C}$ would be needed to show tightness, perhaps along the lines of [13].

11. The two parallel projections are related by the fact that

$$
\begin{aligned}
\Pi_{\mathcal{M}//\mathcal{N}} + \Pi_{\mathcal{N}//\mathcal{M}} &= \Pi_1 \circ H_{P,C} + \Pi_2 \circ H_{P,C} \\
&= \Sigma_{\mathcal{M},\mathcal{N}} \circ H_{P,C} \\
&= I_{\mathcal{W}}
\end{aligned}
$$

so their norms differ by at most 1. In fact, $\|\Pi_{\mathcal{M}//\mathcal{N}}\| = \|\Pi_{\mathcal{N}//\mathcal{M}}\|$ under very general circumstances, e.g. if $\mathcal{W}$ is a Hilbert space, $\|\cdot\|$ denotes the usual induced norm and one of $\mathcal{M}$ or $\mathcal{N}$ is a linear subspace [4].

12. To do this, the following bound can be calculated from (3) and the lemma

$$
\left\| \left(\Sigma_{\mathcal{M}_1, \mathcal{N}} \hat{\Phi} \Sigma_{\mathcal{M}, \mathcal{N}}^{-1} \right)^{-1} \right\|_\Delta < \frac{1}{1 - \delta(\mathcal{G}_P, \mathcal{G}_{P_1}) \|\Pi_{\mathcal{M}//\mathcal{N}}\|_\Delta}
$$

from which an upper bound on $\|\Sigma_{\mathcal{M}_1, \mathcal{N}}^{-1}\|_\Delta$ and thence on $\|\Pi_{\mathcal{N}//\mathcal{M}_1}\|_\Delta$ can be found.

13. An alternative generalization of the gap for nonlinear systems was introduced in [6]. The topology in [6] is a *pseudo-metric* topology and applies to systems with differentiable graphs (modelled as infinite-dimensional manifolds over a Banach space). However, the topology in [6] is not "continuously robust" in the sense of [21]. Two systems are close in the

sense of [6] if they have similar incremental responses, albeit about possibly different input/output operating conditions. Thereby, small values for $\|H_{P,C} - H_{P_1,C}\|_\Delta$ are not required for two open-loop plants P, P_1 to be close. Yet it is interesting to note that the stability of a feedback system is still a robust property with respect to the topology in [6], in spite of the fact that this topology is coarser than the one induced by $\delta(\cdot, \cdot)$, when $\delta(\cdot, \cdot)$ is specialized to systems with differentiable graphs. (A topology $\mathcal{T}$ is coarser than $\mathcal{T}_1$ if $\mathcal{T} \subset \mathcal{T}_1$ [15, p. 47].) We would like to mention that a different metric topology for nonlinear systems is proposed in [22]. Unfortunately, feedback stability is not a robust property for the topology in [22]. Finally it is worth noting that, for the case of linear systems, the question of whether the gap/graph topology ([17]) is the coarsest topology in which stability is a robust property remains open (see [18, page 44], [3]).

References

[1] E. Berkson, Some metrics on the subspaces of a Banach space, *Pac. J. Math.*, **13**, 7–22, 1963.

[2] C.A. Desoer and M. Vidyasagar, *Feedback systems: Input-output properties*, Academic Press: New York, 1975.

[3] F. De Mari and R. Ober, Topological Aspects of Robust Control, *Control of Uncertain Systems*, ed. Hinrichsen and Mårtensson, Birkhauser, pp. 69–82, 1990.

[4] J.C. Doyle, T.T. Georgiou, and M.C. Smith, The parallel projection operators of a nonlinear feedback system, *Systems & Control Letters*, **20**, 79–85, 1993.

[5] C. Foias, T.T. Georgiou, and M.C. Smith, Geometric techniques for robust stabilization of linear time-varying systems, Proceedings of the 1990 IEEE Conf. on Decision and Control, pp. 2868–2873, December 1990; Robust stability of feedback systems: A geometric approach using the gap metric, *SIAM Journal on Control and Optimization*, **31**, 1518–37, 1993.

[6] T.T. Georgiou, Differential stability and robust control of nonlinear systems, *Mathematics of Control, Signals and Systems*, **6**, 289–307, 1993.

[7] T.T. Georgiou and M.C. Smith, Optimal robustness in the gap metric, *IEEE Trans. on Automat. Control*, **35**, 673–686, 1990.

[8] J. Hammer, Nonlinear systems, stabilization, and coprimeness, *Int. J. Control*, **42**, 1–20, 1985.

[9] T. Kato, *Perturbation Theory for Linear Operators*, Springer-Verlag: New York, 1966.

[10] J.L. Massera and J.J. Schäffer, Linear differential equations and functional analysis, I, *Annals of Mathematics*, **67**, 517–573, 1958.

[11] M. Safonov, *Stability and Robustness of Multivariable Feedback System*, MIT Press, 1980.

[12] I.W. Sandberg, On the $\mathcal{L}_2$-boundedness of solutions of nonlinear functional equations, *Bell Sys. Tech. J.*, **43**, 1581–1599, 1964.

[13] J.S. Shamma, The necessity of the small-gain theorem for time-varying and nonlinear systems, *IEEE Trans. on Automat. Control*, **36**, 1138–1147, 1991.

[14] E.D. Sontag, Smooth stabilization implies coprime factorization, *IEEE Trans. on Automat. Control*, **34**, 435–443, 1989.

[15] W.A. Sutherland, *Introduction to Metric and Topological Spaces*, Oxford Science Publications, 1975.

[16] M.S. Verma, Coprime fractional representations and stability of non-linear feedback systems, *Int. J. Control*, **48**, 897–918, 1988.

[17] M. Vidyasagar, H. Schneider and B. Francis, Algebraic and topological aspects of feedback stabilization, *IEEE Trans. on Automat. Control*, **27**, 880–894, 1982.

[18] G. Vinnicombe, Measuring the robustness of feedback systems, Engineering Dept., Cambridge Univ., Ph.D. thesis, December 1992.

[19] G. Zames, On the input-output stability of time-varying nonlinear feedback systems. Part I: Conditions using concepts of loop gain, conicity, and positivity, *IEEE Trans. on Automat. Control*, **11**, 228–238, 1966.

[20] G. Zames, On the input-output stability of time-varying nonlinear feedback systems. Part II: Conditions involving circles in the frequency plane and sector nonlinearities, *IEEE Trans. on Automat. Control*, **11**, 465–476, 1966.

[21] G. Zames and A.K. El-Sakkary, Unstable systems and feedback: The gap metric, Proceedings of the Allerton Conference, pp. 380–385, October 1980.

[22] S.Q. Zhu, The gap topology for nonlinear systems, Proceedings of the American Control Conference, Baltimore, pp. 2114–2118, June 1994.

Acknowledgments. This research has been supported in part by the National Science Foundation and the Air Force Office for Scientific Research, U.S.A. and the SERC, U.K. This paper is based on: Cambridge University, Department of Engineering Technical Report, CUED/F-INFENG/TR 175, April 1994, Metric Uncertainty and Nonlinear Feedback Stabilization, by T.T. Georgiou and M.C. Smith.

Identification in Frequency Domain *

Guoxiang Gu[§] and Pramod P. Khargonekar[†]

In honor of Professor George Zames on his 60th birthday

Abstract

In this paper, we describe some of our recent work on identification in frequency domain.

1 Introduction

System modeling and identification plays a central and critical role in the design of control systems. Some of Zames' most fundamental contributions [47] have focused on the issue of modeling accuracy required to achieve robust control objectives. Zames introduced the concepts of metric complexity and entropy [47] to analyze these questions. These contributions continue to influence current research in system identification.

The classical approach to system identification is from a time-domain stochastic system theory perspective. See the book by Ljung [25] for a comprehensive exposition. In recent years a number of researchers have started to explore alternative (worst-case and/or deterministic and/or frequency domain) approaches [3, 5, 6, 9, 11, 12, 15, 16, 17, 18, 19, 20, 21, 23, 22, 24, 26, 29, 27, 30, 31, 34, 36, 37, 38, 40, 41, 42, 43, 44, 45, 46, 48, 49]. This list is not meant to be complete but just to give a sampling of the literature in this active and growing area.

Among these new research directions, one particular avenue being explored is *identification in frequency domain*. The basic idea is to perform input-output experiments using, for example, sinusoidal excitation, and extract frequency response information on the system under consideration. The resulting (noisy and partial) frequency response information is then mapped into a transfer function model for the system. Such an approach is by no means new — it is a very commonly used procedure. Some early references dealing with this frequency domain approach include [7, 39].

---

*Supported in part by Airforce Office of Scientific Research under contract no. F-49620-93-1-0246DEF and the Army Research Office under grant no. DAAH04-93-G-0012.

§ Dept. of Electrical Engineering, Louisiana State University, Baton Rouge, LA 70803-5901.

† Dept. of Electrical Engineering and Computer Science, The University of Michigan, Ann Arbor, MI 48109-2122.

Frequency response based system identification is again being investigated very actively; see [2, 3, 11, 12, 17, 18, 19, 26, 31, 32, 33, 34, 30, 38] and the references cited there. A canonical problem, known as "identification in $\mathcal{H}_\infty$", was formulated by Helmicki, Jacobson, and Nett [17]. The basic problem is to map noisy and partial frequency response data into a transfer function model for a stable linear system such that the worst case error measured in the $\mathcal{H}_\infty$ norm converges to zero as the number of data points goes to infinity as the noise amplitude goes to zero. Additionally, it is required to produce guaranteed worst case errors on the modeling error. We have been working on this problem for the last few years and have developed quite promising algorithms for solving this problem. In this paper, we will give a brief summary of some of our work on this problem along with a few new results.

Very recently, we have done some work on applying these algorithms to very challenging data sets on flexible systems [8, 13]. These results indicate that our algorithms have the potential of being quite useful in real engineering applications.

2   Problem Formulation

In this paper, we will focus attention on a specific frequency domain identification problem known as "identification in $\mathcal{H}_\infty$". The problem was posed in this particular form by Helmicki, Jacobson and Nett [17]. We briefly describe this problem formulation next.

Suppose that the "true" unknown system to be identified is a linear exponentially stable discrete-time shift-invariant system with transfer function $\hat{h}$. It will be assumed that we have some prior knowledge concerning the transfer function $\hat{h}$. This will be reflected in the assumption that $\hat{h} \in \mathcal{S} \subset \mathcal{A}$, where $\mathcal{A}$ is the disk algebra, i. e.,

$$\mathcal{A} := \{f \in \mathcal{H}_\infty : f \text{ continuous on the unit circle}\}.$$

The set $\mathcal{S}$ reflects the prior knowledge of the properties of the unknown system to be identified.

An important model set $\mathcal{S}$ is the collection of exponentially stable bounded systems

$$\mathcal{H}(D_\rho, M) = \left\{\hat{f} : \hat{f} \text{ analytic in } D_\rho, \; |\hat{f}| \leq M \; \forall z \in D_\rho\right\}$$

where $\rho > 1$ and $D_\rho := \{z : |z| < \rho\}$ is the complex disc of radius ρ.

The experimental data on $\hat{h}$ will be taken to be noisy values of the frequency response of the system at a finite number of frequencies. Such frequency response information can be obtained by applying linear combinations of sinusoids of various frequencies as input excitation to the system to be identified. As this is quite well known in the elementary linear systems literature, we will not dwell on this issue.

Thus, let us suppose we are given a finite number N of possibly noisy experimental frequency response data

$$E_k^N(\hat{h}, \hat{\eta}) := \hat{h}(e^{j2k\pi}) + \hat{\eta}_k, \qquad (2.1)$$

where $k = 0, 1, 2, \ldots, N - 1$ and

$$\hat{\eta} \in B_N(\epsilon) := \left\{ (\hat{\eta}_0, \hat{\eta}_1, \ldots, \hat{\eta}_{N-1}) \in C^N : |\hat{\eta}_k| \leq \epsilon \ \forall k \right\}, \qquad (2.2)$$

is the frequency response measurement noise. The bound ϵ quantifies the uncertainty in the estimation of the frequency response from the time-domain experiment. While extracting a good estimate of the value of the transfer function using the time domain data from a sinusoidal excitation experiment is quite straightforward, the problem of extracting the noise bound ϵ from the experimental data is much harder and will not be dealt with here.

A more general case is to assume a noise bound of the form

$$\|\hat{w}\hat{\eta}\|_\infty \leq \epsilon$$

for some known weighting function $\hat{w}$. As pointed out in [17], the corresponding identification problem can be transformed into the case $\hat{w} \equiv 1$.

The identification problem is to find an algorithm A_N which maps the given experimental information into an identified model $\hat{h}_{id}^N(z) \in A$. The performance of the identification algorithm is measured in a worst-case sense as follows:

$$e_N(\epsilon, \mathcal{S}) := \sup_{\hat{h} \in \mathcal{S}, \hat{\eta} \in B_N(\epsilon)} \|\hat{h} - \hat{h}_{id}^N\|_\infty. \qquad (2.3)$$

An algorithm is called "convergent" if

$$\lim_{\epsilon \to 0, \, N \to \infty} e_N(\epsilon, \mathcal{S}) = 0. \qquad (2.4)$$

If in addition, the convergence of the algorithm is independent of the *a priori* information $\epsilon, \mathcal{S}$ then it is called "robustly convergent and untuned".

There is a close connection between the problem of identification in $\mathcal{H}_\infty$ and ideas from n-width approximation theory [35, 47]. Let $\mathcal{P}_m$ as the collection of all polynomials (or FIR models) of degree no greater than $m - 1$:

$$\mathcal{P}_m = \left\{ \hat{p} : \ \hat{p} = p_0 + p_1 z + \ldots + p_{m-1} z^{m-1} \right\}.$$

We shall adopt following definition for the n-width

$$\delta_n[\mathcal{S}] = \sup_{\hat{h} \in \mathcal{S}} \inf_{\hat{p}_n \in \mathcal{P}_n} \|\hat{h} - \hat{p}_n\|_\infty. \qquad (2.5)$$

For simplicity, we will denote $\delta_n[\mathcal{S}]$ by merely δ_n for the special case $\mathcal{S} = \mathcal{H}(D_\rho, M)$. The following lemma (Theorem 2.1 on page 250 of [35]) will be used extensively for the quantification of the worst-case identification error.

Lemma 2.1 *Let $M > 0, \rho > 1$ and let $S = \mathcal{H}(D_\rho, M)$. Then for $m = 0, 1, \ldots$,*

$$\delta_m = \sup_{\hat{h} \in \mathcal{H}(D_\rho, M)} \inf_{\hat{p} \in \mathcal{P}_m} \|\hat{h} - \hat{p}\|_\infty = M\rho^{-m}.$$

Furthermore for any $\hat{h} \in \mathcal{H}(D_\rho, M)$, its globally optimal approximant is given by

$$\hat{p}_m^*[\hat{h}] := \sum_{k=0}^{m-1} \left(1 - \rho^{2(k-m)}\right) h_k z^k. \tag{2.6}$$

Finally,

$$\sup_{\hat{h} \in \mathcal{H}(D_\rho, M)} \|\hat{h} - \hat{p}_m^*[\hat{h}]\|_\infty = M\rho^{-m}. \tag{2.7}$$

Clearly, certain restrictions on the prior information set S are needed for the problem of identification in $\mathcal{H}_\infty$ to have a solution. We will assume throughout the paper that the model set S is admissible in the following sense.

Definition 2.2 *A subset $S \subset \mathcal{A}$ is said to be admissible if*

$$1.\ M_s := \sup\left\{\|\hat{h}\|_\infty : \hat{h} \in S\right\} < \infty; \quad 2.\ \lim_{n \to \infty} \delta_n[S] = 0.$$

The following proposition reveals an intrinsic property of admissible sets [3].

Proposition 2.3 *A model set $S \subset \mathcal{A}$ is admissible if and only if S is totally bounded in $\mathcal{A}$.*

In what follows we will consider classes of linear algorithms and nonlinear algorithms for identification in $\mathcal{H}_\infty$ and thier engineering applications.

3   Linear Algorithms

For the experimental data in (2.1), define its (inverse) DFT coefficients by

$$f_i(E^N) = \frac{1}{N} \sum_{k=0}^{N-1} E_{k+1}^N e^{j2ik\pi/N}, \tag{3.1}$$

where i is any integer, and $j = \sqrt{-1}$. A class of linear algorithms consists of the identified model based on weighted partial summation

$$\hat{h}_{id}^n := \sum_{k=0}^{n-1} w_{n,k} f_k(E^N) z^k, \tag{3.2}$$

where $n \leq N$ and $w_n = \{w_{n,k}\}_{k=0}^{n-1}$ is a suitable window function, independent of the *a priori* information. It is assumed that n is a function of N and satisfies

$\lim_{N\to\infty} n(N) = \infty$. For simplicity this functional relation will not be shown explicitly.

We associate for each window function a kernel

$$K_n(\omega) = \sum_{k=0}^{n-1} w_{n,k} e^{jk\omega}.$$

Define the *discretized convolution* in the frequency domain by

$$(K_n \circledast \hat{g})(\omega) = \frac{1}{N} \sum_{i=0}^{N-1} \hat{g}(e^{j2k\pi/N}) K_n(\omega - \frac{2i\pi}{N}).$$

Then it follows that

$$\hat{h}_{id}^n(e^{j\omega}) = K_n \circledast \left(\hat{h} + \hat{\eta}\right).$$

The linearity of the algorithm is manifested in the decomposition

$$\hat{h}_{id}^n = \hat{h}_{n,N} + \hat{\eta}_{n,N}, \tag{3.3}$$

where $\hat{h}_{n,N} = K_n \circledast \hat{h}$ and $\hat{\eta}_{n,N} = K_n \circledast \hat{\eta}$. As a simple consequence, we get the inequality

$$\|\hat{h} - \hat{h}_{id}^n\|_\infty \leq \|\hat{h} - \hat{h}_{n,N}\|_\infty + \|\hat{\eta}_{n,N}\|_\infty \tag{3.4}$$

which implies that the identification error for each $\hat{h} \in S$ is bounded by a sum of two components: *the approximation error* $\|\hat{h} - \hat{h}_{n,N}\|_\infty$ and *the noise error* $\|\hat{\eta}_{n,N}\|_\infty$.

Next we give a general procedure for analyzing the worst-case identification error defined in (2.3). This analysis is based on the cocept of n-width [35]. Using the definition in (2.5), each $\hat{h} \in S$ can be written as a sum of a polynomial approximation and a residual as follows:

$$\hat{h} = \hat{p}_n^*[\hat{h}] + \hat{\zeta}[\hat{h}], \quad \hat{p}_n^*[\hat{h}] \in \mathcal{P}_n, \quad \|\hat{\zeta}[\hat{h}]\|_\infty \leq \delta_n[S]. \tag{3.5}$$

For each window function $w_{n,k}$, define

$$C_w^n = \sup_{\omega \in [0,2\pi/N]} \sum_{i=0}^{N-1} |K_n(\omega - 2i\pi/N)|$$

and

$$C_w = \limsup \{C_w^n, n(N) > 0\}.$$

The following result demonstrates the impact of the window function on the worst-case identification error.

Theorem 3.1 *The the worst-case identification error (2.3) for the linear algorithm represented by (3.2) satisfies*

$$e_N(\epsilon, S) \leq (1 + C_w)\delta_n[S] + C_w\epsilon + \sup_{\hat{h} \in S} \|K_n \circledast \hat{p}_n^*[\hat{h}] - \hat{p}_n^*[\hat{h}]\|_\infty. \tag{3.6}$$

This result can be deduced from the inequality (3.4) and the decomposition (3.5) using the triangle inequality.

Now let us consider the rectangular window function

$$w_{n,k} = 1, 0 \le k \le n, \quad \text{and} \quad 0 \quad \text{for} \quad k > n$$

as studied in [10, 30]. Then, it is not difficult to see that $K_n \circledast \hat{p}_n^*[\hat{h}] = \hat{p}_n^*[\hat{h}]$. The next result follows immediately from this observation.

Corollary 3.2 *Let the window function be given by $w_{n,k} = 1$ for $0 \le k \le n-1$ and zero elsewhere. Suppose $S = \mathcal{H}(D_\rho, M)$. Then the linear algorithm in (3.2) gives worst case identification error*

$$e_N(\epsilon, S) \le (1 + C_w)M\rho^{-n} + C_w\epsilon.$$

Next we give some bounds on C_w for the rectangular window. By [50, page 67 of vol. 1], we have that as $N \to \infty$, $C_w = L_n$, a Lebesgue's constant. Combining this result with Gronwall's inequality, we can conclude that [11]

$$C_w \le \alpha \log(n) + \beta, \quad \frac{4}{\pi^2} \le \alpha \le \frac{2}{\pi}$$

where β is an absolute constant. Thus a new error bound for the linear algorithm with rectangular window is given by

$$e_N(\epsilon, S) \le (1 + \alpha \log(n) + \beta) M\rho^{-n} + (\alpha \log(n) + \beta) \epsilon. \tag{3.7}$$

Comparing this against the earlier error bound in [11]:

$$e_N(\epsilon, S) \le \frac{M(\rho^{-n+1} + \rho^{-N+1})}{\rho - 1} + (\alpha \log(n) + \beta) \epsilon, \tag{3.8}$$

we conclude that the error bound in (3.7) is often more attractive than (3.8) especially if ρ is very close to one as in the case of lightly damped system.

A similar result can also be established for the triangular window as studied in [11].

Corollary 3.3 *Let $w_{n,k} = 1 - \frac{k}{n}, k = 0, 1, \ldots, n; w_{n,k} = 0$ for $k \ge n$ be the one-sided triangular window and $S = \mathcal{H}(D_\rho, M)$. Then the linear algorithm in (3.2) gives worst case identification error*

$$e_N(\epsilon, S) \le (1 + C_w)M\rho^{-n} + C_w\epsilon$$
$$+ M \frac{\rho^2(1 - \rho^{-(n+1)}) - (n+1)(\rho - 1)\rho^{-n+1}}{n(\rho - 1)^2},$$

where $C_w = \frac{\log(n)}{\pi} + b$ and b is an absolute constant.

This result is obtained by combining Theorem 3.1 and the results in [11]. It is interesting to compare it with Corollary 3.2. While the noise error is smaller for the triangular window (the coefficient of $\log(n)$ is $1/\pi$ as compared with $2/\pi$), the last term of the worst-case identification error in (3.6) is nonzero thereby increasing the approximation error.

Using a rather clever argument, Partington [32] has shown that there does not exist a robustly convergent linear algorithm. Therefore, we turn our attention next to nonlinear algorithms.

4 Nonlinear Algorithms

A class of nonlinear algorithms for the problem of identification in $\mathcal{H}_\infty$ is based on modifications of the linear algorithm from last section. Let $\{f_i(E^N)\}$ be the inverse DFT coefficients as defined in (3.1). Obviously, this is a periodic sequence with period N. The following type of algorithms will be discussed in this section.

Two-Stage Nonlinear Algorithm:

- Stage 1: Set pre-identified model $\hat{h}_{pi}^n$ by

$$\hat{h}_{pi}^n(\lambda) := \sum_{k=-n+1}^{n-1} w_{n,k} f_k(E^N) z^k \tag{4.1}$$

for some two-sided window function w_n.

- Stage 2: Take identified model $\hat{h}_{id}^n$ as

$$\hat{h}_{id}^n := \text{argmin} \left\{ \|\hat{h}_{pi}^n - \hat{g}\|_\infty : \hat{g} \in \mathcal{H}_\infty(D) \right\}. \tag{4.2}$$

It is noted that Stage 2 involves solving the Nehari best approximation problem. This is necessitated by the fact that the pre-identified model $\hat{h}_{pi}^n$ is not necessarily analytic as it may have nonzero negative Fourier coefficients. Using the standard Nehari theorem [28], the approximation error induced in the second stage is no larger than the Hankel norm of the anticausal part of $\hat{h}_{pi}^n$. Therefore the performance of the two-stage algorithm is critically dependent on the window function in the first stage which is the only *free design parameter* in the two-stage nonlinear algorithm.

We associate for each two-sided window function a kernel

$$K_n(\omega) = \sum_{k=-n+1}^{n-1} w_{n,k} e^{jk\omega}$$

and define C_w as in the previous section. Then, the pre-identified model in Stage 1 can also be written as the discretized convolution:

$$\hat{h}_{pi}^n = K_n \circledast \left(\hat{h} + \hat{\eta} \right).$$

Using the decomposition (3.5) for $\hat{h} \in \mathcal{S}$, we have that

$$\hat{h}_{pi}^n = K_n \circledast \hat{p}_n^*[\hat{h}] + K_n \circledast \left(\hat{\eta} + \hat{\zeta} \right).$$

If $2n \le N + 1$, the function $K_n \circledast \hat{p}_n^*[\hat{h}]$ is analytic, and thus the anticausal component of $\hat{h}_{pi}^n$ comes from the term $K_n \circledast \left(\hat{\eta} + \hat{\zeta} \right)$ and is insignificant if n is large and ϵ is small.

The next result indicates a general procedure for the quantification of the worst-case identification error.

Theorem 4.1 *Let the worst-case identification error at the first stage be defined by*

$$e_N^{pi}(\epsilon, \mathcal{S}) := \sup_{\hat{h} \in \mathcal{S}, \hat{\eta} \in B_N(\epsilon)} \|\hat{h} - \hat{h}_{pi}^n\|_\infty.$$

Let $N + 1 \geq 2n$. Then

$$e_N^{pi}(\epsilon, \mathcal{S}) \leq \delta_n[\mathcal{S}] + C_w \left(\epsilon + \delta_n[\mathcal{S}]\right) + \sup_{\hat{h} \in \mathcal{S}} \|K_n \circledast \hat{p}_n^*[\hat{h}] - \hat{p}_n^*[\hat{h}]\|_\infty.$$

Furthermore, the worst-case identification error for the two-stage algorithm satisfies

$$
\begin{aligned}
e_N(\epsilon, \mathcal{S}) \quad &\leq \quad e_N^{pi}(\epsilon, \mathcal{S}) + C_w \left(\epsilon + \delta_n[\mathcal{S}]\right) \\
&\leq \quad \delta_n[\mathcal{S}] + 2C_w \left(\epsilon + \delta_n[\mathcal{S}]\right) + \sup_{\hat{h} \in \mathcal{S}} \|K_n \circledast \hat{p}_n^*[\hat{h}] - \hat{p}_n^*[\hat{h}]\|_\infty,
\end{aligned}
$$

where C_w is defined as in Theorem 3.1.

It is noted that the worst-case identification error for the class of nonlinear algorithms is similar to the class of linear algorithms except an additional term $C_w \left(\epsilon + \delta_n[\mathcal{S}]\right)$ due to the the anticausal component in $K_n \circledast \left(\hat{\eta} + \hat{\zeta}\right)$.

While C_w^n diverges as $n \to \infty$ for one-sided window functions, it can be made uniformly bounded if the window function is two-sided thereby guaranteeing the convergence of the nonlinear algorithm. In fact such convergence property is robust and independent of the *a priori* information. An important problem is thus the characterization of the robust convergence in terms of the window function. This question was resolved in [12, 14]. Our first result is taken from [12] where robust convergence is characterized in the time-domain.

Theorem 4.2 *Suppose that the window function $w_{n,k}$ is even symmetric with respect to k with $n < N/2$. Then, Stage 1 of the two-stage nonlinear algorithm is robustly convergent if the window function $w_{n,k}$ satisfies*

(i) $\lim_{n \to \infty} \Delta^2 w_{n,k} = 0$ *for $k = 0, 1, ...;$*

(ii) $N_s := \lim \sup \left\{ N_n := n|\Delta w_{n,n-1}| + \sum_{k=0}^{n-2}(k+1)|\Delta^2 w_{n,k}| : n \geq 0 \right\} < \infty;$

(iii) $\lim_{n \to \infty} w_{n,0} = 1.$

Consequently, if the above conditions hold, then the two-stage nonlinear identification algorithm is robustly convergent.

It should be mentioned that for symmetric windows, the inequality $C_w \leq N_s$ holds. For some commonly used window functions, the quantity N_s, an upper bound of C_w, can be explicitly calculated. The following notions of convex and concave windows are useful.

Definition 4.3 *Let the two-side window function be symmetric. The window function $w_{n,k}$ is called convex at k if $\Delta^2 w_{n,k} \geq 0$, and called a convex function if $\Delta^2 w_{n,k} \geq 0$ for all $k \geq 0$; The window function $w_{n,k}$ is called concave at k, if $\Delta^2 w_{n,k} \leq 0$ and called a concave function if $\Delta^2 w_{n,k} \leq 0$ for all $k \geq 0$; The window function $w_{n,k}$ is called non-increasing if $\Delta w_{n,k} \geq 0$ for all $k \geq 0$.*

Theorem 4.4 *Let the window function $w_{n,k}$ used in the two-stage algorithm be even symmetric which is non-increasing and $N > n > 0$. Then the following hold:*

(1) If $w_{n,k}$ is is a convex function, then

$$C_w \leq N_s = w_{n,0};$$

(2) If $w_{n,k}$ is is a concave function, then

$$C_w \leq N_s = 2n w_{n,n-1} - w_{n,0};$$

(3) If $w_{n,k}$ is convex for $k < m(< n)$ and concave for $k \geq m$, then

$$C_w \leq N_s \leq w_{n,0} - 2(m+1) w_{n,m} + 2m w_{n,m+1} + 2n w_{n,n-1};$$

(4) If $w_{n,k}$ is concave for $k < m(< n)$ and convex for $k \geq m$, then

$$C_w \leq N_s \leq 2(m+1) w_{n,m} - 2m w_{n,m+1} - w_{n,0}.$$

We comment that while $C_w \leq N_s$, in some cases even though C_w is finite, N_s is infinity. Hence the conditions in Theorem 4.2 are sufficient but not necessary. The next result characterizes the robust convergence of the two-stage algorithm in frequency domain.

Theorem 4.5 *Suppose that the window function $w_{n,k}$ is even symmetric with respect to k with $n < N/2$. Then, the two-stage nonlinear algorithm is robustly convergent if*

1. $\displaystyle \lim_{N \geq 2n \to \infty} \frac{1}{N} \sum_{i=0}^{N-1} K_n(2i\pi/N) e^{-jk\frac{2i\pi}{N}} = 1$ *for $k = 0, \pm 1, \pm 2, \ldots$;*

2. $C_w < \infty$.

Conversely, if the first stage of the two-stage nonlinear algorithm is robustly convergent then conditions 1 and 2 are satisfied.

The proof of the converse statement involves construction of a noise sequence which is conjugate symmetric [14]. Although the class of window functions characterized in Theorem 4.5 is larger than that in Theorem 4.4, it is difficult to compute or estimate C_w. However for an important class windows — trapezoidal windows — Theorem 4.5 gives a better estimate than Theorem 4.4.

In light of Theorem 4.1, the worst-case identification error bound can be written as

$$e_N(\epsilon, \mathcal{S}) \le 2C_w\epsilon + 2(1 + C_w)\delta_n[\mathcal{S}] + \sup_{\hat{h}\in\mathcal{S}} \|K_n \circledast \hat{p}_n^*[\hat{h}] - \hat{h}\|_\infty. \tag{4.3}$$

The first term is the error induced by measurement noise and the rest of the terms grouped together will be called the approximation error. There appears to be a trade-off between the noise error and approximation error. For the one-sided rectangular window, the last term in the error expression vanishes but the corresponding C_w is not bounded. On the other hand, the two-sided triangular window has the smallest possible value for C_w, namely one, but the corresponding approximation error is $\mathcal{O}(\frac{1}{n})$ [11]. Hence an attractive choice is a *trapezoidal window*: a combination of the triangular and rectangular windows. A special trapezoidal window associated with the de la Vallee kernel [50] was used by Partington [32]. We will employ following parameterized trapezoidal window [12, 13]:

$$w_{n,k} = \begin{cases} 1 & 0 \le k \le m-1, \\ 1 + \frac{k}{n}, & -n \le k \le -1, \\ 1 - \frac{k-m+1}{n} & m \le k \le n+m-1, \\ 0 & \text{elsewhere}, \end{cases} \tag{4.4}$$

where $n + m < N$. Clearly $m = 1$ corresponds to two-sided triangular window while $n = 1$ corresponds to one-sided rectangular window. The above window has two additional advantages. First the free integer parameter m can be used to yield a right trade-off between the two error terms. Second it gives more weighting on the causal part of the pre-identified model. It is noted if m is odd, then the shifted window

$$\tilde{w}_{n,k} = w_{n,k+(m-1)/2}$$

is symmetric about $k = 0$. Because shifting does not change the value of C_w, the worst-case identification error bound in (4.3) remains valid provided that $n + m < N$. Moreover, we have the following result.

Theorem 4.6 *For the trapezoidal window in (4.4), the pre-identified model is interpolatory if $N = n + m - 1$:*

$$\hat{h}_{pi}^n(W_N^k) = E_k^N, \quad k = 0, 1, ..., N-1.$$

Moreover $C_w \le \sqrt{N/n}$ [14]. Thus the resulting worst-case identification error for the two-stage nonlinear algorithm is bounded by

$$e_N(\epsilon, \mathcal{S}) \le 2\epsilon\sqrt{N/n} + \left(1 + 2\sqrt{N/n}\right)\delta_m[\mathcal{S}].$$

In particular if $N = 2n$ and $m = 1 + N - n = 1 + N/2$ are chosen, and $\mathcal{S} = \mathcal{H}(D_\rho, M)$, then

$$e_N(\epsilon, \mathcal{S}) \le 2\sqrt{2}\epsilon + \left(1 + 2\sqrt{2}\right)M\rho^{-(1+N/2)}.$$

Remark 4.7 *It is known that the interpolation algorithm in [5, 15] is optimal within a factor of two. However it can only be shown in [15] that the resulting worst-case identification error for the interpolation algorithm is no larger than twice of $e_N^{pi}(\epsilon, S)$. In light of Theorem 4.1, the resulting identification error for the two-stage algorithm is strictly smaller than twice of $e_N^{pi}(\epsilon, S)$.*

5 Engineering Applications

The results described above indicate that we have computationally feasible algorithms for solving the problem of identification in $\mathcal{H}_\infty$. In [13], we attempted to apply these algorithms to a frequency response data set on a Jet Propulsion Laboratory flexible structure. This data was supplied to us by Dr. David Bayard. The approach taken in [13] was to use a combination of the classical Santhanan-Koerner (SK) iteration followed by the two-stage nonlinear algorithm. SK iteration is an iterative procedure to solve a nonlinear least squares problem of fitting the frequency response data by a parameterized transfer function. In a recent paper [8], two case studies have been presented. One was the same JPL data and the other case was frequency response data of a flexible pointing system test bed at the ARDEC, NJ. This test bed was the subject of two invited sessions at the 1993 American Control Conference. The results in [8] show that the two stage nonl! inear algorithm with the trapezoi

References

[1] V. M. Adamyan, D. Z. Arov and M. G. Krein, "Analytic properties of Schmidt pairs for a Hankel operator and the generalized Schur-Takagi problem," *Math. Sbornik*, vol. 15, pp. 31–73, 1971.

[2] H. Akçay, G. Gu and P. P. Khargonekar, "A class of algorithms for identification in $\mathcal{H}_\infty$: Continuous-time case," *IEEE Transactions on Automatic Control*, vol. 38, pp. 289– 294, 1993.

[3] H. Akçay, G. Gu and P.P. Khargonekar, "Identification in $\mathcal{H}_\infty$ with nonuniformly spaced frequency response measurements," *International J. of Robust and Nonlinear Control*, vol. 4, pp. 613-629, 1994.

[4] E.-W. Bai, "Adaptive quantification of model uncertainties by rational approximation," *IEEE Transactions on Automatic Control*, vol. 36, pp. 441–453, 1991.

[5] J. Chen, C. N. Nett, and M. K. H. Fan, "Worst-case system identification in $\mathcal{H}_\infty$: validation of *a priori* information, essentially optimal algorithms, and error bounds," *Proceedings of American Control Conference*, pp. 251–257, 1992.

[6] J. Chen and C. N. Nett, "The Carathédory-Fejér problem and $\mathcal{H}_\infty$ identification: a time domain approach," preprint.

[7] J. O. Flower and S. C. Forge, "Developments in frequency-response determination using schroeder-phased harmonic signals," *The Radio and Electronic Engineer*, vol. 51, pp. 226–232, 1981.

[8] J. Friedman and P. P. Khargonekar, "Identification of lightly damped systems using frequency domain techniques," *Proc. IFAC Conference on System Identification*, pp. 201–206, 1994. Submitted for publication.

[9] G. C. Goodwin, M. Gevers and B. Ninness, "Quantifying the error in estimated transfer functions with application to model selection," *IEEE Transactions on Automatic Control*, vol. 37, pp. 913–928, 1992.

[10] G. Gu, P. P. Khargonekar and E.B. Lee, "Approximation of infinite dimensional systems," *IEEE Transactions on Automatic Control*, vol. AC-34, pp. 610–618, 1989.

[11] G. Gu and P. P. Khargonekar, "Linear and nonlinear algorithms for identification in $\mathcal{H}_\infty$ with error bounds," *IEEE Transactions on Automatic Control*, vol. 37, pp. 953–963, 1992.

[12] G. Gu and P. P. Khargonekar, "A class of algorithms for identification in $\mathcal{H}_\infty$," *Automatica*, vol. 28, pp. 199–312, 1992.

[13] G. Gu and P. P. Khargonekar, "Frequency domain identification of lightly damped systems: The JPL example," *Proc. of American Control Conference*, pp. 3052–3057, 1993.

[14] G. Gu, P. P. Khargonekar and Y. Li, "Robust convergence of two-stage nonlinear algorithms for system identification in $\mathcal{H}_\infty$," *Systems and Control Letters*, vol. 18, pp. 253–263, 1992.

[15] G. Gu, D. Xiong, and K. Zhou, "Identification in $\mathcal{H}_\infty$ using Pick's interpolation," *Systems and Control Letters*, vol. 20, pp. 263–272, 1993.

[16] R. G. Hakvoort, "Worst-case system identification in $\mathcal{H}_\infty$: Error bounds, interpolation, and optimal models," Internal Report, Delft University of Technology, The NETHERLANDS, 1992.

[17] A. J. Helmicki, C.A. Jacobson and C.N. Nett, "Control oriented system identification: a worst-case/deterministic approach in $\mathcal{H}_\infty$," *IEEE Transactions on Automatic Control*, vol. 36, pp. 1163–1176, 1991.

[18] A. J. Helmicki, C. A. Jacobson and C. N. Nett, "Worst-case/deterministic identification in $\mathcal{H}_\infty$: The continuous-time case," *IEEE Transactions on Automatic Control*, vol. 37, pp. 604–610, 1992.

[19] A.J. Helmicki, C.A. Jacobson, and C.N. Nett, "Identification in $\mathcal{H}_\infty$: linear algorithms," *Proc. American Control Conference*, pp. 2418–2423, 1990

[20] R. Kosut and H. Ailing,"Worst case control design from batch least squares," *Proc. American Control Conference*, pp. 318–322, 1992.

[21] J. M. Krause and P.P. Khargonekar, "Parameter identification in the presence of non-parametric dynamic uncertainty," *Automatica*, vol. 26, pp. 113–124, 1990.

[22] M. Lau, R.L. Kosut, and S. Boyd, "Parameter set identification of systems with uncertain nonparametric dynamics and disturbances," *Proc. 29th IEEE Conference on Decision and Control*, pp. 3162–3167, 1990.

[23] W. Lee, B. D. O. Anderson, and R. Kosut, "On adaptive robust control and control relevant identification," *Proc. American Control Conference*, 1992

[24] L. Lin, L. Wang, and G. Zames,"Uncertainty principles and identification n-widths for LTI and slowly varying systems," *Proc. American Control Conference*, pp. 296–300, 1992.

[25] L. Ljung, *System Identification, Theory for the User*, Prentice-Hall, Englewood Cliffs, NJ, 1987.

[26] P. M. Mäkilä and J. R. Partington, "Robust identification of strongly stabilizable systems," *IEEE Transactions on Automatic Control*, vol. 37, pp. 1709–1716, 1992.

[27] S. H. Mo and J. P. Norton, "Parameter-bounding identification algorithms for bounded-noise records," *IEE Proceedings Part D, Control Theory and Applications*, vol. 135, pp. 127–132, 1988

[28] Z. Nehari, "On bounded linear forms," *Annals of Mathematics*, vol. 65, pp. 153–162, 1957.

[29] B. M. Ninness, "Stochastic and deterministic estimation in $\mathcal{H}_\infty$," *Proc. 33rd IEEE Conference on Decision and Control*, pp. 62–67, 1993.

[30] P. J. Parker and R. R. Bitmead, "Adaptive frequency response identification," *Proc. 28th IEEE Conference on Decision and Control*, pp. 348–353, 1987.

[31] J. R. Partington, "Robust identification in $\mathcal{H}_\infty$," *J. Mathematical Analysis and Applications*, vol. 166, pp. 428–441, 1992

[32] J. R. Partington, "Robust identification and interpolation in $\mathcal{H}_\infty$," *International Journal of Control*, vol. 54, pp. 1281–1290, 1991.

[33] J. R. Partington, "Algorithms for identification in $\mathcal{H}_\infty$ with unequally spaced function measurements", *International Journal of Control*, vol. 58, pp. 21–32, 1993.

[34] J. R. Partington, "Interpolation in normed spaces from the values of linear functionals," *Bulletin of the London Mathematical Society*, vol. 26, pp. 165–170, 1994.

[35] A. Pinkus, *n-Widths in Approximation Theory*, Springer-Verlag, Berlin, 1985.

[36] K. Poolla, P. P. Khargonekar, A. Tikku, J. Krause, and K. Nagpal, "A time-domain approach to model validation," *IEEE Transactions on Automatic Control*, vol. 39, pp. 951–959, 1994.

[37] K. Poolla and A. Tikku, "Time complexity of worst-case identification," *IEEE Transactions on Automatic Control*, vol. 39, pp. 944–950, 1994.

[38] S. Raman and E. W. Bai, "A linear, robust and convergent interpolatory algorithm for quantifying midel uncertainties," *Systems and Control Letters*, vol. 18, pp. 173–178, 1992.

[39] C. K. Sanathanan and J. Koerner, "Transfer function synthesis as a ratio of two complex polynomials," *IEEE Transactions on Automatic Control*, vol. AC-8, pp. 56–58, 1963.

[40] R. J. P. Schrama, "Accurate models for control design: the necessity of an iterative scheme," *IEEE Transactions on Automatic Control*, vol. 37, pp. 991–993, 1992.

[41] R. S. Smith and J. C. Doyle, "Towards a methodology for robust parameter identification," *IEEE Transactions on Automatic Control*, vol. AC-37, pp. 942–952, 1992.

[42] R. Tempo, "IBC: a working tool for robust parameter estimation," *Proc. American Control Conference*, pp. 237–240, 1992.

[43] D. N. C. Tse, M. A. Dahleh, and J. N. Tsitsiklis, "Optimal Asymptotic Identification under Bounded Disturbances", *Proc. American Control Conference*, pp. 1786–1787, 1991.

[44] T. van den Boom, *MIMO-systems identification for $\mathcal{H}_\infty$ robust control: A frequency domain approach with minimum error bounds*, Ph. D. Thesis, Eindhoven University, The NETHERLANDS.

[45] J. C. Willems, "From time series to linear system — Parts 1, 2, 3," *Automatica*, vol. 22–23, 1986-87.

[46] R. C. Younce and C. E. Rohrs, "Identification with non–parametric uncertainty," *IEEE Transactions on Automatic Control*, vol. 37, pp. 715–128, 1992.

[47] G. Zames, "On the metric complexity of causal linear systems: ϵ-entropy and ϵ-dimension for continuous time," *IEEE Transactions on Automatic Control*, vol. AC-24, pp. 222–230, 1979.

[48] Z. Zang, R. Bitmead, and M. Gevers, " $\mathcal{H}_\epsilon$ Iterative model refinement and control robustness enhancement," *Proc. 30th IEEE Conference on Decision and Control*, pp. 279-284, 1991.

[49] T. Zhou and H. Kimura, "Identification for robust control in time domain," *Systems and Control Letters*, vol. 20, pp. 167–178, 1993.

[50] A. Zygmund, *Trigonometric Series*, Cambridge University Press, 1959.

Performance Analysis and Control of Stochastic Discrete Event Systems

R.H. Kwong and L. Zhu
Electrical and Computer Engineering
University of Toronto
Toronto, Ontario
Canada M5S 1A4

Abstract

In this paper, a model of stochastic discrete event system (SDES) based on the Ramadge & Wonham (RW) framework is constructed. The performance analysis of SDES is studied in a systematic fashion. Formulas are provided for calculating the probability of a series of events, and the corresponding expected time and expected cost. Parameter optimization problems are also discussed in this framework. A transfer line example is given which shows the tradeoffs between logical requirements and performance.

1 Introduction

Automata and formal language models for discrete event systems (DES) and the related theory have been initiated by Ramadge and Wonham (RW) [6], [10], and subsequently extended by them and other researchers. Let us define a DES to be a dynamic system with a discrete state space and piecewise constant state trajectories. The time instants at which state transitions occur, as well as the actual transitions, will in general be random. The state transitions of a DES are called *events*, and may be labeled with the elements of some alphabet. These labels usually indicate the physical phenomena that cause the change in state. If the times of occurrence of events are ignored and only the order in which they occur is considered, then this leads to *logical DES models*. In such models a system trajectory is specified simply by listing (in order) the events that occur along the original sample path. The system behaviour can be modelled using formal languages, and the language can be generated by an *automaton*:

$$G = (\Sigma, Q, \delta, q_0, Q_m)$$

where $\Sigma = \{\sigma_1, \sigma_2, ..., \sigma_n\}$ is the *alphabet*, and σ_i is an *event*, $i = 1, 2, ..., n$. Q is called the state set. The partial function $\delta : \Sigma \times Q \longrightarrow Q$ is called the *transition function.* $q_0 \in Q$ is the *initial state*, and $Q_m \subseteq Q$ is called the *marked state set*, which serves to 'mark' the termination of 'tasks' of the underlying physical process that G is intended to model. Often, an automaton is displayed as a directed graph.

Logical models have been successfully used to study the qualitative properties of DES in a variety of applications. In such applications, the formulation and analysis of the model typically proceeds as follows. One first specifies the set of admissible event trajectories, i.e. the physically possible sequences of events. This may be done using some form of state transition structure or other mathematical description. In the cases of interest, the admissible event trajectories form a strict subset of the set of all (mathematically) possible event orderings. Given a property of event sequences, one then seeks to determine if each admissible trajectory has certain desired properties. Or, in a control context, one asks if it is possible to obtain desired trajectories using suitable control action.

Among the mathematical models of discrete event systems, automaton models have provided valuable concepts and insights to serve as guidelines for future work, and have contributed to our understanding of the fundamental issues involved in the analysis and control of DES. Most of the work focus on the logical behaviour of DES [7].

However, in real-time control, the times of occurrence of events are also significant, since they are often related to performance optimization of the DES. In this paper, occurrence times are taken to be as random variables. Using this stochastic framework, this paper studies the performance analysis and control of DES. It provides formulas to compute probabilities of event sequences and to evaluate the expected time and the associated expected cost that a DES might take to reach a certain state. A *Transfer Line* example is given to illustrate the results, and to show that a supervisory synthesis more flexible than conventional RW theory can be done within this framework. Other authors, for example Hoffmann [3] and Lin [4], also studied stochastic aspects of DES within the RW framework, but they treated issues different from those in this paper. Cassandras [1] describes stochastic timed models for discrete event systems, but he only treats distributions related to a single event. We shall only state our main results in this paper. Detailed proofs can be found in [11].

2 SDES Algebra

Let Σ^* be the set of finite sequences of events, including the empty sequence ε. Consider the extension of $\delta : \Sigma^* \times Q \longrightarrow Q$, with

$$\delta(\varepsilon, q) = q$$

and

$$\delta(s\sigma, q) := \delta(\sigma, \delta(s, q))$$

for $q \in Q$, $\sigma \in \Sigma$, $s \in \Sigma^*$, provided $q' := \delta(s, q)$ and $\delta(\sigma, q')$ are defined (denoted by $\delta(s, q)!$ and $\delta(\sigma, q')!$).

Then

$$L(G) := \{s \in \Sigma^* : \delta(s, q_0)!\}$$

and

$$L_m(G) := \{s \in \Sigma^* : \delta(s, q_0) \in Q_m\}$$

are the *closed behavior* and *marked behavior* of G.

Here Σ is partitioned into *controllable events* Σ_c and *uncontrollable events* Σ_u i.e. $\Sigma = \Sigma_c \cup \Sigma_u$. A controllable event can occur only if it is enabled by an external agent whereas an uncontrollable event can always occur. This is the control mechanism in the standard RW framework [6], [10].

We now add stochastic features to the DES dynamics. Given an automaton description for the DES

$$G_{log} = (\Sigma, Q, \delta, q_0, Q_m)$$

the occurrence time t_σ of an event $\sigma \in \Sigma$, once it becomes eligible to occur, is modelled as a random variable. In most cases in this paper, it is assumed to be exponentially distributed [1], [3], [4], with density function

$$f_{t_\sigma}(\tau) = \begin{cases} \lambda_\sigma e^{-\lambda_\sigma \tau} & \text{if } \tau \geq 0 \\ 0 & \text{otherwise} \end{cases}$$

Here λ_σ is called the *parameter* of event σ.

We adopt the following description for SDES. Define $\Lambda_l(s) := \{\lambda_\sigma : s\sigma \in L(G_{log})\}$, which is the set of parameters of events that may be triggered following string s, and $\Lambda_l := \{\Lambda_l(s) : s \in L(G_{log})\}$ A stochastic discrete event system (SDES) G is then written as

$$G = (G_{log}, \Lambda_l)$$

If t_σ is not restricted to be exponentially distributed, then this system is called a *general stochastic discrete event system (GSDES)*. In this case, Λ_l should describe the detailed statistical features of the GSDES in an appropriate way.

The above SDES structure is called a *language model*. We shall also use a *state model*. Let the logical model of a SDES be

$$G_{log} = (\Sigma, Q, \delta, q_0, Q_m)$$

Define $\Lambda_s(q) = \{\lambda_\sigma : \delta(\sigma, q)!\}$, which is the set of parameters of events which may trigger from state q, and $\Lambda_s = \{\Lambda_s(q) : q \in Q\}$. Then the state model of this SDES is

$$G = (G_{log}, \Lambda_s)$$

2.1 Basic Properties of SDES

We first describe some basic statistical properties of SDES. For the SDES

$$G = (G_{log}, \Lambda) \text{ and } G_{log} = (\Sigma, Q, \delta, q_0, Q_m)$$

Let

$$\mathcal{P}_t(s\sigma|s) := \mathcal{P}(s\sigma \text{ occurs before time } t|s \text{ occurred })$$

and

$$\mathcal{P}(s\sigma|s) := \mathcal{P}(s\sigma \text{ eventually occurs } |s \text{ occurred })$$

Let $q = \delta(s, q_0), \Sigma(s) := \{\sigma \in \Sigma : s\sigma \in L(G) \text{ and } \delta(\sigma, q) \neq q\}$ and

$$\mu(s) := \sum_{\sigma \in \Sigma(s)} \lambda_\sigma, \text{ if } \Sigma(s) \neq \emptyset$$

$$\mu(s) := 0, \text{ if } \Sigma(s) = \emptyset$$

If event σ is a selfloop (i.e., $\delta(\sigma, q) = q$), then once the selfloop σ triggers, the system goes back to the original state instantaneously. One cannot tell if σ has occurred or not. Since the selfloop σ contributes nothing to the stochastic dynamics of the system, it can, and will be, ignored.

Proposition 1 *Once string s occurred, the interarrival time of the next event (if possible) is still exponentially distributed, and its parameter is $\mu(s)$. Hence the average interarrival time is $\frac{1}{\mu(s)}$.*

Proposition 2 *If $\sigma \in \Sigma(s)$, then*

$$\mathcal{P}_t(s\sigma|s) = \frac{\lambda_\sigma}{\mu(s)}(1 - e^{-\mu(s)t})$$

$$\mathcal{P}(s\sigma|s) = \frac{\lambda_\sigma}{\mu(s)}$$

2.2 Probabilities

First we examine the probability that the system makes a series of transitions before any given time t. For any string $s \in \{s \in \Sigma^* : \delta(s, q)!\}$, denote

$$\mathcal{P}_t(s|q) := \mathcal{P}(\text{ System generates string } s \text{ before time } t |$$
$$\text{It starts from state } q \text{ at time } 0)$$

The function $\mathcal{P}_t(s|q)$ satisfies an integral equation as described below.

Lemma 1 $$\mathcal{P}_t(s\alpha|q) = \int_0^t \lambda_\alpha e^{-\mu(s)t_\alpha} \mathcal{P}_{t-t_\alpha}(s|q)dt_\alpha$$

and

$$\mathcal{P}_t(\alpha s|q) = \int_0^t \lambda_\alpha e^{-\mu(q)t_\alpha} \mathcal{P}_{t-t_\alpha}(s|q')dt_\alpha$$

where $q' = \delta(\alpha, q)$, t_α is time of α triggering after G starts from state q at time 0.

Using this lemma, the following theorem can be derived.

Theorem 1 *For string $s = \alpha_1\alpha_2\cdots\alpha_n$, assume $s_i = \alpha_1\alpha_2\cdots\alpha_i, i = 1, 2, ..., n-1(n > 1)$ and $s_0 = \varepsilon$, and $\mu(s_0), \mu(s_1), \cdots, \mu(s_{n-1})$ are unequal, then*

$$\mathcal{P}_t(\alpha_1\alpha_2\cdots\alpha_n|q) = (\prod_{r=1}^n \lambda_{\alpha_r})[\prod_{i=0}^{n-1} \frac{1}{\mu(s_i)} - \sum_{i=0}^{n-1} \frac{e^{-\mu(s_i)t}}{\mu(s_i)\prod_{j=1,j\neq i}^{n-1}(\mu(s_j) - \mu(s_i))}]$$

When $n=1$, we assign 1 to the empty product $\prod_{j=1,j\neq i}^0 (\mu(s_j) - \mu(s_i))$.

We can lift the restriction that none of the $\mu(i)$s are the same by taking the limit as they tend to the same value.

Secondly we consider the probability that the system reaches a specified state before any time t once it starts from another state.

Let $Q = \{1, 2, ..., n\}$ and

$$\bar{p}_{ij}(t) := \mathcal{P}(G \text{ first visits state } j \text{ before time } t \mid G \text{ starts from state } i \text{ at time } 0)$$

Define $\Sigma_{ij} := \{\sigma \in \Sigma : \delta(\sigma, i) = j\}$, for $i, j \in Q$, but $i \neq j$ and

$$\lambda_{ij} := \sum_{\sigma \in \Sigma_{ij}} \lambda_\sigma, \text{ if } \Sigma_{ij} \neq \emptyset$$

$$\lambda_{ij} := 0, \text{ if } \Sigma_{ij} = \emptyset$$

For state j , let

$$\bar{p}_j(t) := [\bar{p}_{1j}(t), \bar{p}_{2j}(t), \cdots, \bar{p}_{j-1j}(t), \bar{p}_{j+1j}(t), \cdots, \bar{p}_{nj}(t)]'$$

$$a_j(t) \; := \; [\frac{\lambda_{1j}}{\mu(1)}(1 - e^{-\mu(1)t}), \cdots, \frac{\lambda_{j-1j}}{\mu(j-1)}(1 - e^{-\mu(j-1)t}),$$

$$\frac{\lambda_{j+1j}}{\mu(j+1)}(1 - e^{-\mu(j+1)t}), \cdots, \frac{\lambda_{nj}}{\mu(n)}(1 - e^{-\mu(n)t})]'$$

$$\Lambda_j(\tau) := \begin{bmatrix} 0 & \cdots & \lambda_{1j-1}e^{-\mu(1)\tau} & \lambda_{1j+1}e^{-\mu(1)\tau} & \cdots & \lambda_{1n}e^{-\mu(1)\tau} \\ \lambda_{21}e^{-\mu(2)\tau} & \cdots & \lambda_{2j-1}e^{-\mu(2)\tau} & \lambda_{2j+1}e^{-\mu(2)\tau} & \cdots & \lambda_{2n}e^{-\mu(2)\tau} \\ \cdots & \cdots & \cdots & \cdots & \cdots & \cdots \\ \lambda_{n1}e^{-\mu(n)\tau} & \cdots & \lambda_{nj-1}e^{-\mu(n)\tau} & \lambda_{nj+1}e^{-\mu(n)\tau} & \cdots & 0 \end{bmatrix}$$

We follow the recursive way to compute $\bar{p}_{ij}(t)$. The system moving from one state i to another state j can be realized as follows: The system jumps one step to any other state, then moves from this current state to reach the state j, or else it moves in one step to this target state (if possible). So $\bar{p}_{ij}(t)$ can be represented by $\bar{p}_{kj}(t)(k = 1, 2, ..., n$ but $k \neq i, j)$ and other known information. As Q is finite, $\bar{p}_{ij}(t)$ can be evaluated by dealing with a finite number of recursive equations. These considerations lead to the following result.

Theorem 2 *The probability vector $\bar{p}_j(t)$ satisfies the integral equation*

$$\bar{p}_j(t) = a_j(t) + \int_0^t \Lambda_j(\tau)\bar{p}_j(t - \tau)d\tau.$$

The above equation can often be solved using Laplace transforms.

3 Computation of Expected Transition Time and Expected Cost

In this and the next sections, we study *general stochastic discrete event system* using a state model. Let $x(t)$ be the state of the system G at time t. We construct a discrete time embedded Markov chain $\{x_n\}$ from $\{x(t) : t \geq 0\}$: x_n is the state of the system immediately after its nth transition. The transition matrix of $\{x_n\}$ is denoted by $P = (p_{ij})_{n \times n}$, with $p_{ii} = 0, i = 1, 2, ..., n$, since selfloops are ignored. Additional information on Markov Chain can be found in [1], [2],[5] and [8].

Assume the random occurrence time t_σ of event σ is distributed in $[t_1, t_2]$ with density function $f_\sigma(t_\sigma)$. Then the expectation of t_σ is

$$\bar{t}_\sigma = \int_{t_1}^{t_2} t_\sigma f_\sigma(t_\sigma)dt_\sigma$$

Define $\Sigma_{ij} := \{\sigma \in \Sigma : \delta(\sigma, i) = j\}$, for $i, j \in Q$, but $i \neq j$.

Let $p_{ij}(\sigma)$ be the probability that system makes one step transition from state i to state j via event $\sigma \in \Sigma_{ij}$, and denote the expected time of system's one step transition from state i to state j by e_{ij}. Then the probability that the system makes one step transition via σ with the condition that the system must make one step transition from state i to state j is $p_{ij}(\sigma)/p_{ij}$. So

$$e_{ij} = \sum_{\sigma \in \Sigma_{ij}} p_{ij}(\sigma)\bar{t}_\sigma / p_{ij}$$

We assume that $\{x_n\}$ is irreducible. Then the probability of system eventually hitting state j from state i is 1: $f_{ij} = 1$, for all $i, j = 1, 2, ..., n$ [1], [2], [5], [8].

Let T_{ij} be expected time which G might take when it moves from state i to visit state j for the first time (if possible), where $i, j \in Q$, and let

$$T_k := [T_{1k}, T_{2k}, ..., T_{nk}]'.$$

Define

$$\Delta_k := \begin{bmatrix} -1 & p_{12} & \cdots & p_{1k-1} & 0 & p_{1k+1} & \cdots & p_{1n} \\ p_{21} & -1 & \cdots & p_{2k-1} & 0 & p_{2k+1} & \cdots & p_{2n} \\ \cdot & \cdot & \cdots & \cdot & \cdot & \cdot & \cdots & \cdot \\ p_{k1} & p_{k2} & \cdots & p_{kk-1} & -1 & p_{kk+1} & \cdots & p_{kn} \\ \cdot & \cdot & \cdots & \cdot & \cdot & \cdot & \cdots & \cdot \\ p_{n1} & p_{n2} & \cdots & p_{nk-1} & 0 & p_{nk+1} & \cdots & -1 \end{bmatrix} \tag{1}$$

Let $\eta := [\eta_1, \eta_2, ..., \eta_n]'$, where

$$\eta_i := \sum_{j=1, j \neq i}^{n} e_{ij} p_{ij}, i = 1, 2, ..., n.$$

The following theorem shows that the expected first passage times can be effectively computed.

Theorem 3 *The expected first passage time from any state to state k can be determined through the equation*

$$\Delta_k T_k = -\eta. \tag{2}$$

Besides the timing evolution of the system, one is also concerned with how much cost the system might incur during a series of activities. Mathematically, define an *additive* string cost function $C: \Sigma^* \times \Sigma^* \to \mathbf{R}^1$ as follows:

Suppose for each event $\sigma \in \Sigma$ and each string $s \in \Sigma^*$ such that $s\sigma \in L(G_{log})$, the cost $C(\sigma|s)$ of the system G generating σ after s is given. For any strings $s = \sigma_1\sigma_2 \cdots \sigma_l \in \Sigma^*$ and $s' \in \Sigma^*$ such that $s's \in L(G_{log})$, let $s_i = s_{i-1}\sigma_i, s_0 = s', i = 1, 2, \cdots, l$, then

$$C(s|s') = \sum_{i=1}^{l} C(\sigma_i|s_{i-1}) \text{ with } C(\varepsilon|s') = 0.$$

$\mathcal{C}(s|\varepsilon)$ is also written as $\mathcal{C}(s)$.

Let t_s be the time of system G generating string s and $T(\Sigma^*) = \{t_s : s \in \Sigma^*\}$. We can similarly define an additive time cost function $\mathcal{C}'\colon T(\Sigma^*) \to \mathbf{R}^1$ as follows: When the system generates any string $uv \in L(G_{log})$, it must generate string u, followed by string v. Let the occurrence times of uv, u and v during this performance be t_{uv}, t_u and t_v respectively. They satisfy $t_{uv} = t_u + t_v$. The cost function $\mathcal{C}'$ is required to preserve such an additive property:

$$\mathcal{C}'(t_{uv}) = \mathcal{C}'(t_u) + \mathcal{C}'(t_v),$$

$$\mathcal{C}'(t_\varepsilon) = 0.$$

We have so far described cost functions for the language SDES model. For state SDES models, we can define cost functions associated with state transitions in a completely analogous way. We now show how expected costs associated with transitions can be computed.

Let r_{ij} be the expected cost which G might incur when it moves from state i to visit state j for the first time (if possible), where $i, j \in Q$. And let

$$r_k := [r_{1k}, r_{2k}, ..., r_{nk}]'$$

Let

$$c_{ij} := \sum_{\sigma \in \Sigma_{\imath j}} p_{ij}(\sigma)\mathcal{C}(\sigma)/p_{ij}$$

which is the expected cost the system might incur when making one step transition from state i to state j; and

$$c'_{ij} := \sum_{\sigma \in \Sigma_{\imath j}} p_{ij}(\sigma)\mathcal{E}(\mathcal{C}'(t_\sigma))/p_{ij}$$

which is the expected cost of time as the system performs one step transition from state i to state j, where $\mathcal{E}(\mathcal{C}'(t_\sigma))$ can be evaluated by the following formula:

$$\mathcal{E}(\mathcal{C}'(t_\sigma)) = \int_{t_1}^{t_2} \mathcal{C}'(t_\sigma)f_\sigma(t_\sigma)dt_\sigma.$$

Let $L := [L_1, L_2, ..., L_n]'$, where

$$L_i := \sum_{j=1, j \neq i}^{n} (c_{ij} + c'_{ij})p_{ij}, i = 1, 2, ..., n.$$

Then we obtain following equation to compute the expected costs the system might incur when it moves from each state to visit a specified state k. It can be verified by the same way of Theorem 3.

Theorem 4 *The expected cost vector satisfies the equation*

$$\Delta_k r_k = -L. \tag{3}$$

Δ_k *is given by (1).*

□

Particularly, for SDES $G = (G_{log}, \Lambda)$,

$$p_{ij} = \frac{\lambda_{ij}}{\mu(i)}$$

we can rewrite those corresponding matrices and gain special formulas for the SDES.

4 Control Mechanisms

There are two control mechanisms we can consider. Controllable events can be disabled as in RW theory. This corresponds to setting the transition parameters associated with the disabled event to zero. Another mechanism is to change the parameters and cost assignments of some events. This can be achieved, for example, by replacing components in the system. In the following, we shall focus on the second control mechanism, that of parameter optimization.

The parameter optimization problem is to minimize the average time or average cost for the system to move from one state to another specified state. From results of the previous sections, we see that the equations are of the form:

$$\Delta(u)\gamma(u) + L(u) = 0$$

where u is the vector of rate and cost parameters. For simplicity, assume that each optimization problem involves only one parameter so that u is scalar-valued. The following algorithm, similar to approximation in policy space in dynamic programming, can be used to compute the optimal parameter value.

Algorithm

1. Choose arbitrary u_0. Compute γ_1 satisfying

$$\Delta(u_0)\gamma_1 + L(u_0) = 0$$

Set $k = 1$.

2. Find $u_k = argmin\{\Delta(u)\gamma_k(u) + L(u)\}$

3. Compute γ_{k+1} satisfying

$$\Delta(u_k)\gamma_{k+1} + L(u_k) = 0$$

Set $k = k + 1$. Go to 2.

The following optimality theorem can be proved.

Theorem 5 *If $\Delta(u)$ is nonsingular, for all $u \in \mathcal{U}$ and $\Delta^{-1}L(u)$ has lower bound on compact region $\mathcal{U}$, then for arbitrary $u_0 \in \mathcal{U}$, the sequence $\{\gamma_t\}$ converges to the minimal value:*

$$\lim_{t \to +\infty} \gamma_t = \gamma^*, \ \text{and} \ u^* \in \mathcal{U} : \Delta(u^*)\gamma^* + L(u^*) = 0$$

is an optimal control.

Of course, it is possible to combine logical and performance requirements. For example, one can first design a supervisor to satisfy logical requirements using RW theory. Within the supervisor structure thus obtained, optimize system parameters using the algorithm just described.

We have described in some detail performance analysis and control of stochastic discrete event systems. In the next section, we shall describe the use of these results to study performance tradeoffs.

5 Performance Tradeoffs: A Transfer Line Example

Logical requirements imposed by RW theory can sometimes be overly stringent. For example, to avoid deadlock, RW theory may require disablement of a large number of transitions, leading to a very limited behaviour for the system. If transitions leading to deadlock are very unlikely to occur, it seems to be more practical to allow the transitions to occur. When deadlock does take place, a restart procedure should be followed. By attaching a cost to the restart event, we can study the performance tradeoffs of different supervisors. In this section, we illustrate these ideas with a simple transfer line example.

5.1 Transfer Line System

We consider a 'transfer line' example, taken from [9], consisting of two machines **M1**, **M2** followed by a test unit **TU**, linked by buffers **B1** and **B2**, in the configuration shown (see *Figure 1*). A workpiece tested by **TU** may be accepted or rejected; if accepted, it is released from the system; if rejected, it is returned to **B1** for reprocessing by **M2**. Thus the structure incorporates 'material feedback'. The specification is simply that **B1** and **B2** must be protected against underflow and overflow.

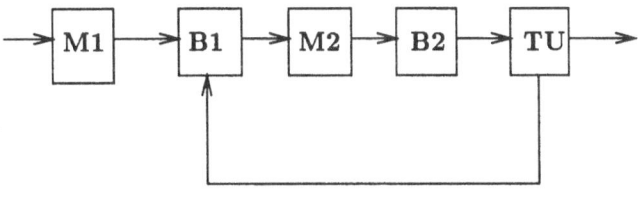

Figure 1

The component DES, displayed below, are taken to be as simple as possible (see *Figure 2*). Event 1 leads **M1** from 'idle' to 'working', picking one workpiece from the input resource, and event 2 brings **M1** back from 'working' to 'idle', feeding the workpiece into **B1**. Event 3 leads **M2** from 'idle' to 'working', picking one workpiece from **B1**, and event 4 brings **M2** back from 'working' to 'idle', processing the workpiece then feeding the finished part into **B2**. Event 5 leads **TU** from 'idle' to 'working', picking one finished part from **B2**. If the finished part is acceptable, then event 8 brings **TU** back from 'working' to 'idle', discharging it from the system. If the finished part is unacceptable, event 6 brings **TU** back from 'working' to 'idle', returning it to **B1** for reprocessing.

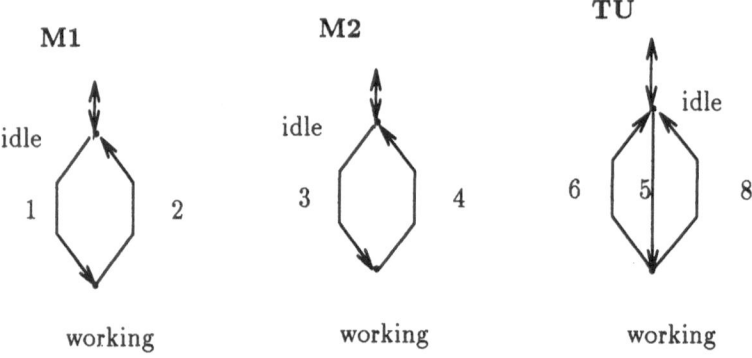

Figure 2

As the three machines work concurrently, the DES representing the transfer line is the shuffle product [9] of these three machines:

TL = shuffle(M1, M2, TU)

The specification is that during manufacturing, **B1** and **B2** must avoid underflow and overflow.

3. Supervisory Control

The control mechanism according to RW theory is:

$\Sigma_c = \{1, 3, 5\}$

Assume the capacities of **B1** and **B2** to be 1. The local specifications for **B1** and **B2** are modelled as **B1SP, B2SP** in lingustic form (see *Figure 3*). Language **B1SP** ensures that **B1** does not underflow and overflow. Once **M1** feeds one workpiece into **B1**, or **TU** rejects one finished part and sends it into **B1**, **M1** and **TU** are no longer allowed to work, until **M2** moves it out of **B1**. Language **B2SP** ensures that **B2** does not underflow and overflow. Once **M2** feeds one finished part into **B2**, **M2** are no longer allowed to work, until **TU** moves it out of **B2**. The intersection or *meet* of **B1SP** and **B2SP** is the global specification.

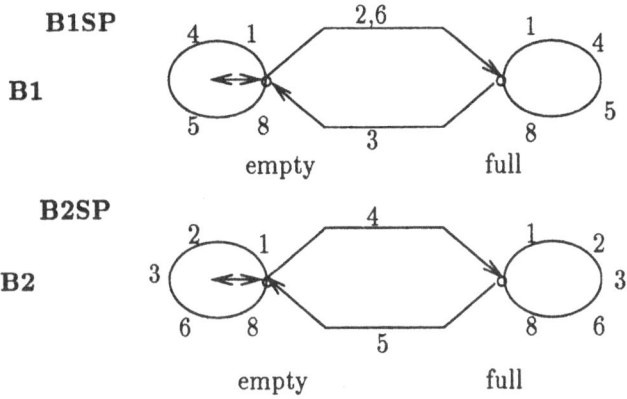

Figure 3

There are several ways to implement the specification. One way is *centralized supervisory control*. The global specification is

BSP = meet(B1SP, B2SP)

and the supervisor is computed in RW theory as

CSUP = supcon(TL, BSP) (see *Figure 4*)

The controlled system works as follows: **M1** picks one workpiece to feed **B1**; then **M2** takes it to process, then releases it into **B2**; next **TU** picks it to test, then either accepts it or returns it to **B1**. If the later case happens, we only permit **M2** to act.

CSUP

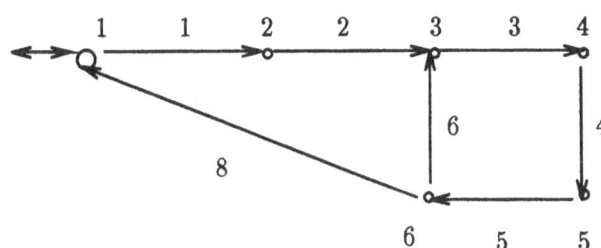

Figure 4

The above supervisor is very conservative. In particular, it never allows the machines to work in parallel. Intuitively, we expect that if M2 and TU work much faster than M1, and the probability of a defective item is small, we should allow M1 to process workpieces in parallel with M2 and TU. To handle the rare occurrence of deadlock (for example when B1 and B2 are both full), we allow a restart event to remove the deadlock. This supervisor is illustrated as a *tree* in Figure 5. Here, let a 5-tuple (M_1, M_2, TU, B_1, B_2) represent the state of M_1, M_2, TU, and the number of slots in B_1, B_2, respectively. The control policy is carried out by considering **B1** and **B2** separately. If **B1** is full and **M1** or **TU** is working, then we shut down the system to clear the troublesome slot in **B1** and keep others unchanged such that the *restart* event leads the system to state $(I, I, W, 0, 0)$ or $(W, I, I, 0, 0)$. If **B2** is full, then **M2** cannot work. We call this modular supervisory control. It is more sophisticated than the naive modular supervisor in [9]. One can ask the question: when will the risk of having to restart the system be outweighed by the increase in efficiency of material processing? Using our stochastic framework, it is now possible to study the performance tradeoffs of the centralized supervisor versus the modular supervisor.

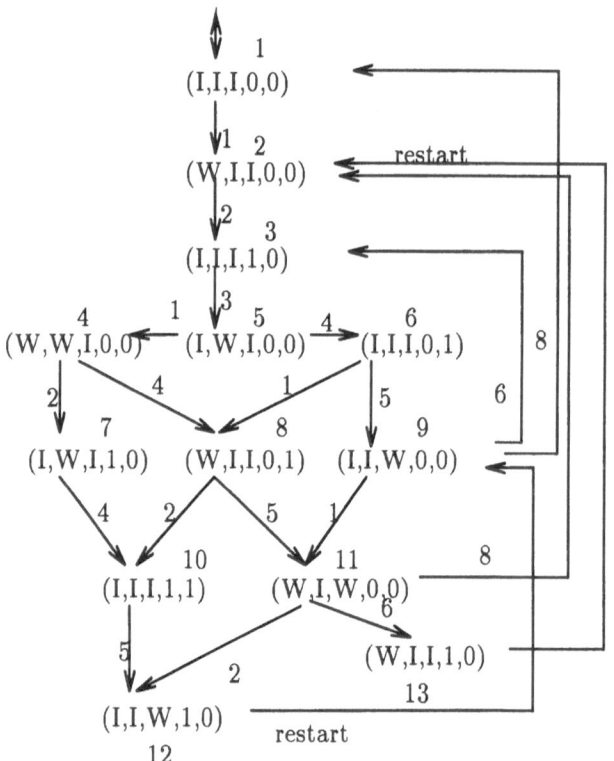

Figure 5

5.2 Numerical Example

We provide numerical details to illustrate the application of our stochastic model. Suppose the occurrence time of each event once it is eligible is an exponentially distributed random variable. The corresponding parameters of events: $1, 2, 3, 4, 5, 6, 8$ and *restart* are $\lambda_1, \lambda_2, \lambda_3, \lambda_4, \lambda_5, \lambda_6, \lambda_8$ and λ_r, respectively.

Set

$$\lambda_1 = \lambda_2 = 1, \lambda_3 = \lambda_4 = \lambda_5 = \lambda_8 = 3, \lambda_6 = 0.01 \text{ and } \lambda_r = 2$$

That λ_6 relatively small can be interpreted to mean that the unacceptable finished part is not frequently checked out by **TU**. λ_r small means once the system is shut down it takes while to restart working over again.

The cost c of each event is assigned according to the following table. Here, the cost of time delay of each event is ignored.

Table 1

event	1	2	3	4	5	6	8	restart
c	1	1	1	1	1	5	−40	5

The performance criterion we shall use is R_w, the fraction of the expected cost C_w of the system's *working cycle* with the expected time θ_w of the system's such process, which is close to *average cost rate*. The *working cycle* is the string which leads the system back to the initial state from the initial state.

Using Theorems 3 and 4, we can determine R_w for both of the two supervisors. For the *centralized supervisor*, $R_w = -10.4815$, and for the *modular supervisor*, $R_w = -12.1758$. Obviously, the *modular supervisor* works more efficiently than the *centralized supervisor* for the above set of parameter values.

5.3 Performance Studies

In this section, we look at performance tradeoffs as a function of parameter values.

(1) Assume $\lambda_r \in [0.1, 4]$ is a control variable. We study how the speed of resetting affects the performances (R_w) of the two supervisors. Figure 6 shows λ_r versus R_w. Intuitively, *modular supervisor* works more naturally, and is not as conservative as the *centralized supervisor*. Indeed, the *modular supervisor* outperforms the *centralized supervisor*, provided the speed of restart operation is fast enough.

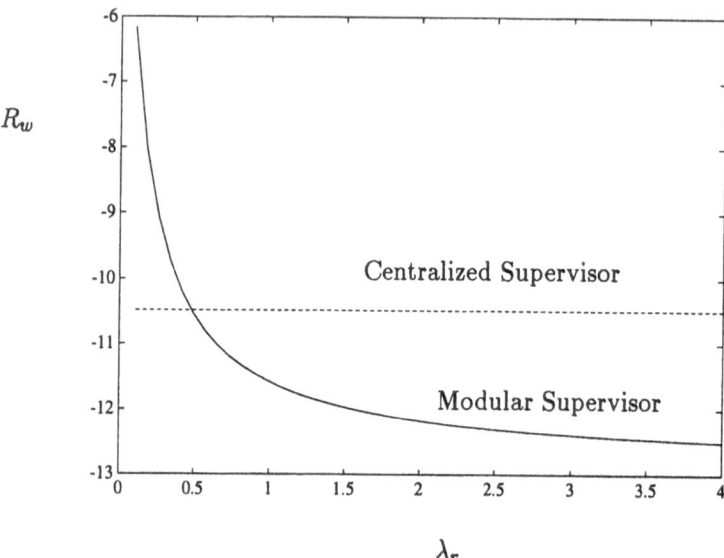

Figure 6: Cost Rate versus Restart Rate

(2) Assume $c_r \in [1, 10]$ is a control variable. We look into how the restart cost affects the performance of the two supervisors. Figure 7 shows c_r versus R_w. From this figure, we see that the *modular supervisor* performs better over the entire range of values of c_r. The reason is exact the same as that for the previous case.

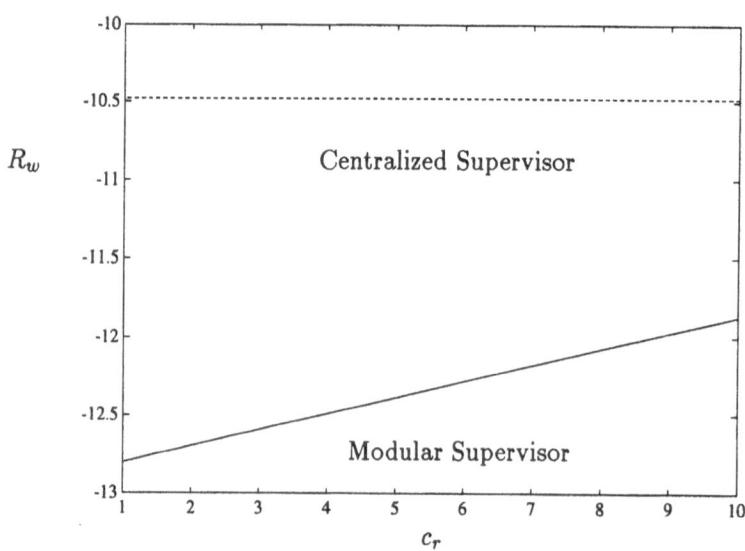

Figure 7: Cost Rate versus Cost of Restart

(3) Assume $\lambda_6 \in [0.01, 1]$ is a control variable. We consider how the probability of a defective product affects the performance of the two supervisors. If TU finds a defective item while M1 is working, a restart will take place. If this rarely happens, the restart operation will occur infrequently. Figure 8 shows λ_6 versus R_w. Again, the modular supervisor outperforms the centralized supervisor over this range of values.

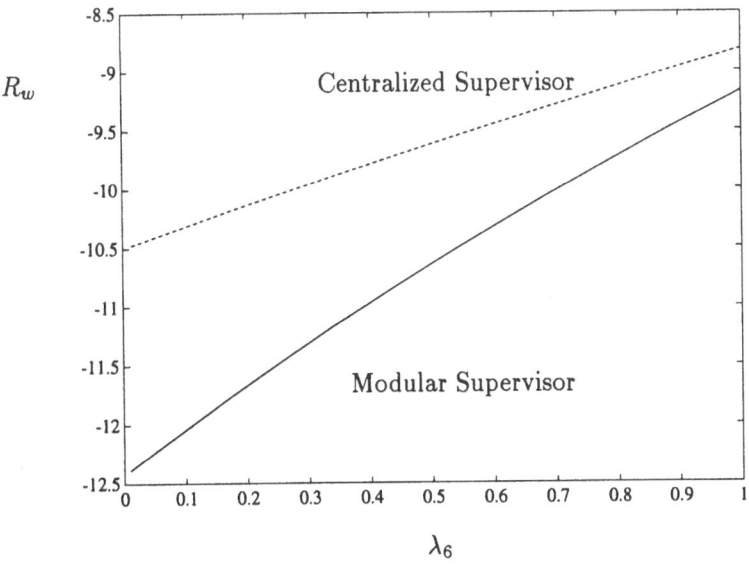

Figure 8: Cost Rate versus Rate of Recycling

6 Conclusion

The stochastic model of discrete event system within RW paradigm has been described. The performance analysis and control of SDES is studied. The analysis can then be used to examine performance tradeoffs between supervisors. These results are applied to a *transfer line* example. It enables us to obtain more tolerant supervisory control than that synthesized by using conventional RW theory.

References

[1]   C.G. Cassandras, *Discrete Event Systems: Modeling and Performance Analysis*, Aksen Associates Inc., Irwin, Homewood, Il, 1993.

[2]   K. L. Chung, *Markov chains with stationary transition probabilities*, Springer, Berlin, 1960.

[3]   G. Hoffmann, G. Franklin and C. Schaper, "Discrete event controller for a rapid thermal multiprocessor", *Proc. of American Control Conference*, Boston, MA, June 1991.

[4]   F. Lin, "A note on optimal supervisory control", *Proc. of 1991 IEEE International Symposium On Intelligent Control*, Washington D. C., August 1991.

[5]   S.M. Ross, *Introduction to probability models*, Academic Press, INC., New York, 1985.

[6]   P. J. Ramadge and W. M. Wonham, "Supervisory control of a class of discrete-event processes", *SIAM J. Control and Optimization*, Vol. 25(1), pp. 206-230, 1987.

[7]   P. J. Ramadge and W. M. Wonham, "The control of discrete event systems", *Proc. IEEE, Special Issue on Discrete Event Dynamic Systems*, Vol. 77(1), pp. 81-97, 1989.

[8]   W. M. Wonham, "Notes on control theory for finite Markov chains", Dept. of Electrical & Computer Engineering, Univ. of Toronto, 1990.

[9]   W.M. Wonham, "Notes on control of discrete-event systems", ELE 1636F/1637S, Dept. of Electrical & Computer Engineering, Univ. of Toronto, 1991.

[10]   W. M. Wonham and P. J. Ramadge, "On the supremal controllable sublanguage of a given language", *SIAM J. Control and Optimization*, Vol. 25(3), pp. 637-659, 1987.

[11]   Ling Zhu and R. H. Kwong, "Performance evaluation and optimal control of stochastic discrete event systems", *Systems Control Group Report No. 9308*, Dept. of Electrical & Computer Engineering, Univ. of Toronto, 1993.

Statistical Validation
for
Uncertainty Models[*]

Dedicated to Professor George Zames
on the occasion of his sixtieth birthday

Lawton H. Lee[†] and Kameshwar Poolla[‡]

Abstract

Statistical model validation is treated for a class of parametric uncertainty models and also for a more general class of nonparametric uncertainty models. We show that, in many cases of interest, this problem reduces to computing relative weighted volumes of convex sets in $\mathbf{R}^N$ (where N is the number of uncertain parameters) for parametric uncertainty models, and to computing the limit of a sequence $(V_k)_1^\infty$ of relative weighted volumes of convex sets in $\mathbf{R}^k$ for nonparametric uncertainty models. We then present and discuss a randomized algorithm based on gas kinetics for probable approximate computation of these volumes. We also review the existing Hit-and-Run family of algorithms for this purpose.

Finally, we introduce the notion of *testability* to describe uncertainty models that can be statistically validated with arbitrary reliability using input-output data records of sufficient (finite) length. It is then shown that some common nonparametric uncertainty models, such as those involving ℓ_1 or $\mathcal{H}_\infty$ norms, do *not* possess this property.

1 Introduction

A major theoretical and technological contribution of modern control has been the development of systematic methods to design and analyze feedback systems that are robust to *a priori* specified modelling uncertainties. This contribution

---

[*]Supported in part by the National Science Foundation under Grant ECS 89-57461, by gifts from Rockwell International, and by the Air Force Office of Scientific Research under a National Defense Science and Engineering Graduate Fellowship.

[†] Dept. of Mechanical Engineering, University of California, Berkeley CA 94720. Tel. (510)642-6152. Email: lawton@jagger.berkeley.edu

[‡] Dept. of Mechanical Engineering, University of California, Berkeley CA 94720. Tel. (510)642-4642. Email: poolla@jagger.berkeley.edu

includes $\mathcal{H}_\infty$, ℓ_1, and μ optimal control (see for example [6, 4, 25]). The principal inspiration and a substantial formulation of this theory was spawned by the seminal paper of Zames [36]. These modern *robust control methodologies* begin with an *uncertainty model* which consists of a nominal model together with a description of the various parametric and dynamic modelling uncertainties attendant with the nominal model.

There has been significant recent research devoted to *control-oriented system identification* (see for example [8, 9, 10, 14, 15, 20, 22, 23, 26, 31, 32, 33, 35] and the references cited therein).

This resurgent interest in system identification has been driven by the recognition that "classical" system identification methods, while eminently successful in producing predictive models, do not marry well with modern robust control design techniques such as $\mathcal{H}_\infty$ or ℓ_1 optimal control in that they do not provide uncertainty models, atleast in a form that is readily digestible by existing robust control design methods.

One segment of this area of research is the *model validation* problem. Here, one is interested in determining whether or not a given *a priori* uncertainty model (describing such factors as unmodelled dynamics and noise) is consistent with an available experimental input-output data record. If the model is consistent with the data, it is said to be *not invalidated*. Otherwise, it is said to be *invalidated*. Model validation problems have been previously addressed in several studies. Ljung [21] discusses model validation in the traditional identification setting. Smith and Doyle [31] address model validation problems in *frequency domain* with structured uncertainty, while Poolla *et. al.* [27] address a closely related problem involving *time-domain* data.

Note that a model can only be *invalidated* by experimental input-output data. Indeed, the results of Popper's [28] analysis of scientific theories would imply that one can never definitively conclude that a given model is valid or that it correctly describes the physical system in question. Consequently, uncertainty models used for control design tend to be conservative and are rarely invalidated by experimental data. This conservatism may at times render impossible the effective design of adequately robust controllers.

Formulating the model validation problem, then, from a *probabilistic* point of view (focusing on the *likelihood* that an uncertainty model is correct) grants the freedom to use more refined uncertainty models that can be "very probably" validated by experimental input-output data. These considerations lead to the following *statistical model validation* problem formulation:

"Given certain prior information about the true plant model, a hypothesized uncertainty model, and experimental time-domain input-output data, determine the *validation probability* that the hypothesized uncertainty model is consistent with both the *a priori* information and the experimental data record."

Statistical model validation can also be posed as a *decision problem* in which one wishes to decide whether or not the hypothesized uncertainty model is consistent with the input-output data and prior information. Equivalently, this is a test of a statistical hypothesis (see [19] for an introduction to hypothesis testing) and uses the aforementioned validation probability as a test statistic.

It is shown in this paper to reduce to a likelihood ratio test having maximum power against a given *a priori* uncertainty model.

For *parametric* uncertainty models, we show that the statistical model validation problem reduces in many cases of interest to computing the relative weighted volume of two convex sets in $\mathbf{R}^N$, where N is the number of uncertain parameters. However, the case of *nonparametric* uncertainty models is more complicated; in this case, statistical model validation reduces instead to computing the *limit* of a *sequence* $(V_k)_1^\infty$ of relative weighted volumes of sets in $\mathbf{R}^k$.

Unfortunately, exact or even approximate volume computation of convex sets in $\mathbf{R}^N$ is known to be #P-complete (see [2, 5]). We therefore examine algorithms for *probable approximate computation* of these volumes. In particular, we present and analyze a randomized algorithm based on gas kinetics as well as the existing Hit-and-Run algorithm.

A further complication in the general problem is the important issue of determining the *kinds* of uncertainty models that can be reliably verifiable (or falsifiable) by statistical methods using finite input-output data records. This property of *testability* is introduced in this paper as the ability to make arbitrarily small the probabilities of failing to accept a hypothesis when it is true or reject it when it is false, given a sufficiently long (but finite) input-output data record. Some common uncertainty models that arise in robust control, such as those involving ℓ_1 or $\mathcal{H}_\infty$ norms, are shown to in fact *lack* this very desirable property.

The remainder of this paper is organized as follows. In Section 2 we establish notation. Following this, in Section 3 we formulate the statistical model validation problem. We discuss general methods of volume computation and develop an algorithm for probable approximate volume computation in Sections 4 and 5. Section 6 introduces the notion of testability and offers a characterization of uncertainty models possessing this property. In Section 7 we summarize our results and discuss topics for further research.

All proofs are omitted from this paper; for details, the interested reader may consult [17] and [18], of which this paper is essentially a summary.

2 Preliminaries

Let $\mathbf{Z}$ denote the set of nonnegative integers. Let S denote the set of one-sided sequences of real numbers, and let $\ell_\infty \subset S$ and $\ell_1 \subset S$ respectively denote Banach spaces of bounded and absolutely summable one-sided sequences of real numbers equipped with the usual norms $\|\cdot\|_\infty$ and $\|\cdot\|_1$, respectively. Let $\pi_L : S \to S$ denote the *L-step truncation operator* with action

$$\pi_L((u_0, u_1, \ldots)) = (u_0, u_1, \cdots, u_{L-1}, 0, 0, \cdots)$$

At times, the range of π_L may be regarded as $\mathbf{R}^L$ via the natural embedding.

Let H be a causal, stable, single-input single-output, linear time-invariant discrete-time system. As is standard, we regard H as a bounded convolution operator $H : \ell_\infty \to \ell_\infty$ represented by its impulse response $h \in \ell_1$ and denote its action on an input sequence u by $Hu = h * u$.

Let $\text{vol}_w A$ denote the volume of the set $A \subseteq \mathbf{R}^N$ weighted by the function $w : \mathbf{R}^N \to \mathbf{R}$, i.e.,

$$\text{vol}_w A = \int_A w(x)\, dx$$

If $w = 1$ everywhere, we omit the subscript and let $\text{vol}\, A$ denote the unweighted volume. Let $\mathcal{B}_A$ denote the family of Borel-measurable subsets of A.

A "probability measure" can be imposed on S as follows. For each integer k, let p_k be a probability density function defined on $\pi_k S = \mathbf{R}^k$. Further assume that the collection $[p] = \{p_k\}_1^\infty$ of these density functions satisfies the recursion

$$\int_{\mathbf{R}^{k+1}} p_{k+1}(x_0, \ldots, x_{k-1}, x_k)\, dx_k \cdots dx_0 = p_k(x_0, \ldots, x_{k-1})$$

Let $\mathcal{A}$ denote the algebra of convex subsets $T \subseteq S$ for which $\pi_k T$ is Borel measurable for every k. Given $T \in \mathcal{A}$, define

$$\Pr_{[p]}\{T\} \quad := \quad \lim_{k \to \infty} \Pr_{p_k}\{\pi_k T\}$$

$$= \quad \lim_{k \to \infty} \int_{\pi_k T} p_k(x_1, \ldots, x_k)\, dx_k \cdots dx_1$$

and denote it by $\text{vol}_{[p]} T$. It can be shown that this limit exists for all $T \subseteq \mathcal{A}$ and that $\text{vol}_{[p]}\{S\} = 1$.

If P is the transition probability operator of an infinite Markov chain on $A \subseteq \mathbf{R}^N$, let $\Pr_m\{x \to B\}$ denote the probability, given $x \in A$, that P sends x into $B \in \mathcal{B}_A$ in m iterations. If $m = 1$, then omit the subscript and let $\Pr\{x \to B\}$ denote the desired probability.

If the random variable x has probability density function p, write $x \sim p$. If x is Gaussian with mean m and variance σ^2 or uniformly distributed on the interval $[a, b]$, denote p by $N(m, \sigma^2)$ or $U[a, b]$, respectively. If x is a jointly Gaussian random *vector* with mean m and covariance matrix Λ, we write $x \sim N(m, \Lambda)$.

3 Problem Formulation

In this section we formulate both the parametric and non-parametric *statistical model validation* problems.

Consider the problem of validating uncertainty models for a *linear time-invariant, causal, stable, single-input single-output discrete-time* plant H. As is standard in identification literature such as [21], assume that the identified plant is relaxed prior to the application of any inputs (i.e., its initial conditions are zero). We depict the situation considered in Fig. 1.

Remark 3.1 Our approach immediately generalizes to multi-input multi-output systems and to noise models of the form $y = Hu + Gn$. For the sake of clarity, however, we will limit our discussion in this paper to single-input single-output systems and additive measurement noise.

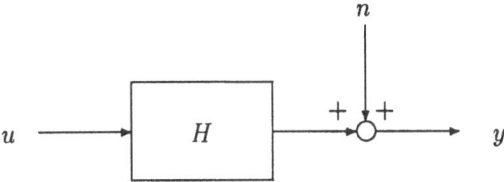

Figure 1: The plant

3.1 Parametric Case

We assume *a priori* that the "true" impulse response $h^\circ = (h_0^\circ, h_1^\circ, \ldots)$ of H lies in some convexly parameterized *model set* $\mathcal{F} \subset \ell_1$. More precisely, h° is assumed to lie in the set

$$\mathcal{F} = \{h(\theta) : \theta \in \Theta_{\mathcal{F}} \subset \mathbf{R}^N\} \subset \ell_1 \tag{3.2}$$

where $\Theta_{\mathcal{F}}$ is a convex set. We also assume that (3.2) is an *identifiable* parameterization, i.e., $h(\theta_1) = h(\theta_2)$ if and only if $\theta_1 = \theta_2$. Let θ° denote the "true" parameters corresponding to h°, or equivalently, $h^\circ = h(\theta^\circ), \theta^\circ \in \Theta_{\mathcal{F}}$. We treat θ° as a fixed but unknown vector and model this uncertainty by regarding θ° as a particular outcome of a *random* vector θ. We incorporate any additional *a priori* knowledge about the plant by specifying a probability density function $p_\theta(\theta)$ on the parameter set $\Theta_{\mathcal{F}}$. In the absence of such information, we may consider θ to be uniformly distributed on $\Theta_{\mathcal{F}}$.

We also assume that the noise $n = (n_0, n_1, \ldots)$ is drawn from the *noise set* $\mathcal{N}$ and that, for any L, the random vector $\pi_L n$ has known joint density $p_{\pi_L n}(\pi_L n)$. We make the reasonable assumption that the parameters θ and noise process n are statistically independent.

Remark 3.3 By stating that θ° represents the "true" parameters and that $\theta^\circ \in \Theta_{\mathcal{F}}$, we mean the following: given a *noisy* input-output record (u, y) produced by the plant, there exists a parameter vector $\theta^\circ \in \Theta_{\mathcal{F}}$ capable of generating that input-output data in a manner consistent with the noise model; in fact, all parameters in some neighborhood of θ° also satisfy this condition. In other words, the prior model set is rich enough to explain "sufficiently well" the finite input-output records produced by the plant, so that unmodelled dynamics need not be a further concern. Note that *in no way* need θ° or the model structure exactly represent the *actual* dynamics of the physical plant. $\square$

We regard this initial model set $\mathcal{F}$ as a conservative description of the true plant. In light of Remark 3.3, this conservatism may be extreme; consequently, it may be impossible to design adequate feedback controllers for the plant that are robust against the model set $\mathcal{F}$.

In view of this difficulty, we *hypothesize* that the true plant h° belongs to the smaller, more manageable model set $\mathcal{G} \subset \mathcal{F}$, convexly parameterized as

$$\mathcal{G} = \{h(\theta) : \theta \in \Theta_{\mathcal{G}} \subset \Theta_{\mathcal{F}}\} \subset \mathcal{F} \tag{3.4}$$

where $\Theta_{\mathcal{G}}$ is a convex set. In effect, we hypothesize that $\theta^\circ \in \Theta_{\mathcal{G}}$ in the sense of Remark 3.3. This hypothesis may, for instance, have arisen from further identification of the plant.

Our objective is to statistically test this hypothesis. In particular, we are given *a posteriori* data acquired from an input-output experiment of finite duration $L > N$ in which we apply an input sequence

$$u = (u_0, u_1, \ldots, u_{L-1}, *, *, \ldots)$$

to the physical system and observe the first L samples of the noisy output

$$y = (y_0, y_1, \ldots, y_{L-1}, *, *, \ldots) = h^\circ * u + n$$

The admissible inputs that we may apply are drawn from the *input set*

$$\mathcal{U} = \{u \in \ell_\infty : \|u\|_\infty \leq 1, u_0 \neq 0\}$$

arising from a reasonable restriction of applying only bounded inputs to the physical system. Without loss of generality, we can take the bound to be one and can assume that $u_0 \neq 0$.

The output y, which is a function of the noise n and the parameters θ, is regarded as a random sequence. In this case, for a fixed input u, the random vectors θ and $\pi_L n$ induce joint densities for $(\theta, \pi_L y) \in \mathbf{R}^{L+N}$ and $\pi_L y$. Since θ and n are independent, these are given by

$$p_{\theta, \pi_L y}(\theta, \pi_L y) = p_\theta(\theta) \, p_{\pi_L n}(\pi_L(y - h(\theta) * u)) \tag{3.5}$$

and

$$p_{\pi_L y}(\pi_L y) = \int_{\mathbf{R}^N} p_{\theta, \pi_L y}(\theta, \pi_L y) \, d\theta \tag{3.6}$$

We now introduce some key definitions.

Definition 3.7 The *a posteriori validation probability* $P_L(u, y)$ is the probability that the hypothesis is true, given both the prior information and the input-output data up to time L. More precisely,

$$P_L(u, y) = \Pr\{\theta \in \Theta_{\mathcal{G}} \mid \theta \in \Theta_{\mathcal{F}}, \pi_L u, \pi_L y\}$$

Note that if a data record $\{\pi_L u, \pi_L y\}$ is given, then $P_L(u, y)$ is a number. For *a priori* analysis, on the other hand, y is a random sequence, hence $P_L(u, y)$ is a random variable.

We use $P_L(u, y)$ in the following way. Given a threshold $\epsilon \in (0, 1/2)$, we say that the hypothesis is *probably true* if $P_L(u, y) > 1 - \epsilon$ and *probably false* if $P_L(u, y) < \epsilon$. In the event $\epsilon \leq P_L(u, y) \leq 1 - \epsilon$, we make no decision regarding the statistical validity of the hypothesis.

Definition 3.8 Given a fixed input-output data record $\{\pi_L u, \pi_L y\}$, the *consistency set* $\mathcal{C}_L(u, y)$ is the set of parameter vectors consistent with the data, i.e.,

$$C_L(u, y) = \{\theta \in \mathbf{R}^N : \pi_L(y - h(\theta) * u) \in \mathcal{N}\}$$

In many situations of interest, including ARX and FIR model structures, $C_L(u, y)$ happens to be *convex*. Note also that we have suppressed the dependence of $C_L(u, y)$ and $P_L(u, y)$ on the *a priori* information and the particular hypothesis, thus restricting the explicit arguments to the *a posteriori* data $\{\pi_L u, \pi_L y\}$.

The following simple result relates the calculation of $P_L(u, y)$ to a volume computation problem.

Lemma 3.9 The validation probability is equal to the weighted relative volume

$$P_L(u, y) = \frac{\text{vol}_w(A_2)}{\text{vol}_w(A_1)} = \frac{\int_{A_2} w(\theta) d\theta}{\int_{A_1} w(\theta) d\theta}$$

where $A_1 = C_L(u, y) \cap \Theta_{\mathcal{F}}$, $A_2 = C_L(u, y) \cap \Theta_{\mathcal{G}}$, and

$$
\begin{aligned}
w(\theta) &= p_{\theta, \pi_L y}(\theta, \pi_L y) \\
&= p_\theta(\theta) \, p_{\pi_L n}(\pi_L(y - h(\theta) * u))
\end{aligned}
$$

Note that the weighting function w has support $A_1 = C_L(u, y) \cap \Theta_{\mathcal{F}}$. The assumptions made in Remark 3.3 assure that the fraction $P_L(u, y)$ is well-defined.

Statistical model validation therefore reduces to computing relative weighted volumes of sets in $\mathbf{R}^N$, where N is the number of uncertain parameters observed in $\mathcal{F}$ and $\mathcal{G}$. These sets are often convex.

3.2 Non-parametric Case

Our development of statistical model validation for nonparametric models closely parallels that of the previous subsection. We begin by assuming *a priori* that the "true" impulse response $h^\circ = (h_0^\circ, h_1^\circ, \ldots)$ of H lies in some convex *model set* $\mathcal{F} \subseteq \ell_1$, $\mathcal{F} \in \mathcal{A}$. The impulse response h° is treated as a (fixed-but-unknown) realization of the random sequence $h \in \mathcal{F}$. As before, any additional prior knowledge about the plant is incorporated by specifying on $\mathcal{F}$ a probability measure $[pf]$ characterized by a family of joint densities $pf_k : \mathbf{R}^k \to \mathbf{R}$, each with support $\pi_k \mathcal{F}$. In the absence of such information, one can use uniform densities.

We hypothesize that the true impulse response h° lies in the smaller, more manageable convex model set $\mathcal{G} \subseteq \mathcal{F}, \mathcal{G} \in \mathcal{A}$. The objective of statistical model validation is to test this hypothesis.

Again, we assume that the random measurement noise sequence $n = (n_0, n_1, \ldots)$ is drawn from the *noise set* $\mathcal{N} \in \mathcal{A}$ and has known probability measure $[pn]$ characterized by joint densities pn_L, each with support $\pi_L \mathcal{N}$. In addition, assume that the sequences n and h° are statistically independent.

To validate our hypothesis that $h° \in \mathcal{G}$, we are given *a posteriori* data acquired from an L-sample input-output experiment in which an input sequence $u = (u_0, u_1, \ldots, u_{L-1}, *, *, \ldots)$ is applied to the physical system and the first L samples of the noisy output

$$y = (y_0, y_1, \ldots, y_{L-1}, *, *, \ldots) = h° * u + n$$

are observed. The admissible inputs are drawn from the the *input set* $\mathcal{U}$ as defined earlier.

Since the output y is a function of the noise n, y can be treated as a random sequence. In this case, for a fixed input u, the random sequences h and n induce joint densities for $(\pi_L y, \pi_k h)$ and for $\pi_L y$. These are given by

$$p_{\pi_L y, \pi_k h}(\pi_L y, \pi_k h) = pf_k(\pi_k h) \cdot pn_L(\pi_L(y - h * u))$$

and

$$py_L(\pi_L y) = \int_{\mathbf{R}^k} p_{\pi_L y, \pi_k h^*}(\pi_L y, \pi_k h) \, dh_0 \cdots dh_{k-1}$$

The *a priori validation probability* P_0 is the probability that the hypothesis is true, given the prior information and *no* input-output data. More precisely,

$$P_0 = \Pr\{h \in \mathcal{G} \,|\, h \in \mathcal{F}\}$$

The *a posteriori validation probability* (or simply *validation probability*) $P_L(u, y)$ is the probability that the hypothesis is true, given both the prior information and the input-output data up to time L. More precisely,

$$P_L(u, y) = \Pr\{h \in \mathcal{G} \,|\, h \in \mathcal{F}, \pi_L u, \pi_L y\}$$

We can define the *consistency set* $C_L(u, y) \in \mathcal{A}$ as before to be the set of impulse responses consistent with the data, i.e.,

$$C_L(u, y) = \{h \in \ell_1 : \pi_L(y - h * u) \in \mathcal{N}\}$$

The following simple result relates the calculation of $P_L(u, y)$ to a volume computation problem.

Lemma 3.10 The validation probability is equal to

$$
\begin{aligned}
P_L(u, y) &= \lim_{k \to \infty} V_k \\
&= \lim_{k \to \infty} \frac{\mathrm{vol}_{w_k} \pi_k [\mathcal{G} \cap C_L(u, y)]}{\mathrm{vol}_{w_k} \pi_k [\mathcal{F} \cap C_L(u, y)]} \\
&= \frac{\mathrm{vol}_{[w]}(\mathcal{G} \cap C_L(u, y))}{\mathrm{vol}_{[w]}(\mathcal{F} \cap C_L(u, y))}
\end{aligned}
$$

where $w_k(\pi_k h) = pf_k(\pi_k h) \cdot pn_L(\pi_L(y - h * u))$.

Statistical model validation therefore reduces to computing the limit of a sequence $(V_k)_1^\infty$ of relative weighted volumes of sets in $\mathbf{R}^k$ that are often convex.

4 Methods of Volume Computation

As discussed in Section 3, determining validation probabilities often reduces to computing relative weighted volumes of convex sets in $\mathbf{R}^N$. In this section we discuss methods of computing these volumes.

Unfortunately, exact or even approximate volume computation of a convex set is very difficult. More precisely, [2, 13] have shown that volume computation for polytopes in $\mathbf{R}^N$ or for general convex sets in $\mathbf{R}^N$ under a separating hyperplane oracle is NP-hard. In fact, the situation is even worse; [5] shows that volume computation is $\#P$-complete (i.e. as hard as determining the number of tours having cost less than γ in the Traveling Salesman Problem on N cities).

In view of these results, we are compelled to abandon exact or even approximate computation of validation probabilities. Instead, we shall be content with *probable approximate computation* (PAC) of these probabilities. There has been extensive research on the PAC model of computation, particularly in the context of learning theory (see [34], for example). Polynomial-time algorithms have been developed in [1] for the problem of volume computation with log-concave weighting functions. In the context of statistical model validation, the weighting functions that arise are often Gaussian densities, which *are* log-concave.

We will focus here on simpler algorithms than those found in [1]; this is reasonable because in statistical model validation we are mainly interested in calculating relative volumes that are close to either zero or one. In particular, we now describe *rejection sampling*, which is the simplest randomized algorithm for volume computation.

Given two sets $A_2 \subset A_1 \subset \mathbf{R}^N$, we can estimate $V = \mathrm{vol}_w(A_2)/\mathrm{vol}_w(A_1)$ by selecting random points in A_1 (drawn according to the distribution w) and invoking an oracle to determine how many of these lie in A_2. Then $\hat{V}$, the fraction of the points drawn from A_1 that lie in A_2, is the rejection sampling estimate of V. The following theorem, which is obtained via Hoeffding bounds, provides an upper bound on the number of samples needed to assure an accurate estimate.

Theorem 4.1 Let $0 < \epsilon \leq 1 - V$, $0 < \delta < 1$. Then computation of V to within an additive accuracy ϵ with confidence $1 - \delta$, i.e.,

$$\Pr\{|\hat{V} - V| < \epsilon\} \geq 1 - \delta$$

is assured by testing $N \geq \frac{1}{2\epsilon^2} \ln(\frac{2}{\delta})$ samples.

The essential difficulty of rejection sampling is in generating points from A_1 distributed according to the weighting function w. The difficulty of this problem stems not from w but from the geometry of the set A_1.

One possible approach to this problem is to circumscribe A_1 by a hypercube $\mathcal{H}$, extend w to $\mathcal{H}$, generate points in $\mathcal{H}$ according to this extended distribution, and retain only those points that happen to lie in A_1. Generating points uniformly on $\mathcal{H}$ is easy. The shortcoming of this "Box & Throw" method is that it tends to be inefficient: exponentially many (in N) points are rejected.

Another approach is to impose a fine grid on A_1, select an initial point in A_1, and generate subsequent points by an appropriate random walk on this grid. Generating the initial point is easily done via linear programming. The shortcoming of this "Grid & Walk" approach is that the resulting Markov chain may not converge rapidly enough to the target distribution on A_1. As a result, producing a collection of points that is faithfully distributed as desired requires a very long walk. Using "Rook's move" or "Queen's move" walks as in [1] can accelerate convergence, but the points generated are still selected from only a discrete grid and not from a continuum.

In our work, we have developed alternative Markov chain-based scheme (based on gas kinetics) for generating points *uniformly* drawn from the bounded, convex set $A_1 \subset \mathbf{R}^N$.

5   Gas Kinetics Point Generation

In this section we present the randomized Gas Kinetics Point Generator (GKPG) algorithm for generating uniformly distributed points used in rejection sampling (we will suppress the weighting function for now). We begin by giving an algorithm description. Following this, we analyze statistical properties of our algorithm and summarize our simulation experience. Finally, we discuss generalizations of GKPG and previous work on random point generation.

5.1   Algorithm Description

We wish to generate a collection $\{\theta(k)\}_0^{K-1}$ of K points nearly uniformly distributed on the bounded, convex set $A \subset \mathbf{R}^N$. In the context of statistical model validation, A is often a polytope defined by P hyperplanes, where P equals the number of hyperplanes defining $\Theta_{\mathcal{F}}$ ($2N$ for a box) plus the number of hyperplanes defining $\mathcal{C}(u, y, L)$ (typically $2L$ if noise is bounded and zero otherwise). We will discuss the extension of GKPG to general bounded (measurable) sets in Section 5.4.

Imagine that we fill this closed volume A with a large number of hot gas particles. We mark one of these particles and give it an initial position $\theta(0) \in A$. This requires finding a point in A and can be done rapidly using linear programming. We make the following assumptions on the subsequent motion of this particle:

1. The marked particle will collide with other particles only at times $t = T, 2T, \ldots, (K-1)T$ (without loss of generality, we can set $T = 1$).

2. Immediately after the k-th such collision, the particle's velocity is a random vector $v(k) \in \mathbf{R}^N$. Moreover, the random sequence $\{v(k)\}_0^\infty$ is i.i.d. with density function $p_{v(k)}(v) = \phi(v))$ (for example, $\phi = N(0, \sigma^2 I)$).

3. Any collisions with the boundary ∂A are assumed to be elastic, and these may occur at any time.

To avoid confusion, we will hereafter refer to collisions with the boundary as *reflections* and to collisions with particles simply as *collisions*.

Each iteration generates a position $\theta(k)$ corresponding to the kth collision occurring at time kT. The algorithm performs $K - 1$ iterations to provide the points $\{\theta(k)\}_0^{K-1}$ for use in rejection sampling. The computational burden of this algorithm is modest: it requires $(2NP + P + 3N + 2)r + (2NP + P + N)$ flops to compute each point $\theta(i)$ where r is the number of reflections.

5.2   Statistical Properties

It is clear that the algorithm above defines a discrete-time Markov chain on the continuous state-space A. We now present an analysis of the statistical properties of this Markov chain. As is well-known, the steady-state distribution of this chain is determined by the eigenfunction associated with the largest eigenvalue λ_1 of the one-step probability transition operator P, and the rate of convergence to this steady-state distribution is governed by the magnitude of the second-largest eigenvalue λ_2; see [24] for details.

In the case where A is the hyperrectangle $[0, a_1] \times [0, a_2] \times \ldots \times [0, a_N]$ we can compute the eigenvalues and eigenfunctions of the probability transition operator P *explicitly*. This is somewhat surprising and we are able to do compute the eigenstructure of P using elementary Fourier methods. More precisely, we have the following:

Theorem 5.1 Let $A = [0, a_1] \times \ldots \times [0, a_N]$. Suppose the velocity distribution $\phi(v)$ is a function of $\|v\|_2$ alone. Let $\Phi(\omega)$ be the moment generating function of

$$\Phi(\omega) = E(e^{j\omega v})$$

Then, the steady-state distribution of the GKPG Markov chain is the uniform distribution on A. In addition, the eigenvalues of P are

$$\left\{ \prod_{n=1}^{N} \Phi\left(\frac{m_n \pi}{a_n}\right) : m_1, \ldots, m_N \in \mathbf{Z} \right\}$$

The second largest eigenvalue of P is therefore $\lambda_2(P) = \Phi\left(\frac{\pi}{a_{max}}\right)$ where a_{max} is the length of the largest side of A.

Example 5.2 Let $A = [0, a_1] \times \ldots \times [0, a_N]$, and let $v(k) \sim N(0, \sigma^2 I)$ for all k. Then by Theorem 5.1 the steady-state distribution is uniform and $\lambda_2 = e^{-\pi^2 \sigma^2 / 2a_{max}^2}$ where a_{max} is the largest side of the hyperrectangle A. Observe that λ_2 decreases as σ increases or a_{max} decreases. This is consistent with our intuition that the rate of convergence is faster as velocity increases and/or the dimensions of A decrease. □

In the general case where A is a bounded, convex set in $\mathbf{R}^N$, we can establish the following result.

Theorem 5.3 The probability transition operator P is Hermitian if and only if the velocity distribution $\phi(v)$ is a function of $\|v\|_2$ alone. In this case, the steady-state distribution of the GKPG Markov chain is the uniform distribution on A.

In order to characterize the rate of convergence, we can use the coupling arguments of Diaconis (see [24]) to obtain a loose upper bound for $|\lambda_2|$ in the general case.

Proposition 5.4 Let A be a bounded, convex subset of $\mathbf{R}^N$. Then λ_2 is bounded by

$$|\lambda_2| \leq 1 - \text{vol}(A) \cdot \inf_{x \in A} \inf_{B \subseteq A} \inf_{y \in B} \phi(y - x)$$

Example 5.5 Let A be a bounded, convex subset of $\mathbf{R}^N$, and let the velocity jointly Gaussian with zero mean and variance σ^2 in each coordinate, i.e., $v(k) \sim N(0, \sigma^2 I_n)$ for all k. Then

$$|\lambda_2| \leq 1 - \frac{\text{vol}(A)}{(\sigma\sqrt{2\pi})^N} \exp\left[-\frac{(\text{diam } A)^2}{2\sigma^2}\right]$$

where diam $A = \sup_{x,y \in A} \|y - x\|_2$. $\qquad\qquad\qquad\qquad\qquad\qquad\Box$

The general case resists a more accurate analysis. Indeed, $A_2 \subseteq A_1$ does *not*, for example, imply that $|\lambda_2(P)|_{A=A_2} \leq |\lambda_2(P)|_{A=A_1}$. Nevertheless, both our analysis here and simulations in MATLAB suggest that the convergence of this Markov chain is faster than for random walks and is accelerated by higher velocities. This makes sense intuitively, considering the physical gas kinetics model on which the algorithm is based: the current location of a gas particle in a closed volume becomes less correlated to its past locations as the gas temperature (and hence its *rms* velocity) increases.

5.3   Simulation Results

Here we discuss our comparison by simulation of the GKPG algorithm's ability to approximate the uniform distribution on sets of the form $A = [-a, a]^N$ (hypercubes of size $2a$ in $\mathbf{R}^N$, centered at the origin) with that of MATLAB's random number generator *rand*. The velocity distribution used is $v(k) \sim N(0, \sigma^2 I)$. We primarily observe the effect of the variance σ^2 of the velocity on the algorithm's rate of convergence in K (the number of generated points) to the uniform distribution. Dependence of the algorithm's performance on N and the geometry of A is not straightforward.

We consider two performance criteria for the uniformity of the point distributions: (*a*) the squared norm of the mean, and (*b*) the approximated L_2 norm of the difference between the sample density and the (desired) uniform density. The sample density is computed from a histogram of generated points. We find that for sufficiently large σ the algorithm recovers the convergence rate (as measured by criteria (a) and (b)) of the random number generator; roughly speaking, the two criteria vary as $1/K$.

We obviously cannot make the velocity very large, because of the cost incurred by computing too many reflections. To navigate this tradeoff between convergence rate and computation time, then, we need to choose σ and K suitably to minimize the computational cost J_c, subject to an upper limit on some cost J_e associated with the errors (deviations from uniformity) in the distribution.

In the general case, the relationship between $\bar{r}$ (the average frequency of reflection) and $\sigma_1, \ldots, \sigma_N$ is highly dependent on the geometry of A and hence is rarely known *a priori*. One possible way to compensate for this is through *adaptive* selection of $\sigma_1, \ldots, \sigma_N$, based on observing the number of reflections $r(k)$ of each iteration and specifying a desired frequency of reflection (such as $\bar{r} = 1$ or $\bar{r} = N$). such a strategy is complicated, though, in geometries for which $r(k)$ varies widely; moreover, there is no guarantee that the variances will converge to advantageous values.

5.4 Discussion

In this section, we consider ways to use GKPG with non-uniform weighting functions and with sets that are not convex or have nonlinear constraints.

Suppose A is a bounded but possibly non-convex set of the form

$$A = \{\theta \in \mathbf{R}^N : g(\theta) < c\}$$

where $c \in \mathbf{R}^P$ and $g : \mathbf{R}^N \to \mathbf{R}^P$ is continuous but nonlinear. Assume A is also connected (if the actual set of interest is not connected, treat A as one of its connected components). Then GKPG readily extends to this case, provided that:

1. The constraints are *invertible*, i.e., given a point $\theta_0 \in A$, displacement $x \in \mathbf{R}^N$, and index i, we can determine all (if any) intersection points of the line $\{\theta_0 + ax : a \in \mathbf{R}\}$ with the ith hypersurface $g_i(\theta) = c_i$.

2. The $(N-1)$-dimensional hypersurfaces $g_i(\theta) = c_i$ are differentiable.

These assumptions allow us to calculate impact points and tangent planes for reflections in step 3(c) (see Figure 2). If this transition probability operator P is Hermitian, then by Theorem 5.3 the steady-state distribution is uniform.

Suppose the weighting function w is not uniform. Then the point generation problem is equivalent to generating uniformly distributed points in the set

$$A_+ = \left\{ \begin{bmatrix} \theta \\ \eta \end{bmatrix} \in \mathbf{R}^{N+1} : \theta \in A, \eta \in \mathbf{R}, \eta < w(\theta) \right\} \subset \mathbf{R}^{N+1}$$

(which is the region under the graph of w), then projecting the points onto A. This assumes that the graph of w (as a hypersurface in $\mathbf{R}^{N+1}$) satisfies the preceding conditions 1 and 2 on nonlinear constraints.

To avoid increasing the dimension and introducing a nonlinear constraint, we can instead generate points $\{\theta(k)\}_0^{K-1}$ on A as before but use *weighted* sums to calculate the relative volume of A_2 and A_1, i.e.,

$$\frac{\text{vol}_w(A_2)}{\text{vol}_w(A_1)} = \frac{\sum_{\theta(k) \in A_2} w(\theta(k))}{\sum_{k=0}^{K-1} w(\theta(k))}$$

However, not only does Theorem 4.1 no longer apply here, but numerical problems may result, as when w is concentrated about one point and so few points are generated that we fail to capture the cluster point.

5.5 Hit-and-Run Algorithms

We have seen previously that using GKPG does impose some limitations on volume computation. For example, the lack of prior knowledge of the geometry of A can cause a conservative choice of ϕ to multiply the computation time unnecessarily. Also, any nonlinear constraints must be both invertible and differentiable (as hypersurfaces). An alternative approach to point generation is the *Hit-and-Run* family of mixing algorithms, that is similar to GKPG but does not share these computational and analytical shortcomings. We offer a general discussion of this method here (see [3, 11, 30] for more detailed analyses).

We wish to generate a sequence $\{x_n\}_0^\infty$ of points in the bounded open set $A \subseteq \mathbf{R}^N$ approximating the almost everywhere continuous joint density function w (the *target density*). Let $d \in \mathbf{R}^N$ be a random vector on the N-dimensional unit sphere with joint density ν (the *direction density*). We can describe the general Hit-and-Run algorithm as follows.

1. Choose a starting point $x_0 \in A$

2. For $k = 1, 2, \ldots$, generate point $x_k \in A$

    (a) Choose $d_k \in \mathbf{R}^N$ according to the density ν

    (b) Choose $\alpha_k \in \mathbf{R}$ from the set

    $$\Lambda_k = \{\alpha \in \mathbf{R} : x_k + \alpha d_k \in A\}$$

    according to the density

    $$w_k(\alpha) = \frac{w(x_k + \alpha d_k)}{\int_{\beta \in \Lambda_k} w(x_k + \beta d_k) \, d\beta}$$

    (c) Set $x_k = x_{k-1} + \alpha_k d_k$

Each density w_k represents a conditionalization of the target density w (a normalized "slice" of w) along a line.

Definition 5.6 The direction density ν is *full dimensional* if its support spans $\mathbf{R}^N$. The connected components of A are ν-*communicating* if, for all components $A_1, A_2 \subset A$ there exists $x_0 \in A_1$ and $k \geq 1$ for which $\Pr_k\{x_0 \to A_2\} > 0$.

According to [3], if ν is full dimensional and the components of A are ν-communicating, then the Hit-and-Run algorithm converges to the target density w in *total variation*, i.e.,

$$\lim_{k \to \infty} \Pr_k\{x_0 \to B\} = \text{vol}_w(B)$$

uniformly in $B \in \mathcal{B}_A$ for any starting point $x_0 \in A$.

Consider the special case where ν is uniform on the unit sphere and w is uniform on A (hence, f_k also is uniform). Smith [30] shows that, for all $K \geq 1$, starting points $x_0 \in A$, and subsets $B \in \mathcal{B}_A$, the total variation of this *Hypersphere Directions* algorithm (which closely resembles GKPG) is bounded by

$$|\Pr_k\{x_0 \to B\} - \text{vol}_w(B)| < \left(1 - \frac{\gamma}{N2^{N-1}}\right)^{k-1}$$

where γ is the (unweighted) relative volume of A and the smallest sphere containing A.

The Hit-and-Run algorithms clearly have several computational advantages over GKPG. For example, they require no prior knowledge of the dimensions of A; we need not worry about choosing a proper density for velocities. The P defining constraints, if nonlinear, need only be invertible (as defined in Section 5.4). If A is a polytope and w is uniform, then each point requires only $2NP + N$ time to generate; compare this with the $2NP + P$ time per *reflection* required by GKPG.

Forming the conditional densities f_k may be complicated at times (unless, for example, w is Gaussian or uniform), but just as with GKPG we can instead generate uniformly distributed points in the region $A_+ \subset \mathbf{R}^{N+1}$ under the graph of w and projecting onto A.

6 Testability

This section addresses the issue of decision errors in statistical model validation. Testability is introduced as a property of hypotheses for which, based on the decision rule in the previous section, the probability of making an incorrect decision can be made arbitrarily small by choice of input and experiment length. The testability (or lack thereof) of a particular class of uncertainty models is then examined.

Definition 6.1 For a given input $u \in \mathcal{U}$, experiment length L, and threshold $\epsilon \in (0, 1/2)$, the *error plus indecision probability sum* is the sum of the probability of deciding that $h^\circ \in \mathcal{G}$ when in fact $h^\circ \notin \mathcal{G}$, the probability of deciding that $h^\circ \notin \mathcal{G}$ when in fact $h^\circ \in \mathcal{G}$, and the probability of making no decision. More precisely,

$$\begin{aligned} E(u, L, \epsilon) \; = \; & \Pr\{P_L(u, h^\circ * u + n) > \epsilon \,|\, h^\circ \notin \mathcal{G}\} + \\ & \Pr\{P_L(u, h^\circ * u + n) < 1 - \epsilon \,|\, h^\circ \in \mathcal{G}\} \end{aligned}$$

which may exceed 1, as these events are not mutually exclusive.

One important question to ask when testing a hypothesized uncertainty model is whether or not one can, given a sufficient amount of input-output data, determine the probable truth or falsity of the hypothesis with arbitrary reliability (i.e., make $E(u, L, \epsilon)$ arbitrarily small). The following definition introduces this notion of *testability*.

Definition 6.2 $\mathcal{G}$ is *testable* relative to $\mathcal{F}$ if there exists an input $u° \in \mathcal{U}$ for which

$$\text{plim}_{L \to \infty} P_L(u°, h° * u + n) = \begin{cases} 1 \text{ if } h° \in \mathcal{G} \\ 0 \text{ if } h° \notin \mathcal{G} \end{cases}$$

which is equivalent to $\lim_{L \to \infty} E(u°, L, \epsilon) = 0$.

Clearly, this is a very desirable property for an uncertainty model to possess, for if an uncertainty model is *not* testable, one must question whether or not input-output experimentation is a suitable approach to validating such models.

The following proposition provides a necessary condition for testability that is independent of the noise statistic.

Proposition 6.3 If $\mathcal{G}$ is testable relative to $\mathcal{F}$, then $0 < \text{vol}_{[pf]}\mathcal{G} < 1$.

It follows that an uncertainty model composed of ball in ℓ_1, $\mathcal{H}_\infty$, or some other system or induced norm is in general *not* testable relative to larger ball in the same norm. The following corollary demonstrates this.

Corollary 6.4 Let the prior model set be $\mathcal{F} = \{h \in \ell_1 : \|h\| \leq \gamma\}$ (where $\|\cdot\|$ is some system or induced norm) with uniform finite joint densities

$$pf_k(h_0, \ldots, h_{k-1}) = \begin{cases} \frac{1}{\text{vol}_{\pi_k \mathcal{F}}} & \pi_k h \in \pi_k \mathcal{F} \\ 0 & \text{otherwise} \end{cases}$$

and let the hypothesized model set be $\mathcal{G} = \{h \in \ell_1 : \|h\| \leq \beta < \gamma\}$. Then $\mathcal{G}$ is not testable relative to $\mathcal{F}$.

This result can be further interpreted as follows:
"It is impossible to ensure arbitrary reliability when using *finite* amounts of input-output data to test uncertainty models for which *arbitrarily long delays* (the impulse response tail) account for the difference between the hypothesized and prior model sets."

In essence, these uncertainty models are not testable because the distinction between model sets cannot be captured (even asymptotically) by a finite collection of input-output data.

7   Conclusions

In this paper, we have formulated the statistical model validation problem for a class of uncertainty models and shown that, in many cases of interest, it reduces to computing relative weighted volumes of convex sets in $\mathbf{R}^N$. We have also discussed a randomized algorithm based on gas kinetics for probable approximate computation of these volumes. The related decision problem has been shown to reduce to a multihypothesis likelihood ratio test having maximum power over a class of statistical tests. The property of *testability* has been introduced to characterize the ability of an uncertainty model to be statistically validated with arbitrary reliability using input-output data collected over a finite (though potentially long) time interval. Some common uncertainty models, such as those involving ℓ_1 or $\mathcal{H}_\infty$ norms, have been shown to be *not* testable and therefore should not be validated with input-output data.

References

[1] D. Applegate and R. Kannan, "Sample and Integration of Near Logconcave Functions," Comp. Science report CMU, 1990.

[2] I. Barany and Z. Furedi, "Computing the Volume Is Difficult," *Proc. of the 18th Annual Symp. on Theory of Computing*, pp. 442-447, 1986.

[3] C. J. P. Bélisle, H. E. Romejin, and R. L. Smith, "Hit-and-Run Algorithms for Generating Multivariate Distributions," *Mathematics of Operations Research*, **18**, No. 2, 255-266, 1993.

[4] M. Dahleh and J. B. Pearson, "ℓ^1-Optimal controllers for MIMO discrete-time Systems," *IEEE Trans. Automatic Control*, vol. AC-32, pp. 314-323, 1987.

[5] M. E. Dyer and A. M. Frieze, "On the Complexity of Computing the Volume of a Polyhedron," *SIAM J. Computing*, vol. 17, pp. 967-974, 1988.

[6] B. A. Francis, *A Course in $\mathcal{H}_\infty$ Control Theory*, Springer-Verlag, 1987.

[7] B. K. Ghosh, *Sequential Tests of Statistical Hypotheses*, Addison-Wesley, Reading, 1970.

[8] G. C. Goodwin, B. Ninness, and M. E. Salgado, "Quantification of Uncertainty in Estimation," *Proc. of the 1990 ACC*, pp. 2400-2405, 1990.

[9] G. Gu and P. P. Khargonekar, "Linear and Nonlinear Algorithms for Identification in H_∞ with Error Bounds," to appear in *IEEE TAC*, 1994.

[10] A. J. Helmicki, C. A. Jacobson, and C. N. Nett, "Identification in H_∞: a Robustly Convergent Nonlinear Algorithm," *Proc. of the 1990 ACC*, pp. 386-391, 1990.

[11] M. H. Karwan, ed., *Redundancy in Mathematical Programming*, Springer-Verlag, New York, 1983.

[12] M. J. Kearns and U. V. Vazirani, *Topics in Computational Learning Theory*, preprint, 1993.

[13] L. G. Khachiyan, "On the Complexity of Computing the Volume of a Polytope," *Izvestia Acad. Nauk. SSSR Engr. Cybernetics*, vol. 3, pp. 216-217, 1988. (in Russian)

[14] R. L. Kosut, M. Lau and S. Boyd, "Identification of Systems with Parametric and Non-parametric Uncertainty," *Proc. of the 1990 ACC*, pp. 2412-2417, 1990.

[15] J. M. Krause and P. P. Khargonekar, "Parameter Identification in the Presence of Non-parametric Dynamic Uncertainty," *Automatica*, vol. 26, pp. 113-124, Jan. 1990.

[16] L. Lee and K. Poolla, "Statistical Testability of Uncertainty Models," *Proc. of the 1994 IFAC Symposium on System Identification*, Copenhagen, 1994.

[17] L. H. Lee and K. Poolla, "On Statistical Model Validation," submitted to *ASME Journal of Dynamic Systems, Measurement and Control.*

[18] L. H. Lee and K. Poolla, "Statistical Validation and Testability of Uncertainty Models," in preparation.

[19] E. L. Lehmann, *Testing Statistical Hypotheses*, Wiley, New York, 1986.

[20] L. Lin, L. Y. Wang and G. Zames, "Uncertainty Principles and Identification n-widths for LTI and Slowly Varying Systems," *Proc. of the 1992 ACC*, pp. 296-300, 1992.

[21] L. Ljung, *System Identification, Theory for the User*, Prentice-Hall, Englewood Cliffs, New Jersey, 1987.

[22] L. Ljung and Z-D Yuan, "Asymptotic Properties of Black-box Identification of Transfer Functions," *IEEE TAC*, vol. 30, pp. 514-530, June 1985.

[23] P. M. Mäkilä and J. R. Partington, "Robust Approximation and Identification in H_∞," *Proc. of the 1991 ACC*, pp. 70-76, 1991.

[24] S. P. Meyn and R. L. Tweedie, *Markov Chains and Stochastic Stability*, Springer-Verlag, New York, 1993.

[25] A. Packard and J.C. Doyle, "The complex structured singular value," *Automatica*, vol. 29, no. 9, pp:71–110, 1993.

[26] J. R. Partington, "Robust Identification and Interpolation in H_∞," *Int. J. Contr.*, **54**, pp. 1281-1290, 1991.

[27] K. Poolla, P. P. Khargonekar, A. Tikku, J. Krause, and K. M. Nagpal, "A Time-Domain Approach to Model Validation," *IEEE TAC*, vol. 39, no. 5, pp. 951-959, 1994.

[28] K. Popper, *Conjectures and Refutations*, Harper and Row, 1963.

[29] M.G. Safonov, "Stability margins of diagonally perturbed multivariable feedback systems," *IEE Proceedings*, Part D, vol. 129, no. 6, pp:251–256, 1982.

[30] R. L. Smith, "Efficient Monte Carlo Procedures for Generating Points Uniformly Distributed over Bounded Regions," *Operations Research*, **32**, No. 6, pp. 1296-1308, 1984.

[31] R. Smith and J. C. Doyle, "Towards a Methodology for Robust Parameter Identification," *Proc. of the 1990 ACC*, pp. 2394-2399, 1990.

[32] R. Tempo and G. Wasilkowski, "Maximum Likelihood Estimators and Worst Case Optimal Algorithms for System Identification," *Systems and Control Letters*, vol. 10, pp. 265-270, 1988.

[33] D. N. C. Tse, M. A. Dahleh, and J. N. Tsitsiklis, "Optimal Asymptotic Identification under Bounded Disturbances", *Proc. of the 1991 ACC*, pp. 1786-1787, 1991.

[34] L. G. Valiant, "A Theory of the Learnable," *Communications of the ACM*, vol. 27, pp. 1134-1142, 1984.

[35] R. C. Younce and C. E. Rohrs, "Identification with Non-parametric Uncertainty," *Proc. of the 1990 CDC*, pp. 3154-3161, 1990.

[36] G. Zames, "On the metric complexity of causal linear systems: ϵ-entropy and ϵ-dimension for continuous time" *IEEE Trans. on Auto. Contr.*, vol. AC-24, pp. 222–230, 1979.

[37] T. Zhou and H. Kimura, "Time domain identification for robust control," Systems and Control Letters, vol. 20, pp. 167-178, 1993.

An Experimental Comparison of $\mathcal{H}_2$ and $\mathcal{H}_\infty$ Designs for an Interferometer Testbed

Leonard Lublin and Michael Athans *

Space Engineering Research Center
Massachusetts Institute of Technology
Cambridge, MA 02139
USA

Abstract

A comparison between an $\mathcal{H}_\infty$ and $\mathcal{H}_2$ multivariable controller designed for and implemented on the MIT SERC Interferometer testbed is presented. The testbed is a modally rich, lightly damped structure that exhibits non-minimum phase behavior and has a disturbance rejection performance metric. Both controllers were 38 state two-input two-output digital compensators implemented at a sampling frequency of 2 kHz, and both were designed using a frequency weighted loop shaping approach. The details of how the various design variables and their values were chosen for both methods are compared. Comparing the methods and the resulting control designs illustrates that the $\mathcal{H}_\infty$ and $\mathcal{H}_2$ controllers are quite similar. The similarity is in part due to the fact that both controllers nearly cancel all the minimum phase transmission zeros of the plant. We also analyze performance robustness properties using structured singular value (μ) methods together with experimental frequency response data.

1 Introduction

Research in the field of controlled structures technology over the past few decades has been vast [1, 2]. The application of multivariable control to structural systems is an especially challenging area of research that brings up a broad range of issues which are yet to be resolved. In this paper we are specifically concerned with the issues involved in rejecting an independent disturbance

---

*This work was carried out at the MIT Space Engineering Research Center with support provided by NASA grant NAGW-1335.

source in a modally dense, lightly damped, non-minimum phase system. For such a system, the dynamics can be represented in the state space as

$$\begin{aligned}
\dot{x}(t) &= Ax(t) + Bu(t) + Ld(t) \\
y(t) &= Cx(t) + Du(t)
\end{aligned} \tag{1}$$

and in the frequency domain as

$$\begin{aligned}
y(s) &= G_1(s)d(s) + G_2(s)u(s) \\
G_2(s) &= C(sI - A)^{-1}B + D \\
G_1(s) &= C(sI - A)^{-1}L
\end{aligned} \tag{2}$$

Note that the independent disturbance source, d, propagates through the system dynamics in a very specific way and, for performance reasons, can not be simply treated as entering at the inputs or outputs of the system. Even though many methods for synthesizing multivariable controllers to provide disturbance rejection exist, we lack an empirical and experimental understanding of what the methods accomplish in the MIMO case. This paper is intended to address these issues.

Specifically the paper presents a comparison between $\mathcal{H}_\infty$ and $\mathcal{H}_2$ controllers designed using a loop shaping method for a complex, modally dense, lightly damped, non-minimum phase structural testbed. In presenting a side by side development of how to shape specific and distinct loops using $\mathcal{H}_\infty$ and $\mathcal{H}_2$ synthesis, we bring to light the strong physical and mathematical connections between the two commonly used design techniques in an effort to move toward a more systematic design procedure for such complex systems. In doing so we demonstrate, with experimental verification, that it is possible to obtain good disturbance rejection over the entire range of modeled dynamics while being robust to unmodeled dynamics just beyond the range of modeled dynamics. Further, by comparing the results of two distinct synthesis frameworks for a common complex problem, we can begin to illicit a fundamental understanding of the inherent mechanisms needed to achieve performance and robustness in such systems. In particular for the designs to be compared, we found that the $\mathcal{H}_\infty$ and $\mathcal{H}_2$ controllers deliver very similar results and that both of them, as expected, nearly cancel all of the minimum phase transmission zeros of $G_2(s)$.

The loop shaping carried out here employs frequency weights in the $\mathcal{H}_\infty$ and $\mathcal{H}_2$ synthesis to simultaneously shape closed loop performance and robustness transfer functions. While in themselves such techniques have been employed in the past [3, 4], we develop both of them in parallel to demonstrate the physical motivation one can use to select the design variables and their values in the $\mathcal{H}_\infty$ and $\mathcal{H}_2$ (or LQG) frameworks. It is important to emphasize that the loop shaping method presented here focuses on shaping specific closed loop transfer functions. More classical loop shaping methods for multivariable systems rely on shaping the open loop, loop transfer matrix $G_2(s)K(s)$ [5, 6, 7]. These methods are not particularly transparent in the context of rejecting independent disturbance sources in multivariable systems with lightly damped dynamics because it is not clear how to shape $G_2(s)K(s)$ to provide desirable shapes of $[I + G_2(s)K(s)]^{-1} G_1(s)$, the closed loop transfer function from the disturbances to the outputs. This is especially true of the loop transfer recovery (LTR) type methods that use the disturbance influence matrix, L of (1), as a design variable.

Comparisons of various design techniques can also be found in the literature [7, 8]. All too often though the comparisons are not fair in that they are biased to highlight one method over another. Indeed, comparison of various design approaches is a difficult business. Comparisons based on specific designs for specific systems, like the ones presented here, may lead to conclusions that do not generalize and may thus be misleading. Likewise, comparisons based on pure mathematics, like the one in [9], lack the iterative and engineering judgment factors that are an integral part of designing feedback controllers. As a result of these factors, we do not make any claims about the general nature of the $\mathcal{H}_\infty$ and $\mathcal{H}_2$ synthesis methods based on our comparisons between them. Rather we use the comparisons to gain some fundamental understanding of what existing multivariable design methods yield when applied to disturbance rejection for complex flexible structures.

The paper is outlined as follows. In the next section a brief description of the testbed for which the controllers were designed is presented along with the models used to design the controllers. In doing so, the desired performance goals are stated in terms of achieving certain shapes of specific closed loop transfer functions. This then becomes the focus of the $\mathcal{H}_\infty$ and $\mathcal{H}_2$ synthesis methods which are described and compared in Section 3. The comparisons of the optimal controllers and experimental results, as well as the insights gained from them, are given in Sections 4 and 5 respectively.

As for notation, $\overline{\sigma}[G(\jmath\omega)]$ will denote the maximum singular value of the matrix $G(\jmath\omega)$, $\|G\|_\infty$ the $\mathcal{H}_\infty$ norm of the system $G(s)$, and $\|G\|_2$ the $\mathcal{H}_2$ norm of the system $G(s)$.

2　System and Model Descriptions

The $\mathcal{H}_2$ and $\mathcal{H}_\infty$ control designs which we shall be comparing were carried out on the M.I.T. Space Engineering Research Center's Interferometer testbed. The testbed is a large scale laboratory experiment that allows various approaches to controlled structures technology to be validated in a realistic setting. As discussed in detail in [10], the Interferometer testbed is a scaled version of a proposed spaced based, stellar interferometer telescope. A schematic drawing of the tetrahedral testbed is shown in Figure 1. The performance requirement for the testbed is to maintain the internal pathlength errors between multiple points on the tetrahedron to within stringent tolerances in the presence of a disturbance shaker that causes the structure to vibrate.

As seen in the schematic drawing in Figure 1, the Interferometer structure is tetrahedral in configuration, has 3.5 meter long truss–beam legs, and has mounted to it various plates that house the optics for an internal laser metrology system. At the time of the control experiment, 50 constrained layer

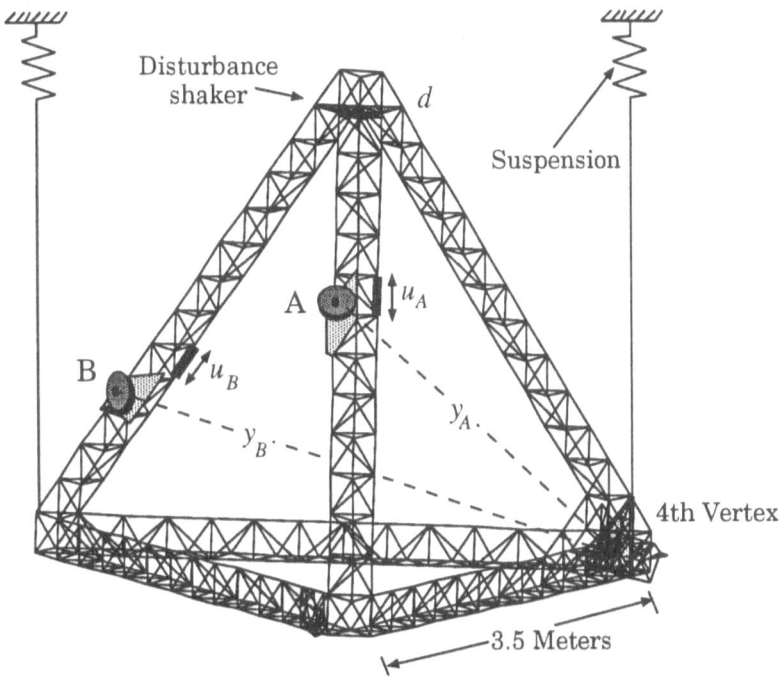

Figure 1: Schematic drawing of the Interferometer testbed highlighting the topology used.

viscoelastic struts were incorporated in the structure to provided a damping level of about one percent in the primary structural modes [11]. Even though the Interferometer testbed is a scaled version of a proposed spacecraft, it does not contain an attitude control system. Rather, it contains a disturbance shaker located at the top vertex of the structure to simulate the effect that an attitude control system and other characteristic disturbance sources have on the structural dynamics of such spacecraft. To this end, the disturbance shaker is driven with a filtered broad band signal which is characterized by a Power Spectral Density (PSD) profile that overbounds a typical spectrum produced by spacecraft specific disturbances.

The disturbance shaker induces vibrations in the structure that are measured by a laser metrology system which serves as the sensor for the control experiments. The laser independently measures the absolute pathlengths between the fourth vertex and points A and B on the structure, denoted by y_A and y_B, as shown in Figure 1. Since points A and B represent the position of the testbed's mock light collecting locations, the differential pathlength

$$\delta_{AB}(t) = y_A(t) - y_B(t) \qquad (3)$$

must be controlled with micro-precision for the purposes of imaging interferometry [10]. The performance objective of the controllers was thus focused on minimizing the Root Mean Square (RMS) motion in δ_{AB} by rejecting the effect of the disturbance source on it over the bandwidth consistent with the digital control computer limitations.

The authority to control δ_{AB} was provided by two active strut actuators [12] located near points A and B in the truss, as seen in Figure 1 and denoted as u_A and u_B. As a result of their location, the actuators induce bending in the truss–beam legs that produces a pistoning motion of points A and B towards the fourth vertex. By using the laser to measure y_A and y_B which make up δ_{AB} via (3), the two-input two-output control system is then able to sense and influence the degradation in δ_{AB} caused by the disturbance source. Controllers to do so are implemented digitally at a sampling frequency of 2 kHz using a real time computer that contains D/A and A/D boards as well as an array processor to facilitate the controller computations.

In order to achieve satisfactory stability and performance from $\mathcal{H}_2$ and $\mathcal{H}_\infty$ controllers, the design model upon which the controllers are based must accurately predict the input/output dynamics of the system and the effect of the disturbance source on the dynamics. Since prior studies [13] have shown that finite element models are not sufficiently accurate for the high frequency modes of the testbed structure, a system identification technique based on measured transfer function data was used to arrive at an accurate model of the Interferometer's dynamics. Transfer function measurements between u_A, u_B, and d (the broad band signal sent to the disturbance source filter) to y_A and y_B were taken through the real time computer that was used to run the controllers and the low pass analog filters used to smooth the outputs of the D/A's. This data, which included the time delays associated with the low pass filters and digital computer running at 2 kHz, was used to synthesize the design model, to evaluate the fidelity of the model, and to predict the closed loop stability and performance properties of the controllers. It should be noted that since the laser is a very accurate sensor and since the transfer functions were averaged, the set of identification data was essentially noise free. Further, since we experimentally computed the transfer functions from the broadband random signal which drove the disturbance source, $d(t)$, to the sensor measurements and incorporated them into the state space model, it was simple to model the disturbance's effect on the state dynamics and pathlengths.

To arrive at a design model that would result in compensators that could be implemented by the available real time computer, we chose to use the data to identify a reduced order model that was only accurate over a portion, 10-85 Hz, of the desired performance frequency range. The identification procedure, which is extensively detailed in [14], consisted of a combination of Ho/Kalman realization [15] and log least squares curve fitting techniques [16]. The resulting model, which had 49 states, was cast into the standard state space description of (1) with

$$y^T = \begin{bmatrix} y_A & y_B \end{bmatrix} \quad \text{and} \quad u^T = \begin{bmatrix} u_A & u_B \end{bmatrix}.$$

Since

$$\delta_{AB} = C_p y(t) \quad \text{where} \quad C_p = \begin{bmatrix} 1 & -1 \end{bmatrix} \quad (4)$$

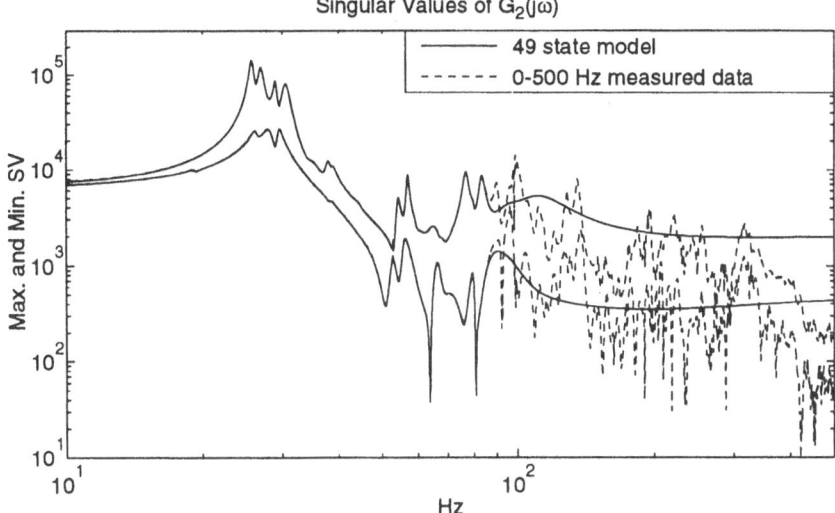

Figure 2: Comparison of the actuator to sensor singular values between the 49 state design model and the measured data.

it should be noted that the performance measure, $d(t)$ influencing $\delta_{AB}(t)$, is
scalar while the input/output topology, $u(t)$ to $y(t)$, is multivariable.

The fidelity of the 49 state design model can be seen in Figures 2 and 3 that
compare the model's frequency response to 10-500 Hz measured data. Figure 2
compares the Singular Values (SV) of the data and of the actuator to sensor
transfer function matrix $G_2(s)$ from (2). Similarly, Figure 3 shows a comparison
of the magnitudes for the disturbance source to performance variable transfer
functions, $|G_p(\jmath\omega)|$, where

$$\frac{\delta_{AB}(\jmath\omega)}{d(\jmath\omega)} = G_p(\jmath\omega) = C_p G_1(\jmath\omega) \tag{5}$$

and $G_1(s)$ is the disturbance to sensor transfer function matrix of (2).

The accuracy of the model over 10-85 Hz is apparent from Figure 2. Any
control design based on this model will only be able to improve the RMS values
of δ_{AB} up to 85 Hz, because there is a large range of prevalent unmodeled
dynamics just beyond there.

To deal with the unmodeled dynamics, an unstructured multiplicative error
robustness criterion assuming that the errors are reflected to the output of the
design model was used [17]. Letting $\Delta_m(s)$ denote the multiplicative error at
the output of the model, $K(s)$ the compensator, and

$$C(s) = [I + G_2(s)K(s)]^{-1} G_2(s)K(s) \tag{6}$$

the complementary sensitivity function, the stability robustness of the controllers based on the 49 state model can be assessed by comparing $C(s)$ and
$\Delta_m(s)$ via the well known stability robustness criterion that

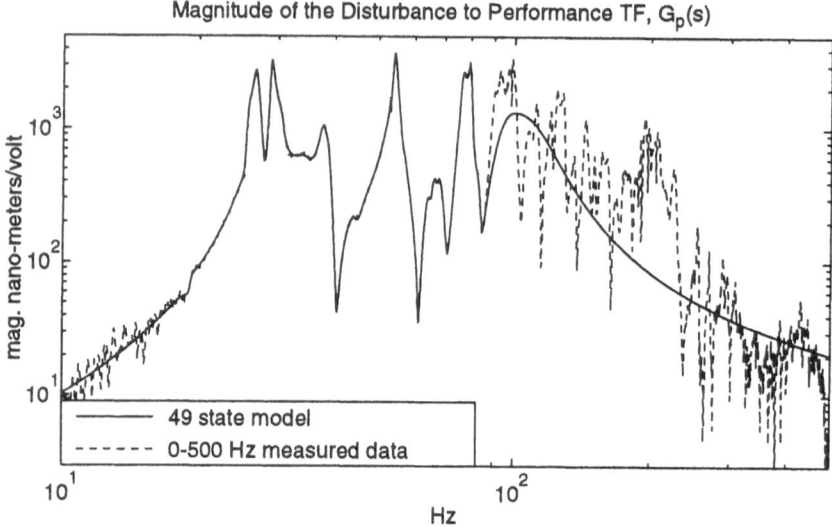

Figure 3: Comparison of $|G_p(\jmath\omega)|$ from (5) between the data and the design model.

$$\bar{\sigma}[C(\jmath\omega)] < \frac{1}{\bar{\sigma}[\Delta_m(\jmath\omega)]} \quad \forall\omega \tag{7}$$

to guarantee closed loop stability robustness. Assembling the measured input/output transfer function data into a transfer function matrix and denoting the result as $\tilde{G}_2(\jmath\omega)$, an estimate of the multiplicative error at the output of the plant was computed as

$$\Delta_m(\jmath\omega) = \left[\tilde{G}_2(\jmath\omega) - G_2(\jmath\omega)\right] G_2(\jmath\omega)^{-1} \tag{8}$$

by carrying out the matrix manipulations at each frequency point of the available data. Tests for stability robustness via (7) for specific controllers using this estimate of the multiplicative error will be discussed in the sequel.

There is another feature of the system's dynamics that is relevant to this work. Namely, the extremely lightly damped zeros seen in the actuator to sensor singular values of Figure 2 at 63 Hz and 80 Hz are non-minimum phase transmission zeros. Unlike the non-minimum phase transmission zeros contained in the model in the vicinity of 120 Hz that help account for the time delays in the system, it is believed that the 63 Hz and 80 Hz non-minimum phase zeros physically arise from the non-collocation between the active strut actuators and the plates to which the optics of the laser sensors are mounted [18].

Given the performance objectives of the testbed and the limited model accuracy, two main design goals are evident:

Goal 1: *Minimize the RMS degradation in δ_{AB} caused by the disturbance source.*

Goal 2: *Maintain stability robustness in the presence of the high frequency unmodeled dynamics beyond 85 Hz.*

Realize that **Goal 1** can be assessed from the closed loop transfer function of the disturbance source, d, to the performance variable of interest, δ_{AB},

$$G_p^{cl}(s) = C_p S(s) G_1(s) \quad \text{where} \quad S(s) = [I + G_2(s)K(s)]^{-1}. \tag{9}$$

Specifically, we would like $|G_p^{cl}(\jmath w)|$ to look like a flattened version of $|G_p(\jmath w)|$, seen in Figure 3, over the entire range where the model is accurate, since the RMS degradation in δ_{AB} is directly related to the area under the curve of $|G_p(\jmath w)|$. On the other hand, **Goal 2** can be assessed from the complementary sensitivity function, $C(s)$ via (7). Given the desire to flatten the peaks of $|G_p^{cl}(\jmath w)|$ over the entire range where the model is precise, the constraint (7) along with the unmodeled dynamics implies that $\overline{\sigma}[C(\jmath w)]$ should be shaped to have a bandwidth near 85 Hz and a very steep roll off beyond that frequency to ensure that the controller is gain stabilized to the unmodeled dynamics. As one can see, the design goals can thus be achieved by synthesizing a compensator, $K(s)$, that shapes the loops of $G_p^{cl}(s)$ and $C(s)$ in the manner just described. As will be detailed in the following section, the $\mathcal{H}_2$ and $\mathcal{H}_\infty$ controller synthesis was set up to do exactly this. Note that implicit in these design goals is a somewhat skewed mixed sensitivity optimal control problem. The difference here being that we must trade off the loop shapes of complementary sensitivity, $C(s)$, with a weighted sensitivity, $C_p S(s) G_1(s)$, rather than the strict sensitivity $S(s)$.

3   Controller Synthesis

The actual synthesis was carried out in the modern paradigm. In this paradigm one synthesizes a stabilizing controller, $K(s)$, to minimize either the $\mathcal{H}_2$ or $\mathcal{H}_\infty$ norm of $T_{zw}(s)$, the closed loop transfer function between the exogenous disturbance signals, w, and the performance variables, z, of the system shown in Figure 4.

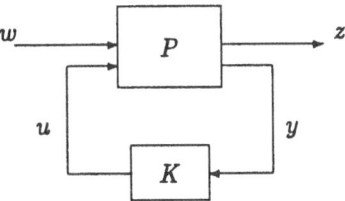

Figure 4: Feedback system description of the modern paradigm.

To shape the desired closed loop transfer functions, the open loop transfer function matrix, P, was defined in such a way that both $G_p^{cl}(s)$ and $C(s)$ multiplied by suitable design variables appeared directly in $T_{zw}(s)$. While there are many ways to define a $P(s)$ to do so, the system interconnection for the $P(s)$ we chose is shown in Figure 5.

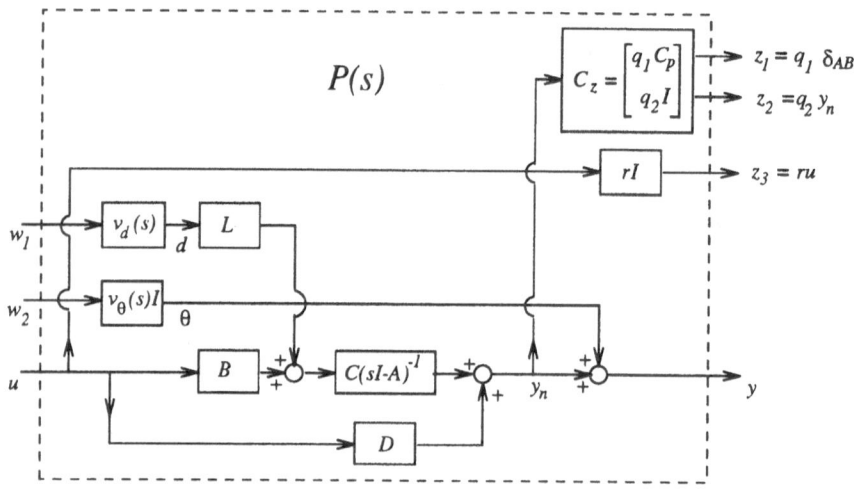

Figure 5: Block diagram interconnection for $P(s)$

All the quantities in the interconnection have been previously defined except for the noiseless output, $y_n(s)$, the scalar frequency dependent design variables $v_d(s)$ and $v_\theta(s)$, and the positive, constant design variables q_1, q_2 and r. The defined performance variables, exogenous disturbances, and interconnection structure for $P(s)$ make good mathematical and physical sense. Mathematically, the $P(s)$ shown in Figure 5 leads to the following $T_{zw}(s)$.

$$T_{zw}(s) = \begin{bmatrix} q_1 v_d(s) G_p^{cl}(s) & -q_1 v_\theta(s) C_p C(s) \\ q_2 v_d(s) S(s) G_1(s) & -q_2 v_\theta(s) C(s) \\ r v_d(s) K(s) S(s) G_1(s) & r v_\theta(s) K(s) S(s) \end{bmatrix} \tag{10}$$

The above partition of $T_{zw}(s)$ is in accordance with $z^T = \begin{bmatrix} q_1 \delta_{AB} & q_2 y_n & ru \end{bmatrix}$ and $w^T = \begin{bmatrix} w_1 & w_2 \end{bmatrix}$ as specified in Figure 5. Notice that the loops of interest we wish to shape, $G_p^{cl}(s)$ and $C(s)$, appear directly in (10) and are directly influenced by the design variables. The fact that both $G_p^{cl}(s)$ and $C(s)$ are multiplied by the frequency weights $v_d(s)$ and $v_\theta(s)$ is deliberate, as previous designs for the Interferometer testbed showed the necessity of using frequency dependent weights [19].

The choice of $P(s)$ makes good sense in terms of the physics of the design objectives. $P(s)$ includes the effects of limited sensor accuracy via the sensor noise term $\theta = v_\theta(s)w_2$ and the knowledge of the actual disturbances that degrade the errors in δ_{AB} via $d = v_d(s)w_1$. Assuming that θ and d are truly present in the physics of the system, $v_d(s)$ and $v_\theta(s)$ can be viewed as design variables that capture the spectral content of the sensor noise and of the disturbance source. The choice of z also makes good physical sense in that it reflects our desire to make errors in δ_{AB}, which is a linear combination of the sensor measurements contained in y_n, small without using unrealistic levels of control u. We chose to include both δ_{AB} and y_n in z because previous designs in which only δ_{AB} was weighted lead to controllers with many unstable modes while ones in which y_n was weighted as well diminished this tendency.

Given that synthesizing $\mathcal{H}_\infty$ and $\mathcal{H}_2$ controllers is straightforward once all of the values in $P(s)$ are fixed [20], all that remains to discuss is how to assign values to the design variables to achieve the desired loop shapes. Here is where comparison of the two design methods begins, as the interpretations of how the design variables influence what the controller does take on different flavors.

In the $\mathcal{H}_\infty$ framework it is possible to attempt to directly shape the loops of any of the transfer function matrices that appear in the partitioned $T_{zw}(s)$ of (10). This is due to the fact that

$$\|T_{zw}\|_\infty < \gamma \Rightarrow \|(T_{zw})_{ij}\|_\infty < \gamma \quad \forall i,j \tag{11}$$

where $(T_{zw})_{ij}$ denotes the closed loop transfer function matrix between exogenous disturbance w_j and performance variable v_i. As a result of (11) and the use of scalar design variables, an $\mathcal{H}_\infty$ controller that achieves a specific value of γ ensures that

$$|G_p^{cl}(\jmath\omega)| < \frac{\gamma}{q_1|v_d(\jmath\omega)|} \quad \forall\omega \tag{12}$$

$$\bar{\sigma}[C(\jmath\omega)] < \frac{\gamma}{q_2|v_\theta(\jmath\omega)|} \quad \forall\omega. \tag{13}$$

Similar bounds will also hold for the other $(T_{zw})_{ij}$ in (10). Now if the values of q_1, q_2, r, $v_d(s)$, and $v_\theta(s)$ are chosen assuming $\gamma = 1$ to provide desirable bounds on $G_p^{cl}(s)$ and $C(s)$ via (12) and (13) and the resulting $\mathcal{H}_\infty$ controller achieves $\|T_{zw}\|_\infty \approx 1$, then the desired loops will in fact be shaped to satisfy (12) and (13). This is how one can choose the values of the design variables to shape the loops of interest in an $\mathcal{H}_\infty$ design.

In using this method of weight selection there are a few issues the designer must keep in mind. First of all, realize that the bounds implied by (11) and exemplified by (12) are not necessarily tight over all frequencies. As a result it helps to graphically inspect all the constraints implicit in the choice of $T_{zw}(s)$ as one iterates through the values of the design variables. More importantly, simply assuming $\gamma = 1$ when the values of the weights are chosen does not ensure an $\mathcal{H}_\infty$ controller that achieves $\|T_{zw}\|_\infty \approx 1$. In fact, when $\|T_{zw}\|_\infty \gg 1$ it is a strong indication that the values of the design variables impose unrealistic constraints on the system's dynamics. One can not choose $v_d(s)$ and $v_\theta(s)$ arbitrarily. They must complement each other.

Another reason why the design variables can not be chosen arbitrarily involves the fact that $\|(T_{zw})_{ij}\|_\infty < \gamma \quad \forall i,j$. Choosing $v_d(s)$ and $v_\theta(s)$ to constrain $G_p^{cl}(s)$ and $C(s)$ via (12) and (13) also imposes constraints on the other loops in $T_{zw}(s)$, such as $C_p C(s)$. While these loops are not of primary interest, they do influence the overall performance of the controller. This is why the constant scalars q_1 and q_2 were included in $P(s)$. They provide additional freedom in shaping $G_p^{cl}(s)$ and $C(s)$, while helping to reduce placing unreasonable constraints on the remaining loops within $T_{zw}(s)$. Even with this additional freedom, the values of the weights could not be chosen arbitrarily.

This is evident when considering how the values of the DC gains for the weights that bound the sensitivity and complimentary sensitivity transfer functions in $(10) - q_1|v_d(0)|$, $q_1|v_\theta(0)|$, $q_2|v_d(0)|$, $q_2|v_\theta(0)|$ – were chosen. These gains influence the desired level of disturbance rejection achieved by the controller and its bandwidth, as will be explicitly shown in the sequel. To meet the design goals for the Interferometer, we require that $q_1|v_d(0)| > q_2|v_d(0)|$ to insure that degradations in differential pathlength, $\delta_{AB} = G_p^{cl}d$, are penalized heavier than degradations in absolute pathlength, $y_n = SG_1d$; see (10). Likewise we require both $q_1|v_\theta(0)| < 1$ and $q_2|v_\theta(0)| < 1$ to facilitate an acceptable bandwidth. Combining these three physically motivated inequality constraints yields the constraint that

$$q_2 < q_1 < \frac{1}{|v_\theta(0)|} \tag{14}$$

if the values of the weights are chosen assuming $\gamma = 1$. Designs in which the values of q_2, q_1, and $|v_\theta(0)|$ violated (14) led to controllers with $\|T_{zw}\|_\infty > 10$ that did not achieved the desired performance. The point here is that in using this method of selecting the design variables of the $\mathcal{H}_\infty$ controller to shape the primary loops of interest, one needs to concentrate on all the loops effected by the constraints imposed by (11).

Turning now to the $\mathcal{H}_2$ case, there is no direct way to shape the loops, as in the $\mathcal{H}_\infty$ framework, because $\| \cdot \|_2$ is not a true norm. Rather, in the $\mathcal{H}_2$ case the loops are shaped by reflecting the desired control system performance into the frequency weights that appear in $P(s)$. The actual values of the design variables are chosen by assuming a stochastic framework for the $\mathcal{H}_2$ control problem and then using the insights afforded by the resulting LQG problem. As such, the exogenous disturbance is considered to be a unit intensity white noise signal in which w_1 and w_2 are independent. Then $v_d(s)$ acts as a filter, or frequency weight, that specifies over what frequencies and how strongly the disturbances impact the system. As a consequence, the $|v_d(\jmath w)|$ should be large over 10-85 Hz where disturbance rejection performance is important, and small beyond 85 Hz where it is not achievable. So while there is really no direct way to shape $G_p^{cl}(s)$ using $\mathcal{H}_2$ synthesis, the design variable $v_d(s)$ can be chosen in the aforementioned way to strongly influence it.

Similarly, $v_\theta(s)$ can be chosen to influence $C(s)$. Here $v_\theta(s)$ specifies over what frequencies the sensors are able to provide accurate measurements. Large values of $|v_\theta(\jmath w)|$ indicate very noisy measurements while small values indicate clean ones. Choosing $|v_\theta(\jmath w)|$ to be very large over the region of unmodeled dynamics beyond 85 Hz and small elsewhere will result in a controller that does not exert significant energy in the region of unmodeled dynamics.

Realize that the manner in which the filters and intensities for the process noise, d, and sensor noise, θ, are chosen is not necessarily consistent with their actual physical values. Rather they are chosen to reflect in the optimization problem our desire to appropriately shape the loops of $G_p^{cl}(s)$ and $C(s)$. In the general case where either $d(t)$ or $\theta(t)$ are colored noises with specific power spectral densities, $v_d(s)$ and $v_\theta(s)$ should be viewed as additional design vari-

Controller	q_1	q_2	r	$|v_d(0)|$	$|v_\theta(0)|$	$\|T_{zw}\|_\infty$
$\mathcal{H}_\infty$	1/2	1/6	1	0.0031	1	0.975
$\mathcal{H}_2$	1/2	1/6	100	0.005	1	2.97

Table 1: Values of the various design variables for the $\mathcal{H}_\infty$ and $\mathcal{H}_2$ controllers.

ables that over emphasize in what frequency ranges the impact of the noises is most vital.

As far as q_1, q_2 and r are concerned, q_1 is chosen to be larger than q_2 to reflect in the cost the fact that we care more about minimizing the RMS energy in δ_{AB} than in y_n, while r is chosen in the typical LQG fashion to influence the level of control authority. This then is how the values of the performance variables can be chosen to shape the desired loops.

The previous discussion alludes to some differences and similarities between the $\mathcal{H}_2$ and $\mathcal{H}_\infty$ controller synthesis methods. Both frameworks require an in depth understanding of the system's open loop dynamics to arrive at a successful set of weights, since the inherent trade off between performance and stability robustness is present no matter what the mathematics indicate. Furthermore given the complexity of the system involved, it is also necessary to iterate over the choice of frequency weights in both frameworks to arrive at a successful design. As for differences, while the $\mathcal{H}_\infty$ framework allows one to more directly shape the loops, thanks to (11), the $\mathcal{H}_2$ synthesis directly minimizes the performance variable of interest; namely, the RMS energy in δ_{AB}. Even though the loop shaping $\mathcal{H}_\infty$ controllers do not directly minimize RMS values, it is important to realize that the framework provides an *ad hoc* means to achieving good $\mathcal{H}_2$ performance. Given the RMS specification on δ_{AB}, if $q_1 v_d(s)$ is chosen to flatten out $|G_p(\jmath w)|$ then the $\mathcal{H}_\infty$ design will actually obtain good $\mathcal{H}_2$ performance. This is because all of the RMS energy is contained in the lightly damped modes of $G_p(s)$, and flattening these modes directly reduces the RMS energy in δ_{AB}. As a final similarity, realize that the optimal value of the $\mathcal{H}_\infty$ or $\mathcal{H}_2$ norm of $T_{zw}(s)$ obtained by the controllers is of little consequence. It is the loop shapes of $G_p^{cl}(s)$ and $C(s)$ that matter here, which have very little to do with the scalar numbers $\|T_{zw}\|_\infty$ or $\|T_{zw}\|_2$. In summary, the differences between the $\mathcal{H}_2$ and $\mathcal{H}_\infty$ synthesis methods are slight given the loop shaping methodology pursued here. This is a direct result of the manner in which the physics of the design problem have been incorporated into the mathematics of the synthesis. The similarities will be further demonstrated by the design examples and experimental results in the following sections.

4   Comparison of Optimal Designs

The intent of this section is to compare the results of an $\mathcal{H}_\infty$ and an $\mathcal{H}_2$ optimal controller, both of which achieved the desired loop shapes. In doing so, the results will verify that the loop shaping design approach works in this case,

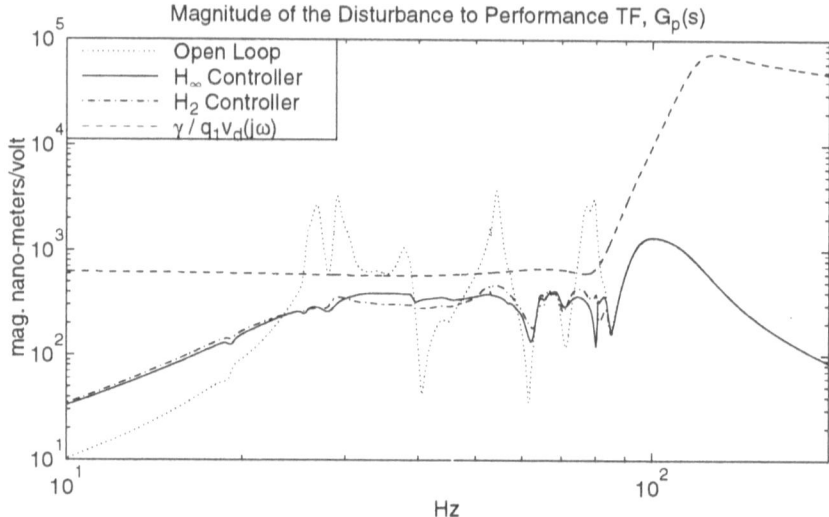

Figure 6: Comparison of $|G_p^{cl}(\jmath\omega)|$ between the $\mathcal{H}_\infty$ and $\mathcal{H}_2$ optimal controllers and the frequency weight, $\gamma/q_1|v_d(\jmath\omega)|$, used to shape $G_p^{cl}(s)$.

and that the differences in the controllers is minimal.

The controllers were synthesized using the methods outlined in Section 3, and analyzed using the methods detailed in [19]. Briefly, the analysis methods consisted of applying the frequency response of the controller to the actual, measured, transfer function frequency response data. In doing so, the closed loop performance and stability properties of the controllers when implemented on the actual testbed can be accurately predicted. This in turn provides a realistic indication of how well the loops of $G_p^{cl}(s)$ and $C(s)$ were shaped.

To emphasize the strong connection between the physical motivation behind selecting $v_d(s)$ and $v_\theta(s)$ in the $\mathcal{H}_2$ framework and the mathematical motivation in the $\mathcal{H}_\infty$ framework, the designs to be compared were arrived at by using the same values of $v_d(s)$ and of $v_\theta(s)$ in both designs. However, because the controllers are arrived at by distinct optimization criterion, different values for the DC gains of $v_d(s)$ and $v_\theta(s)$ and the control weight r were used. Table 1 shows the various values chosen for the design variables along with the values of $\|T_{zw}\|_\infty$, and Figures 6 and 7 show the magnitudes for the values of $v_d(s)$ and $v_\theta(s)$. Since $v_d(s)$ contained six states and $v_\theta(s)$ ten states, the optimal $\mathcal{H}_2$ and $\mathcal{H}_\infty$ controllers for the 49 state model had 75 states. This explosion of controller state dimension is the price one pays for using frequency weights to shape the loops of interest. Fortunately, as the experimental results in the next section will show, existing controller reduction techniques are able to significantly reduce the number of states in the controllers (from 75 to 38 states) without sacrificing much performance.

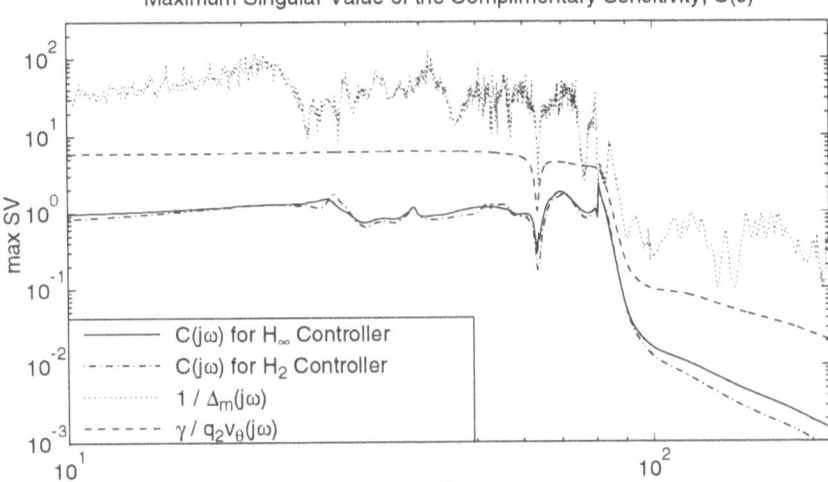

Figure 7: Comparison of $\bar{\sigma}[C(\jmath\omega)]$ between the $\mathcal{H}_\infty$ and $\mathcal{H}_2$ optimal controllers and $1/\bar{\sigma}[\Delta_m(\jmath\omega)]$ using the estimate of $\Delta_m(\jmath\omega)$ from (8). The frequency weight $\gamma/q_2|v_\theta(\jmath\omega)|$ used to shape $C(s)$ is also shown.

Figures 6 and 7 show that the loops of interest have in fact been shaped in accordance with the design goals outlined in Section 2. As seen in Figure 6, **Goal 1** is clearly achieved by both designs, since they have flattened the peaks of $G_p(s)$ up to 85 Hz. The degradation in performance below 25 Hz is not a cause for concern, since the majority of the RMS energy is contributed by the lightly damped modes. In this figure, $\gamma/q_1|v_d(\jmath\omega)|$ is also shown to illustrate that the method described in Section 3 for choosing $v_d(s)$ to shape $G_p(s)$ worked. As discussed, the $|v_d(\jmath\omega)|$ is large up to 85 Hz and $\gamma/q_1|v_d(\jmath\omega)|$ appropriately bounds $|G_p^{cl}(\jmath\omega)|$.

Similarly, Figure 7 illustrates that **Goal 2** is also met by both controllers. The figure graphically shows the stability robustness criterion of (7) using the estimate of the multiplicative error in (8). Clearly both designs achieve the desired loop shapes for $C(s)$. Again, the figure also illustrates that the method outlined for choosing $v_\theta(s)$ works. As can be seen from $\gamma/q_2|v_\theta(\jmath\omega)|$, the weights force the controller to roll off rapidly beyond 85 Hz to ensure stability robustness to the unmodeled dynamics. At 63 Hz, note the notch in the chosen $v_\theta(s)$. This notch was placed in the weight to prevent the synthesis from producing controllers that had an unstable pole near the very lightly damped and somewhat uncertain non-minimum phase zero in $G_2(s)$ at 63 Hz.

Notice that Figures 6 and 7 not only demonstrate that the design goals have been met, but they also illustrate that the $\mathcal{H}_\infty$ and $\mathcal{H}_2$ designs are very similar. In those figures, both controllers achieve very similar shapes of $G_p^{cl}(s)$ and $C(s)$. Further evidence of the similarity between the controllers can be seen

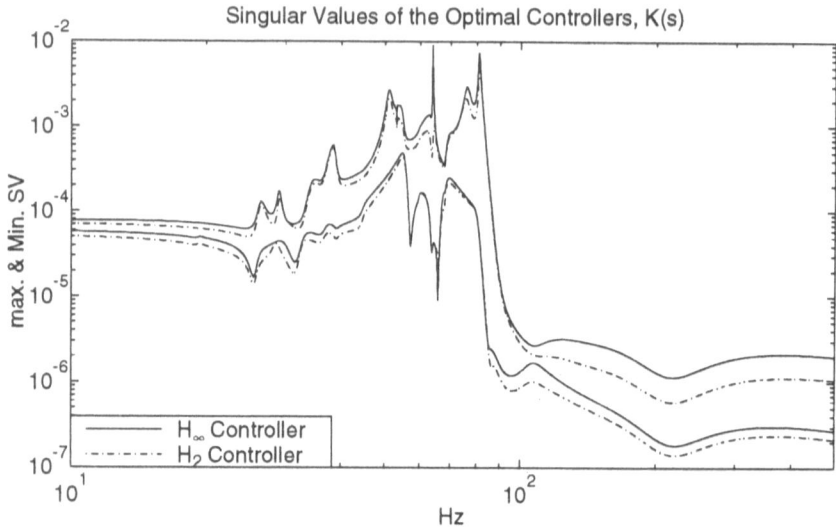

Figure 8: Comparison of the singular values for the optimal $\mathcal{H}_\infty$ and $\mathcal{H}_2$ controllers.

in Figure 8 that compares the singular values of the $\mathcal{H}_\infty$ and $\mathcal{H}_2$ controllers. Clearly, the dynamics are very similar. In fact, in comparing the numerical values of the poles and transmission zeros of the two controllers, we found a significant number of them to be nearly identical.

There is a clear reason why the controllers are so similar even though the optimization problems used to arrive at them are distinct. Essentially, once the desired control system performance is incorporated into $P(s)$ via the design variables, the task of minimizing the $\mathcal{H}_2$ norm of $T_{zw}(s)$ becomes nearly identical to the task of minimizing the $\mathcal{H}_\infty$ norm of $T_{zw}(s)$. This can be seen in Figure 9 that compares the values of $\overline{\sigma}[T_{zw}(\jmath\omega)]$ and $\overline{\sigma}[P_{zw}(\jmath\omega)]$ for the two controllers. Here $P_{zw}(s)$ denotes the open loop transfer function matrix between z and w of $P(s)$, where all of the values of the design variables are included in $P(s)$. As such, $\overline{\sigma}[P_{zw}(\jmath\omega)]$ is an indication of the nominal cost that the controllers seek to minimize. Specifically, to minimize the $\mathcal{H}_\infty$ norm of $T_{zw}(s)$ for this design, all of the spikes seen in the open loop plot of $\overline{\sigma}[P_{zw}(\jmath\omega)]$ must be flattened or damped. This is also the case for minimizing the $\mathcal{H}_2$ norm of $T_{zw}(s)$ which is related to the area under $\overline{\sigma}[P_{zw}(\jmath\omega)]$. While the optimization problems are distinct, the manner in which the costs are minimized is similar. A controller that does a good job minimizing the $\mathcal{H}_\infty$ norm here is also one that does a good job minimizing the $\mathcal{H}_2$ norm. Thus the similarities in the $\mathcal{H}_\infty$ and $\mathcal{H}_2$ controllers for this system with its particular design goals.

There is an interesting difference between the controllers that is worth discussing. Namely, while both controllers achieve the nominal performance defined by (12) and the stability robustness defined by (13) as seen in Figures 6 and 7, the $\mathcal{H}_\infty$ controller guarantees robust performance where as the $\mathcal{H}_2$ one

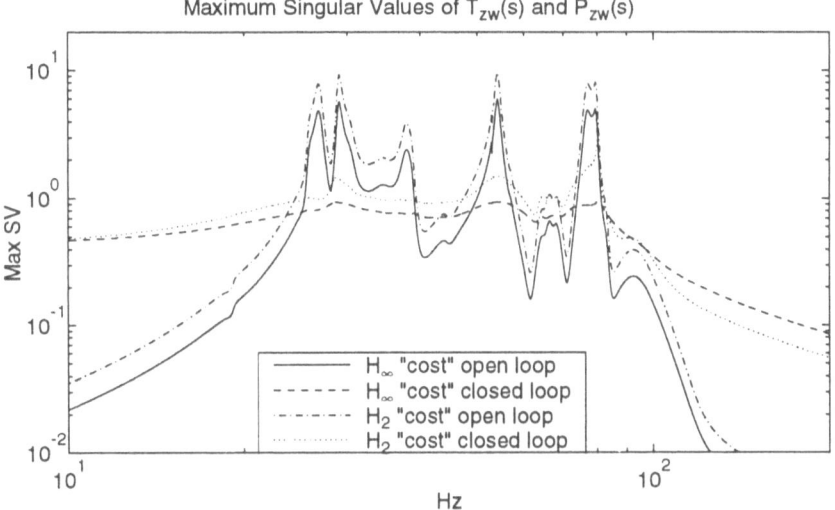

Figure 9: Comparison of the open loop "cost", $\overline{\sigma}[P_{zw}(\jmath w)]$, and
the closed loop "cost", $\overline{\sigma}[T_{zw}(\jmath w)]$, for the $\mathcal{H}_\infty$ and
$\mathcal{H}_2$ controllers.

does not. Here robust performance indicates that the optimal controller applied
to the design model achieves the level of performance defined by the bound on
$G_p^{cl}(s)$ from (12) with $\gamma = 1$ in the presence of any output multiplicative error
bounded in magnitude by $q_2|v_\theta(\jmath w)|$ as per (13) with $\gamma = 1$ [21]. More precisely,
given the aforementioned stability robustness condition and performance goal,
we can define a structured uncertainty Δ such that $w = \Delta\tilde{z}$ where

$$\tilde{z}^T = \begin{bmatrix} z_1 & z_2 \end{bmatrix}, \quad \Delta = \begin{bmatrix} \Delta_1 & 0 \\ 0 & \Delta_2 \end{bmatrix}, \quad \|\Delta_i\|_\infty < 1, \qquad (15)$$

and Δ_i is complex. Then the robust performance properties of the $\mathcal{H}_\infty$ and
$\mathcal{H}_2$ controllers can be analyzed by evaluating the structured singular value
of $T_{\tilde{z}w}(\jmath w)$, denoted as $\mu[T_{\tilde{z}w}(\jmath w)]$, and checking if $\mu[T_{\tilde{z}w}(\jmath w)] < 1 \; \forall w$ [22].
Note that $T_{\tilde{z}w}$ is given by (10) with the third row removed. Since the uncer-
tainty structure consists of only two complex blocks, the computation of μ is
exact and provides a non-conservative indication of the performance robustness to the defined uncertainties here. For the defined uncertainties, it should
be clear that the $\mathcal{H}_\infty$ controller automatically guarantees robust performance
since $\mu[T_{\tilde{z}w}(\jmath w)] < \overline{\sigma}[T_{\tilde{z}w}(\jmath w)] < 0.975 \; \forall w$. This was accomplished through
scaling the values of the design variables for the $\mathcal{H}_\infty$ design in such a way
that achieving a controller with $\|T_{zw}\|_\infty < 1$ satisfied the design goals. Given
the illustrated similarities between the $\mathcal{H}_\infty$ and $\mathcal{H}_2$ controllers, one might suspect that the $\mathcal{H}_2$ design would guarantee the same performance robustness as
the $\mathcal{H}_\infty$ design. This is not the case, as seen in Figure 10 which shows that
$\mu[T_{\tilde{z}w}(\jmath w)] > 1$ at various frequencies for the $\mathcal{H}_2$ controller. In retrospect,

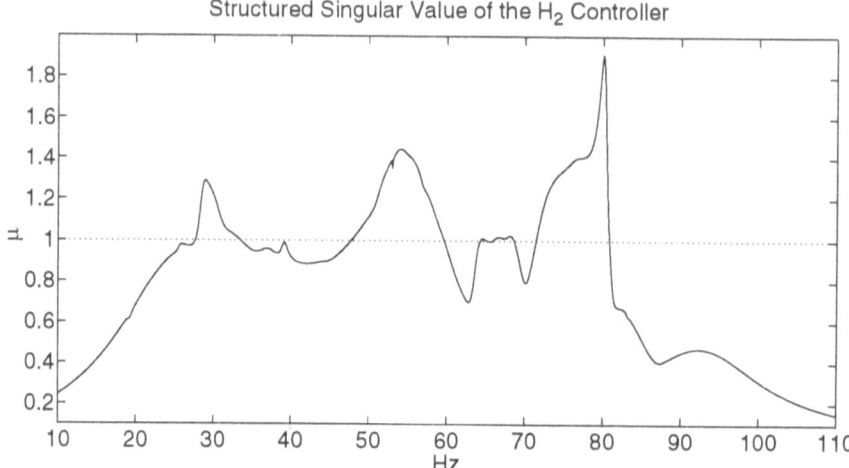

Figure 10: The structured singular value for the $\mathcal{H}_2$ controller, $\mu\left[T_{\tilde{z}w}(\jmath\omega)\right]$. Note the linear scales.

it makes sense that the $\mathcal{H}_2$ controller is not as robust with respect to performance. The $\mathcal{H}_2$ synthesis is not concerned with uniformly minimizing the peaks of $\bar{\sigma}[T_{zw}(\jmath\omega)]$, nor do we incorporate any scaling in the methodology for choosing the values of the design variables that would lead to a performance robust controller. In essence, it is the scaling of the design variables that leads to the guaranteed performance robustness in the $\mathcal{H}_\infty$ design, and the lack of scaling which produced a $\mathcal{H}_2$ controller that is not performance robust. This is further supported by the differences in the singular values of the controller seen in Figure 8. While the poles of the controllers are similar, their zeros and gains are slightly different. Such differences are consistent with a frequency dependent scaling between the two controllers. It is such a difference that leads to the disparate guarantees of robust performance for the otherwise very similar controllers.

To gain some insight into what the design methods accomplish, it is important to go beyond the comparisons of the controllers and understand how they achieve the specific performance objectives. In analyzing the poles and zeros of the compensators and comparing them to those of the system, we found that both the $\mathcal{H}_2$ and $\mathcal{H}_\infty$ controllers nearly cancel all of the minimum phase transmission zeros of $G_2(s)$ from (2) and substitute some of the dynamics from the frequency weights into $G_2(s)K(s)$. Realize that this cancellation occurs even though there are non-minimum phase zeros in $G_2(s)$ at 63 Hz and 80 Hz, which is within the bandwidth achieved by the controllers. While the compensators did not directly cancel the non-minimum phase zeros, they both contained a slightly unstable mode near the 80 Hz non-minimum phase zero, and without the notch in $v_\theta(s)$ they would have contained an unstable mode near the 63 Hz zero as well. This near cancellation of the zeros makes sense. Given an undesirable open loop plant, $G_2(s)$, it is necessary to choose a $K(s)$ that appropriately shapes the loop transfer function $G_2(s)K(s)$ to produce desirable loop shapes in $G_p^{cl}(s)$ and $C(s)$. Appreciating that these high performance design approaches produce controllers that cancel plant zeros is vital. For one thing, it highlights the importance of accurately modeling the transmission zeros of a structural system. If the zeros of the model are in any way uncertain, then this method should be used with caution, as canceling lightly damped uncertain dynamics lacks robustness.

It is interesting to note that while most loop shaping methods are based on directly shaping $G_2(s)K(s)$ [5, 6], the one presented here indirectly shapes $G_2(s)K(s)$ to achieve the desired loop shapes of $G_p^{cl}(s)$ and $C(s)$. Even though this method has the disadvantage endemic to any loop shaping method of canceling the system's dynamics, we believe it has a distinct advantage over the others: it allows one to directly shape the loops of interest, rather than only being able to shape $G_2(s)K(s)$ to influence loops such as (9).

5 Comparison of Experimental Results

The previous section illustrated that the loop shaping design method produced controllers which were able to shape the loops of the mathematical design model. As this section will illustrate, these design techniques also lead to controllers that when implemented on the actual system achieved the desired performance.

In order to be implemented on the testbed, the optimal $\mathcal{H}_2$ and $\mathcal{H}_\infty$ controllers had to be reduced to facilitate the limited computational power of the available real time computer. Without going into details, which appear in [14], the order reduction technique consisted of factoring the unstable compensator into a stable and unstable portion, using balancing and truncation techniques to reduce the stable portion, recombining the reduced stable portion with the unstable one, and using the log least squares curve fitting algorithm of [16] to tune the reduced order controller's dynamics to fit the optimal order controller's frequency response. Using these methods, the 75 state optimal controllers were reduced to 38 states and discretized using a bi-linear transform to run a sampling frequency of 2 kHz. To negate the frequency warping inherent in the bi-linear transform, the dynamics of the digital controllers were further tuned to fit the frequency response of the continuous time reduced order controllers.

In carrying out the reduction and discretization, the controllers were no longer "truly optimal". This fact is of no consequence as long as the reduced controllers provide the desired performance when implemented on the Interferometer testbed. To ensure that the reduced order controllers would not destabilize the testbed, their frequency response was applied to the measured data in the entire 10-500 Hz region in order to evaluate the multivariable Nyquist criterion [19]. The Nyquist plots predicted a stable closed loop, and in fact both controllers were successfully implemented on the testbed. The fact that both of the unstable controllers produced a stable closed loop further verifies that

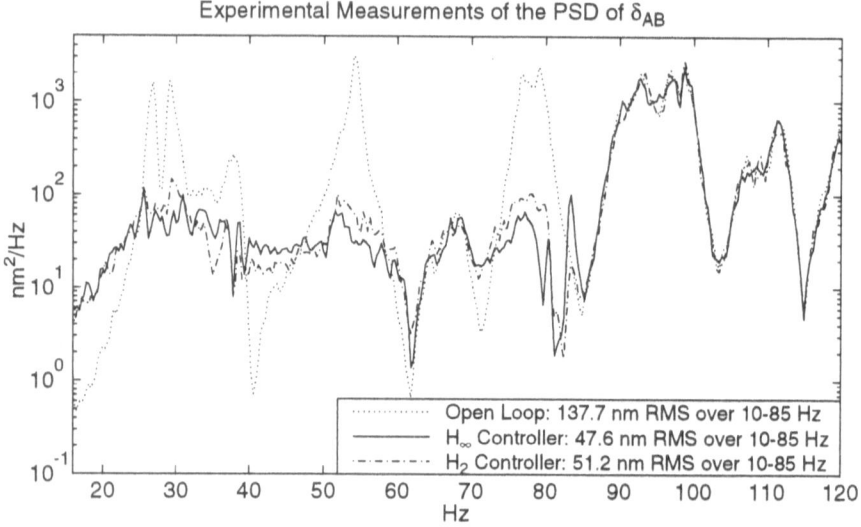

Figure 11: Comparison of the experimentally measured PSD of δ_{AB} shown with the RMS values of δ_{AB} over 10-85 Hz.

the loop shaping of $C(s)$ to guarantee stability robustness to the unmodeled dynamics was successful.

The results of the implementation can be seen in Figure 11 that shows a comparison of the experimentally measured Power Spectral Density (PSD) of δ_{AB}, the performance variable of interest, for the actual closed loop systems. As seen in the figure, the performance actually achieved by the $\mathcal{H}_\infty$ and $\mathcal{H}_2$ controllers is quite similar, which is to be expected given the analysis of the optimal controllers. These experimental measurements also verify that the loop shaping of $G_p^{cl}(s)$ was successful. Notice that both controllers flatten out the peaks in the PSD of δ_{AB} up to 85 Hz where the model is accurate, but they do not degrade the performance in the region of unmodeled dynamics beyond 85 Hz. This striking ability to achieve good performance over the entire region of modeled dynamics while being robust to the unmodeled dynamics a few Hz beyond the bandwidth of the control system is a testament to the ability of the frequency weights to shape the loops of $C(s)$ and $G_p^{cl}(s)$ rather than to the specific design methodology ($\mathcal{H}_\infty$ or $\mathcal{H}_2$) employed. Also realize that the successful experimental results confirm our ability to achieve the desired performance even though the design method leads to an explosion of controller dimension.

A few words about the performance achieved by the controllers is in order here. First of all, it is worth noting that the amount of performance attained by these design methods was limited by the bandwidth allowed by the digital control computer, not the methods themselves, as is evident from the results reported by Jacques in [14]. Secondly, notice that over 10-85 Hz the RMS

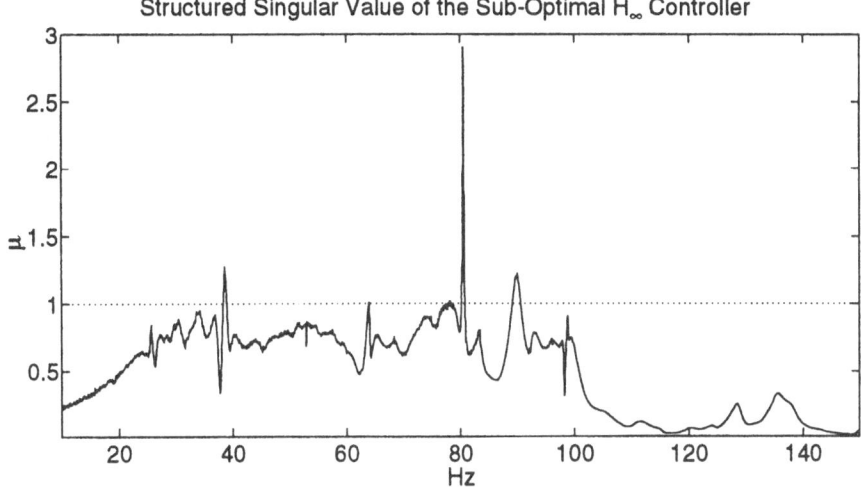

Figure 12: The structured singular value for the reduced order, discrete $\mathcal{H}_\infty$ controller computed from the measured data. Note the linear scales.

performance of the $\mathcal{H}_\infty$ controller is slightly better than that of the $\mathcal{H}_2$ one. Although this seems counterintuitive, recall that different weights were used in the $\mathcal{H}_2$ and $\mathcal{H}_\infty$ designs. Also, this is to be expected given the slightly higher gain of the $\mathcal{H}_\infty$ controller seen in Figure 8. It would be wrong to conclude from this single set of designs for this specific system that $\mathcal{H}_\infty$ controllers are thus better than $\mathcal{H}_2$ ones. A set of designs for which an $\mathcal{H}_2$ controller achieved more performance than the $\mathcal{H}_\infty$ one could have readily been presented here instead. The message of the results is not which controller is better, but simply that both are similar.

As for the guaranteed robustness properties of the controllers that were implemented, realize that since they are reduced versions of the optimal ones, the robustness guarantees discussed in the previous section no longer apply. In particular, even though the optimal $\mathcal{H}_\infty$ controller was guaranteed to be performance robust when applied to the design model, it does not mean the sub-optimal $\mathcal{H}_\infty$ controller that was implemented is also performance robust. To assess whether or not the sub-optimal $\mathcal{H}_\infty$ controller guarantees robust performance, we applied the reduced order digital controller to the measured frequency response data to form an empirical estimate of $T_{\tilde{z}w}(\jmath\omega)$ and then computed its associated structured singular value, as shown in Figure 12. Such a computation is not only viable, as μ can be evaluated for any complex matrix, but it also provides a very accurate indication of what happens when the controller is applied to the actual system. From Figure 12 it is clear that the sub-optimal $\mathcal{H}_\infty$ controller that was implemented is not performance robust. Further, the analysis predicts that only a small perturbation is required at 80 Hz to destroy the guaranteed performance robustness. This is not surprising given that the controller has an unstable mode near the very lightly damped, non-minimum phase zero in the plant at 80 Hz. In summary, this exposition highlights the need to build a better empirical understanding of what current synthesis methods yield in the multivariable case, as there are clearly differences between the conclusions one can draw from the analysis of an optimal controller applied to a design model and a sub-optimal controller applied to an actual system.

6 Conclusions

The objective of this paper has been to present a comparison of $\mathcal{H}_\infty$ and $\mathcal{H}_2$ controllers designed for and implemented on a complex, lightly damped structure with non-minimum phase behavior. In the process, a loop shaping design methodology that uses $\mathcal{H}_\infty$ and $\mathcal{H}_2$ synthesis as means to shape specific performance and robustness loops of interest was detailed. The experimental results presented verify that this methodology does in fact produce implementable high order digital controllers that are able to satisfy the rigid performance requirements of the Interferometer testbed.

Comparison of the designs allowed us to gain some understanding of the mechanisms used by the synthesis methods to achieve the desired performance. In particular, we found that both controllers nearly cancel all of the minimum phase transmission zeros of the plant to achieve the desired performance. The fact that the cancellations are carried out in the presence of non-minimum phase zeros in the control system bandwidth highlights the need to better understand the fundamental mechanisms behind the cancellations. That this is an issue which warrants future research should be clear given the relative lack of robustness inherent in controllers which partially invert a system's dynamics.

References

[1] Hyland, D. C., Junkins, J. L., and Longman, R., "Active Control Technology for Large Space Structures," *Journal of Guidance, Control, and Dynamics*, Vol. 16, Sept-Oct 1993, pp. 801–821.

[2] Sparks, D. and Juang, J.-N., "Survey of Experiments and Experimental Facilities for Control of Flexible Structures," *Journal of Guidance, Control, and Dynamics*, Vol. 15, Jul.-Aug. 1992, pp. 801–816.

[3] Anderson, B. D. O. and Mingori, D. L., "Use of Frequency Dependence in Linear Quadratic Control Problems to Frequency–Shape Robustness," *Journal of Guidance, Control, and Dynamics*, Vol. 8, No. 3, 1985, pp. 397–401.

[4] Safonov, M. and Chiang, R., "CACSD Using the State Space L_∞ Theory - A design Example," *IEEE Transactions on Automatic Control*, Vol. 33, May 1988, pp. 477–479.

[5] Stein, G. and Athans, M., "The LQG/LTR Procedure for Multivariable Feedback Control Design," *IEEE Transactions on Automatic Control*, Vol. 32, Feb. 1987, pp. 105–114.

[6] McFarlane, D. and Glover, K., "A Loop Shaping Design Procedure Using $\mathcal{H}_\infty$ Synthesis," *IEEE Transactions on Automatic Control*, Vol. 37, June 1992, pp. 759–769.

[7] Sparks, A., Banda, S., and Yeh, H.-H., "A Comparison of Loop Shaping Techniques," in *Proceedings of the 30th Conference on Decision and Control*, (Brighton, England), Dec. 1991.

[8] Lim, K., Maghami, P., and Joshi, S., "A Comparison of Controller Designs for an Experimental Flexible Structure," in *Proceedings of the American Control Conference*, (Boston, MA), June 1991.

[9] Zhou, K., "Comparison Between $\mathcal{H}_2$ and $\mathcal{H}_\infty$ Controllers," *IEEE Transactions on Automatic Control*, Vol. 37, Aug. 1992, pp. 1261–1265.

[10] Blackwood, G. H., Jacques, R. N., and Miller, D. W., "The MIT multipoint alignment testbed: Technology development for optical interferometry," in *Proceedings of the SPIE Conference on Active and Adaptive Optical Systems*, (San Diego, CA), July 1991.

[11] Anderson, E., Blackwood, G., and How, J., "Passive Damping in the MIT SERC Controlled Structures Testbed," in *Proceedings of the International Symposium on Active Materials and Adaptive Structures*, Nov. 1991.

[12] Anderson, E., Moore, D., and Fanson, J., "Development of an Active Member Using Piezoelectric and Electrostrictive Actuators for Control of Precision Structures," in *Proceedings of the 31st AIAA/ASME/ASCE/AHS Structures, Structural Dynamics, and Materials Conference*, 1990.

[13] Balmès, E. and Crawley, E., "Designing and Modeling Structural Dynamics for Control. Applications to the MIT/SERC Interferometer Testbed," in *1st SMAC Conference*, (Nice, France), Dec. 1992.

[14] Jacques, R., *On-Line System Identification and Control Design for Flexible Structures*, PhD thesis, Massachusetts Institute of Technology, 1994. SERC report #7-94.

[15] Liu, K., Jacques, R., and Miller, D., "Frequency Domain Structural System Identification by Observability Range Space Extraction," in *Proceedings of the American Control Conference*, (Baltimore, MDS), June 1994.

[16] Jacques, R. and Miller, D., "Multivariable Model Identification from Frequency Response Data," in *Proceedings of the 32nd Conference on Decision and Control*, 1993.

[17] Lehtomaki, N., Castanon, D., Levy, B., Stein, G., Sandell, Jr., N., and Athans, M., "Robustness and Modeling Error Characterization," *IEEE Transactions on Automatic Control*, Vol. AC–29, Mar. 1984, pp. 212–220.

[18] Fleming, F., "The Effect of Structure, Actuator, and Sensor on the Zeros of Controlled Structures," Master's thesis, Massachusetts Institute of Technology, 1991. MIT SERC report # 18-90.

[19] Lublin, L. and Athans, M., "Application of $\mathcal{H}_2$ Control to the MIT SERC Interferometer Testbed," in *Proceedings of the American Control Conference*, (San Francisco, CA), June 1993.

[20] Doyle, J., Glover, K., Khargonekar, P., and Francis, B., "State Space Solutions to Standard $\mathcal{H}_2 / \mathcal{H}_\infty$ Control Problems," *IEEE Transactions on Automatic Control*, Vol. 34, Aug. 1989, pp. 831–847.

[21] Doyle, J., Francis, B., and Tannenbaum, A., *Feedback Control Theory*, New York: Macmillan Publishing Company, 1992.

[22] Packard, A. and Doyle, J., "The Complex Structured Singular Value," *Automatica*, Vol. 29, No. 1, 1993, pp. 71–109.

LOGIC-BASED SWITCHING and CONTROL *

A. S. Morse[†]
Department of Electrical Engineering
Yale University
New Haven, Connecticut, 06520-208267

Dedicated to George Zames on the occasion of his 60th birthday.

Abstract

Recently a new concept called "cyclic switching" was shown provide a viable solution to the long standing certainty equivalence stabilizability problem which arises in the design of estimator-based parameter adaptive controls because of the existence of points in parameter space where the estimated model upon which certainty equivalence control is based, looses stabilizability [1]. Still more recently a simple, 'high-level' controller called a 'supervisor' was proposed for the purpose of coordinating the switching into feedback with a siso process, of a sequence of linear positioning or set-point controllers from a family of candidate controllers, so as to cause the output of the process to approach and track a constant reference input [2]. The purpose of this paper is to demonstrate that cyclic switching and supervisory control are compatible concepts.

*A full-length version of this paper including proofs will appear at a later date. This research was supported by the National Science Foundation under Grant No. ECS-9206021, by the U. S. Air Force Office of Scientific Research under Grant No. F49620-94-I-0181.

[†]Part of the author's research was done while he was visiting the Control and Dynamical Systems Group at the California Institute of Technology.

1   Introduction

A problem of continuing interest is that of developing control methods for processes modeled by linear time invariant systems with uncertain parameters. For processes of this type which are truly stationary, standard off-line identification and time invariant linear control techniques are usually applicable. On the other hand, for processes whose parameters are not fixed, but instead change "slowly" with time in an unpredictable manner, control with linear time-invariant compensators may not be possible because the range of admissible parameter variations is too large to ensure satisfactory performance. It is for such "quasi-stationary" processes that the use of some form of self-adjusting control makes sense.

Perhaps the most promising types of algorithms for controlling processes with such modeling uncertainties are those which are estimator based. Estimator - based algorithms generate control signals in accordance with the idea of certainty equivalence; i.e., at each instant of time the controller in feedback with the process is based of a current estimate of what the process model is; such estimates are selected from a suitably defined admissible model set $\mathcal{M}$ which may or may not include a copy of the actual process model. $\mathcal{M}$ is typically chosen to best satisfy a number of conflicting requirements. For example, $\mathcal{M}$'s set of "nominal" models $\mathcal{M}_{nom}$ should be a subset of a finite dimensional linear space. In addition, $\mathcal{M}_{nom}$ should be "big" enough to ensure that $\mathcal{M}$ includes a model which closely approximates that of the process. If $\mathcal{M}_{nom}$ contains a continuum of models, then for on-line model estimation to be tractable, $\mathcal{M}_{nom}$ should be convex or at least the union of a finite number of convex sets. If each model in $\mathcal{M}$ is considered to be a candidate process model, then each such model should be at least stabilizable.

It is not very difficult to see that these are conflicting requirements. In particular, stabilizability, convexity and largeness of $\mathcal{M}_{nom}$ are at odds. If stabilizability and largeness are required, then convexity and consequently tractability must be sacrificed. If convexity and stabilizability are required, then $\mathcal{M}_{nom}$ must be "small."

A third possibility is to require largeness and convexity of $\mathcal{M}_{nom}$, but not necessarily stabilizability. Naturally those models in $\mathcal{M}_{nom}$ and therefore $\mathcal{M}$ which are not stabilizable cannot be candidate process models. Nevertheless, because of the tractability issue it is useful to consider such models to be admissible for estimation purposes. Therefore an alternative to certainty equivalence is needed for selecting controllers when such models are encountered during the on-line estimation process. Such an alternative, based on the concept of "cyclic switching," has recently been proposed in [1, 3]. These references explain how to use cyclic switching to effectively solve the estimated model stabilizability problem in the context of parameter adaptive control. The purpose of this paper is to demonstrate that the concept can be employed to the same end in a supervisory control system [2]. A 'supervisor' is a high-level, hybrid controller whose purpose is to coordinate the switching into feedback with a process, of a sequence of fixed controllers from a family of candidate controllers, so as to cause the output of the process to behave in a prescribed matter. Using the idea of cyclic switching, this paper extends the concept of a supervisor to encompass those situations in which $\mathcal{M}_{nom}$ contains models which are not stabilizable.

The overall problem of interest is to construct a supervisory control system capable of transferring to and holding at a prescribed set-point, the output of a siso process modeled by a dynamical system with large scale parametric uncertainty. The process is assumed to modeled by a linear system whose transfer function is in the union of a continuum of subclasses $\mathcal{C}(p)$, p being a parameter vector taking values in some set $\mathcal{P}_P$ over which the union is taken §2. $\mathcal{P}_P$ is closed, bounded subset of a finite dimensional linear space. Each subclass $\mathcal{C}(p)$ contains a *nominal process model* transfer function ν_p about which the subclass is "centered." Although ν_p is assumed to depend linearly on p, neither $\mathcal{P}_P$ nor the class of nominal process model transfer functions need be convex.

The tractability issue is dealt with by embedding $\mathcal{P}_P$ in a conveniently chosen closed, bounded, convex subset $\mathcal{P}$ on which ν_p is linearly defined. All points $p \in \mathcal{P}$ at which the unreduced transfer function $\frac{1}{s}\nu_p$ is not stabilizable are taken to be contained in a closed subset $\mathcal{S} \subset \mathcal{P}$ called a *singular region*. The set of candidate process model transfer functions and the set $\{\nu_p : p \in \mathcal{S}\}$ are required to be disjoint.

Presumed given is a set of loop controller transfer functions κ_p, $p \in \mathcal{P} - \mathcal{S}$ with the property that κ_p would solve the positioning control problem were the process model transfer function to be either ν_p or any member of $\mathcal{C}(p)$ if p were in $\mathcal{P}_P$. Also taken as given is a finite set of constant controller gains $\{\kappa_1, \kappa_2, \ldots, \kappa_{n_S}\}$ which satisfy a version of the "observation requirement" defined in [3].

Each nominal model transfer function, controller transfer function pair (ν_p, κ_q) $p \in \mathcal{P}$, $q \in (\mathcal{P} - \mathcal{S}) \cup \{1, 2, \ldots, n_S\}$ determines an "estimator-based" controller Σ_{pq} with two outputs - one a candidate feedback control signal corresponding to κ_q and the other an "output estimation error" based on nominal model transfer function ν_p §3. The problem of having to realize as many estimator-based controllers Σ_{pq} as there are points in $\mathcal{P} \times ((\mathcal{P} - \mathcal{S}) \cup \{1, 2, \ldots, n_S\})$ is dealt with by requiring all such controllers to share the same state x_C. Σ_{pq}'s two outputs are then generated as parameter-dependent outputs of a single, finite dimensional linear system Σ_C with state x_C §3.

The supervisor is a specially structured hybrid dynamical system Σ_S whose output is a switching signal σ_S taking values in $(\mathcal{P} - \mathcal{S}) \cup \{1, 2, \ldots, n_S\}$ and whose inputs are x_C, and the process output y §4. The value σ_S at each instant of time, specifies which controller is in feedback with the process. The supervisor sets this value in accordance with the idea of certainty equivalence whenever it is possible to do so. In particular, whenever the supervisor picks an estimated model transfer function ν_p with $p \in (\mathcal{P} - \mathcal{S})$, a controller with transfer function κ_p is placed in the feedback loop and held there for at least τ_D time units, τ_D being a prespecified positive number called a dwell time. On the other hand, when the supervisor picks an estimated model transfer function ν_p with $p \in \mathcal{S}$, instead of attempting to use certainty equivalence, a "switching cyclic" is executed using the controller gains $\{\kappa_1, \kappa_2, \ldots, \kappa_{n_S}\}$; i.e., each such gain, one after another, is placed in the feedback loop for τ_S time units, τ_S being a prespecified positive number called a cycle dwell time. At the end of the switching cycle, after $n_S \tau_S$ time units have elapsed, the supervisor either applies a certainty equivalence control or executes another switching cycle depending on whether the parameter p associated with the current estimated nominal model transfer function ν_p is in $\mathcal{P} - \mathcal{S}$ or $\mathcal{S}$.

Closed-loop system behavior is analyzed in §5 and §6 subject to the assumption that the transfer function of the process to be controlled is equal to one of the nominal process model transfer functions; i.e., unmodelled dynamics are not taken into account in this paper. It is shown that the proposed supervisor performs just like a standard setpoint-controller in a linear feedback system; i.e., a zero steady state tracking error is attained for every setpoint input and every system initialization, even if process disturbances are present, provided they are bounded and constant.

Notation: In the sequel prime denotes transpose. $\mathbb{R}^{n \times m}$ is the linear space of real $n \times m$ matrices. The norm of $M \in \mathbb{R}^{n \times m}$, written $|M|$ is the sum of the magnitudes of its entries. If $f : [0, \infty) \to \mathbb{R}^n$ and $g : [0, \infty) \to \mathbb{R}^n$ are piecewise-continuous time functions we sometimes write $f \to g$ if normed difference $|f - g|$ goes to zero. $\mathbb{R}_S$ denotes the linear space of all real, rational, proper, stable transfer functions. For $\alpha \in \mathbb{R}_S$, $||\alpha||_\infty$ denotes the norm $||\alpha||_\infty = \sup_{\omega \in \mathbb{R}} |\alpha(j\omega)|$.

2   Problem Formulation

The objective of this paper is to outline a procedure for constructing a hybrid control system capable of setting and holding at a prescribed set-point, the output of a process modeled by a dynamical system with large scale parametric uncertainty. The process model is a siso controllable, observable linear system Σ_P perturbed by an unknown vector d of bounded step disturbance signals; i.e.

$$\dot{x}_P = A_P x_P + b_P u + d$$
$$y = c_P x_P \tag{1}$$

It is assumed that Σ_P's transfer function is a member of a known class of candidate transfer functions of the form

$$C_P = \bigcup_{p \in \mathcal{P}_P} C(p)$$

where $\mathcal{P}_P$ is a closed, bounded subset of a real, finite-dimensional linear space. For each fixed $p \in \mathcal{P}_P$, $C(p)$ denotes the subclass

$$C(p) = \left\{ \frac{\alpha_p}{\beta_p} + \delta_p : \|\delta_p\|_\infty \le \epsilon_p \right\}$$

where

$$\nu_p \triangleq \frac{\alpha_p}{\beta_p}$$

is a preselected, strictly proper *nominal transfer function*, ϵ_p is a real non-negative number and $\delta_p \in \mathbb{R}_S$ is strictly proper norm-bounded perturbation representing unmodelled dynamics of the additive type. It is assumed for each $p \in \mathcal{P}_P$, that β_p is monic, that α_p and β_p are coprime, and that the allowable values of $\delta_p \in \mathbb{R}_S$ exclude transfer functions for which $\nu_p + \delta_p$ has unstable poles and zeros in common. All transfer functions in C_P are thus strictly proper, but not necessarily stable rational functions. Prompted by the requirements of set-point control, it is assumed that the numerator of each transfer in C_P is nonzero at $s = 0$. For each $p \in \mathcal{P}_P$, $\frac{1}{s}C(p)$ is required be at least small enough so that it can be robustly stabilized with a single, fixed-parameter, linear controller. Of course $\frac{1}{s}C_P$ need not have this property.

This paper is concerned specifically with the case when the set of nominal process model transfer functions

$$\mathcal{N}_P = \{\nu_p : p \in \mathcal{P}_P\}$$

contains a continuum of elements. For simplicity it is assumed that each such transfer function has the same McMillan degree n_P. This means that $\mathcal{N}_P$ can be viewed as a subset of the $2n_P$-dimensional linear space of strictly proper {unreduced} rational functions whose denominators are monic and of degree n_P.

In the sequel it is assumed without loss of generality that $\mathcal{P}_P$ is a subset of a finite dimensional linear space and that $p \longmapsto \alpha_p(s)$ and $p \longmapsto \beta_p(s)$ are affine linear functions. We are interested in the case when $\mathcal{P}_P$ is not necessarily convex since convexity of $\mathcal{P}_P$ would imply convexity of $\mathcal{N}_P$. To ensure a tractable model estimation problem, we presume that $\mathcal{P}_P$ has been embedded in a conveniently chosen, closed, bounded convex subset $\mathcal{P}$ {e.g., the convex hull of $\mathcal{P}_P$} and that $\mathcal{M}_{nom} \triangleq \{\nu_p : p \in \mathcal{P}\}$ and $\mathcal{M} \triangleq C_P \cup \mathcal{M}_{nom}$. Thus $\mathcal{M}_{nom}$ is convex, $\mathcal{N}_P \subset \mathcal{M}_{nom}$ and $C_P \subset \mathcal{M}$. We shall assume that all of the points $p \in \mathcal{P}$ {if any} at which α_p and β_p are not coprime are in the interior of a specified closed set $\mathcal{S} \subset \mathcal{P}$, called a *singular region*. We shall require C_P

and $\{\nu_p : p \in S\}$ to be disjoint. As our aim is to use integral control to achieve set-point tracking, we shall also require α_p to be nonzero at $s = 0$ for all values of $p \in \mathcal{P}$ except possible those in the interior of S. We summarize.

Assumption 1

a. $\mathcal{P}$ is a closed, bounded, convex subset[1] of a finite dimensional linear space.

b. $p \longmapsto \alpha_p$ and $p \longmapsto \beta_p$ are affine linear functions on this space.

c. $\mathcal{P}_P$ is a closed, bounded subset of $\mathcal{P}$ on which each nominal transfer function ν_p has McMillan degree n_P.

d. S is a closed, bounded subset of $\mathcal{P}$ whose interior contains all values of $p \in \mathcal{P}$ if any, at which either α_p has a zero at $s = 0$ or α_p and β_p are not coprime.

e. $\mathcal{C}_P \cap \{\nu_p : p \in S\} =$ empty set

Remark 1 In the sequel we write $\mathcal{R}$ for the closure of the complement of S in $\mathcal{P}$. Note that for each $p \in \mathcal{R}$, α_p and β_p are coprime and α_p is nonzero at $s = 0$.

Example 1: More often than not, the selection of $\mathcal{P}$ is likely to precede that of $\mathcal{P}_P$ and S. Consider for example, the two-parameter family of transfer functions

$$\nu_p = \frac{s - p_1}{s^2 + (p_1 + p_2)s + 1}$$

defined on the rectangle

$$\mathcal{P} = \left\{ \begin{bmatrix} p_1 \\ p_2 \end{bmatrix} : 0.1 \le p_i \le 2, \ i \in \{1, 2\} \right\}$$

in $\mathbb{R}^2$. This transfer function has an unstable pole-zero cancellation an each point in $\mathcal{P}$ on the algebraic curve $p_1 p_2 = 1$. Thus one might define S to be the band of points $S \triangleq \{p : |p_1 p_2 - 1| \le \epsilon, p \in \mathcal{P}\}$. where ϵ is small number. In this case $\mathcal{P}_P$ could be taken as $\mathcal{P}_P \triangleq \{p : |p_1 p_2 - 1| \ge \bar{\epsilon}, p \in \mathcal{P}\}$ where $\bar{\epsilon}$ is a number greater than ϵ.

---

[1] This assumption can be replaced with the weaker assumption that $\mathcal{P}$ is a finite union of closed, bounded, convex subsets.

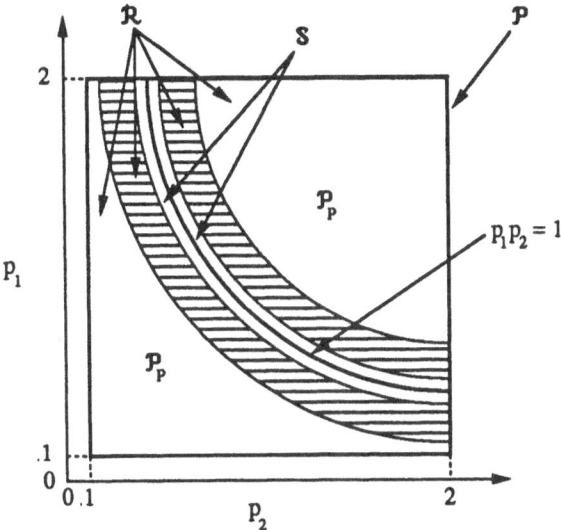

Figure 1: Singular Curve $p_1p_2 = 1$ in $\mathcal{P}$

We now turn to the main objective of the paper which is to construct a positioning or set-point control system capable of causing process output y to approach and track any constant reference input r. Towards this end we first define a *tracking error*

$$\epsilon_T \stackrel{\Delta}{=} r - y \tag{2}$$

and introduce an integrating subsystem to drive u; i.e.,

$$\dot{u} = v \tag{3}$$

Here v is a control signal which will be defined in the sequel. We take as given a parameterized family of proper, "loop-controller" or "internal regulator" transfer functions of the form

$$\kappa_p = \frac{\gamma_p}{\rho_p} \tag{4}$$

where for each $p \in \mathcal{R}$, γ_p and ρ_p are coprime and ρ_p is monic. For each $p \in \mathcal{R}$, κ_p is presumed to have been chosen to endow the feedback interconnection shown in Figure 2 with desired properties. In particular, it is assumed that for each $p \in \mathcal{R}$, the system shown in Figure 2 is at least stable with "stability margin" λ_S. More precisely κ_p is required to satisfy the following.

Property 1 There is a positive number λ_S with the property that for each $p \in \mathcal{R}$, the real parts of all zeros of

$$s\beta_p\rho_p + \alpha_p\gamma_p$$

are no greater than $-\lambda_S$.

As is well-known, the existence of such transfer functions is assured by the properties of α_p and β_p noted in Remark 1.

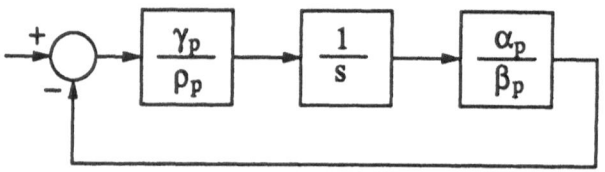

Figure 2: Feedback Interconnection

We have already stipulated that the assignments $p \longmapsto \nu_p$ be as affine linear function. The restriction of this function to $\mathcal{R}$ is therefore continuous. We require the same of κ_p:

Assumption 2 The function $p \longmapsto \kappa_p$ is continuous on $\mathcal{R}$

We shall not demand that the transfer functions in $\mathcal{N} = \{\nu_p : p \in \mathcal{R}\}$ be distinct. We shall however require there to be exactly one controller transfer function in $\mathcal{K} \triangleq \{\kappa_p : p \in \mathcal{R}\}$ for all nominal process model transfer functions in $\mathcal{N}$ which are the same:

Assumption 3 $\alpha_p, \beta_p, \rho_p$, and γ_p, have the property that $\gamma_p = \gamma_q$, and $\rho_p = \rho_q$ for each pair of points $p, q \in \mathcal{R}$ at which $\alpha_p = \alpha_q$ and $\beta_p = \beta_q$.

Assumption 3 implies that the assignment $\nu_p \longmapsto \kappa_p$, $p \in \mathcal{R}$ is a well-defined function from $\mathcal{N}$ to $\mathcal{K}$. We shall require this function to be smooth:

Assumption 4 The aforementioned function is continuous on $\mathcal{N}$.

In the sequel we shall develop a supervisory logic for switching into the feedback loop between ϵ_T and v a sequence of internal regulators so as to achieve desired closed-loop performance. The "supervisor" which performs this function, does this in accordance with the idea of certainty equivalence whenever it is possible to do so. In particular, whenever the supervisor picks an estimated model transfer function ν_p with $p \in (\mathcal{P} - \mathcal{S})$, an internal regulator with transfer function κ_p is connected between ϵ_T and v. On the other hand, when the supervisor picks an estimated model transfer function ν_p with $p \in \mathcal{S}$, instead of attempting to use certainty equivalence which may not be feasible[2], a form of

---

[2] This is because ν_p may not be stabilizable or may have a zero at $s = 0$.

cyclic switching is employed similar to that proposed in [1, 3]. Cyclic switching requires a set of real gains $\{g_1, g_2, \ldots, g_{n_S}\}$ which satisfies the following

Observation Requirement: For each $l \in \mathcal{P}_P$ and each $p \in \mathcal{S}$, there is a value of $q \in \{1, 2, \ldots, n_S\}$ for which the feedforward interconnection of canonical realizations of ν_p and ν_l shown in Figure 3 is observable through e_{ff}.

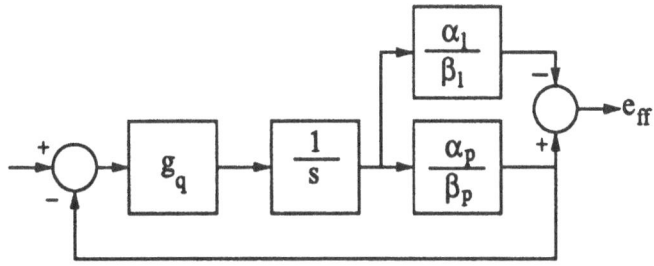

Figure 3: Feedforward Interconnection

The following proposition is a simple consequence of Assumptions 1c to 1e. The proposition's proof is similar to that of Proposition 1 of [3].

Proposition 1 *The observation requirement is satisfied by any set of n_S distinct gains provided $n_S \geq 2n_P + 1$.*

In the sequel we assume that $\{\gamma_1, \gamma_2, \ldots, \gamma_{n_S}\}$ is a set of real distinct gains chosen to satisfy the observation requirement. In addition we adopt the notation $\mathcal{I} \triangleq \{1, 2, \ldots, n_S\}$ and write $\mathcal{Q}$ for the disjoint union $\mathcal{Q} \triangleq (\mathcal{P} - \mathcal{S}) \cup \mathcal{I}$.

3 State Space Systems

To deal with the assumption Σ_P's transfer function is in $\mathcal{C}_P$, but is otherwise unknown, we shall employ a parameterized family of estimator-based controllers $\{\Sigma_{pq} : p \in \mathcal{P}, q \in \mathcal{Q}\}$. Each Σ_{pq} is a stable linear system with inputs y, v, and ϵ_T, and outputs e_p and v_q; e_p is an "output estimation error" and v_q is a candidate signal for v. The Σ_{pq} are defined on $\mathcal{P} \times (\mathcal{R} \cup \mathcal{I})$ as follows. First pick monic, stable polynomials ω_P and ω_R of degrees $n_P + 1$ and n_R respectively, n_R being an upper bound on the McMillan degree of κ_q, $q \in \mathcal{R}$; choose these polynomials so that all of their roots have real parts no greater than $-\lambda_S$. Second, for each $q \in \mathcal{R}$ pick a monic polynomial ψ_q of degree $n_R - \deg \rho_q$ whose roots also have real parts no greater than $-\lambda_S$. Third, set $\psi_q = 1$ and $\rho_q = 1$ for $q \in \mathcal{I}$. For $p \in \mathcal{P}$ and $q \in \mathcal{R} \cup \mathcal{I}$ the transfer matrix of Σ_{pq} from $[y \quad v \quad \epsilon_T]'$ to $[e_p \quad v_q]'$ is specified to be

$$
\begin{bmatrix}
-\frac{s\beta_p}{\omega_P} & \frac{\alpha_p}{\omega_P} & 0 \\[2mm]
0 & 1 - \frac{\psi_q \rho_q}{\omega_R} & -\frac{\psi_q \gamma_q}{\omega_R}
\end{bmatrix}
\tag{5}
$$

This definition implies that no matter how Σ_{pq} is realized, the following are true:

- If the transfer function of Σ_P is the same as the nominal process model transfer function ν_p and d is a constant disturbance input, then no matter how v is defined, $e_p \rightarrow 0$ as fast as $e^{-\lambda_S t}$.

- If v is set equal to v_q then the transfer function from ϵ_T to v is

$$
\frac{\gamma_q}{\rho_q}
$$

after cancellation of the stable common factor $\psi_q \omega_R$.

We shall require all estimator-based controllers to share the same state x_C. In other words, e_p and v_q will be generated as parameter-dependent outputs of a single, stable, finite dimensional linear system Σ_C with state x_C and inputs y, v and ϵ_T. Σ_C is defined as follows: First pick any three single-input controllable pairs $(\tilde{A}_P, \tilde{b}_P)$, $(\tilde{A}_{PR}, \tilde{b}_{PR})$ and $(\tilde{A}_R, \tilde{b}_R)$ in such a way that ω_P, ω_{PR} and ω_R are the characteristic polynomials of $\tilde{A}_P$, $\tilde{A}_{PR}$ and $\tilde{A}_R$ respectively where ω_{PR} is the least common multiple of ω_P and ω_R. Second define $A_C = $ block diag.$\{\tilde{A}_P, \tilde{A}_{PR}, \tilde{A}_R, \}$, $d_C = $ block diag.$\{\tilde{b}_P, 0, 0\}$, $b_C = $ block diag.$\{0, \tilde{b}_{PR}, 0\}$, and $h_C = $ block diag.$\{0, 0, \tilde{b}_R\}$. Third, for $p \in \mathcal{P}$ and $q \in \mathcal{R} \cup \mathcal{I}$ define c_p, f_q, and g_q to be the unique solution to the equation

$$
\begin{bmatrix} c_p \\ f_q \end{bmatrix} (sI - A_C)^{-1} [\, d_C \quad b_C \quad h_C \,] +
\begin{bmatrix} 0 & 0 & 0 \\ 0 & 0 & g_q \end{bmatrix}
$$

$$
= \begin{bmatrix}
1 - \frac{s\beta_p}{\omega_P} & \frac{\alpha_p}{\omega_P} & 0 \\[2mm]
0 & 1 - \frac{\psi_q \rho_q}{\omega_R} & -\frac{\psi_q \gamma_q}{\omega_R}
\end{bmatrix}
\tag{6}
$$

Finally, define Σ_C to be the parameter dependent system

$$
\begin{aligned}
\dot{x}_C &= A_C x_C + d_C y + b_C v + h_C \epsilon_T \\
v_q &= f_q x_C + g_q \epsilon_T \\
e_p &= c_p x_C - y
\end{aligned}
\tag{7}
$$

It is straightforward to verify that the transfer matrix of Σ_C from $[\,y \quad v \quad \epsilon_T\,]'$ to $[\,e_p \quad v_q\,]'$ is the transfer matrix of Σ_{pq} specified in (5). It is also not difficult see that if Σ_P's transfer function were the same as nominal process model transfer function ν_p and d were a constant disturbance, then e_p would be of the form

$$e_p = c_p e^{A c t}(x_C(0) - M_1 x_P(0) - M_2 u(0)) \tag{8}$$

where M_1 and M_2 are time-independent matrices depending only on Σ_P and p.

The feedback control to the integrator driving u can now be written as

$$v = v_\sigma \tag{9}$$

where σ is a piecewise-constant, "switching signal" taking values in $\mathcal{Q}$. Closing the "supervisory loop" means setting

$$\sigma = \sigma_S \tag{10}$$

where σ_S is the output of a suitably defined "supervisory control system" whose structure we discuss in the next section.

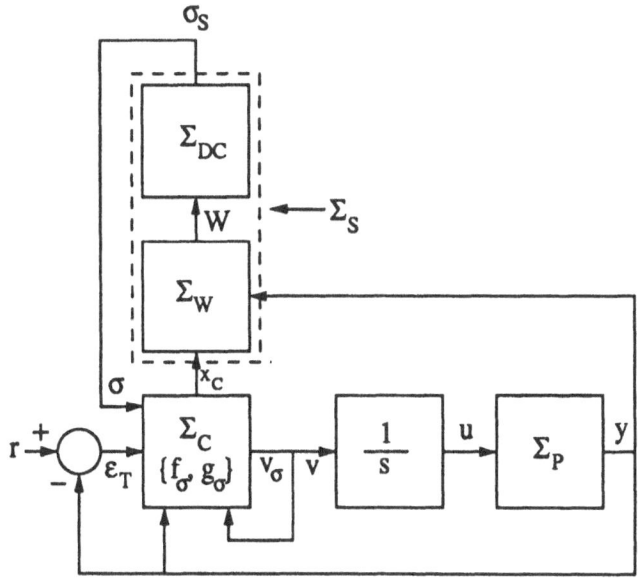

Figure 4: Closed-Loop Supervisory Control System

Remark 2 The uniqueness of parameter vectors f_q and g_q defined by (6) together with Assumption 2 ensures that f_q and g_q are continuous on $\mathcal{R}$. Note in addition that Assumption 1b implies that $p \longmapsto c_p$ is an affine linear function.

4   Supervisor

By a supervisor is meant a specially structured causal dynamical system Σ_S whose output σ_S is a switching signal taking values in $\mathcal{Q}$ and whose inputs are x_C and y. Internally a supervisor consists of two subsystems, one a *performance weight generator* Σ_W and the other a *dwell-time/cyclic switching logic* Σ_{DC}. Σ_W is a causal dynamical system whose inputs are x_C and y and whose state and output W is a "weighting matrix" which takes values in a linear space $\mathcal{W}$. W together with a suitably defined *performance function* $\Pi : \mathcal{W} \times \mathcal{P} \to \mathbb{R}$ determine, for each $p \in \mathcal{P}$, a scalar-valued *performance signal* of the form

$$\pi_p = \Pi(W, p), \ p \in \mathcal{P} \tag{11}$$

which is viewed by the supervisor as a measure of the expected closed-loop performance of controller p. Σ_W and Π are defined by the equations

$$\dot{W} = -\lambda W + \begin{bmatrix} x_C \\ y \end{bmatrix} \begin{bmatrix} x_C \\ y \end{bmatrix}' \tag{12}$$

and

$$\Pi(W, p) = \begin{bmatrix} c_p & -1 \end{bmatrix} W \begin{bmatrix} c_p & -1 \end{bmatrix}' \tag{13}$$

respectively where λ is a nonnegative number. In the light of (7) and (11) it is easy to see that these definitions imply that

$$\dot{\pi}_p = -\lambda \pi_p + e_p^2, \quad p \in \mathcal{P} \tag{14}$$

The supervisor's other subsystem, called a *dwell-time/cyclic switching logic* Σ_{DC}, is a combined version of the switching logics of [3] and [2]. It is a hybrid dynamical system whose input and output are W and σ_S respectively, and whose state is the ordered quintuple $\{X, \hat{p}, \tau, \beta, \sigma_S\}$. Here X is a discrete-time matrix which takes on sampled values of W, $\hat{p}$ is a discrete-time variable taking values in $\mathcal{P}$, τ is a continuous-time variable called a *timing signal*, and β is a logic variable taking values in $\{0, 1\}$. τ takes values in the closed interval $[0, \max\{n_S \tau_S, \tau_D\}]$, where τ_D and τ_S are a prespecified positive numbers called a *dwell time* and a *cycle dwell time* respectively. It is assumed that there is a second prespecified nonnegative number $\tau_C \leq \max\{n_S \tau_S, \tau_D\}$ called a *computation time*, which bounds from above for any $X \in \mathcal{W}$, the time it would take the supervisor to compute a value $p = p_X \in \mathcal{P}$ which minimizes $\Pi(X, p)$. Between "event times" τ is generated by a reset integrator according to the rule

$$\dot{\tau} = 1$$

Such event times occur for $\beta \in \{0,1\}$, when the value of τ reaches either $T(\beta) - \tau_C$ or $T(\beta)$ where $T(0) \triangleq \tau_D$ and $T(1) \triangleq n_S \tau_S$; at such times τ is reset to either 0 or $T(\beta) - \tau_C$ depending on the value of Σ_{DC}'s state. Σ_{DC}'s internal logic is defined by the computer diagram shown in Figure 5 where p_X denotes a value of $p \in \mathcal{P}$ which minimizes $\Pi(X,p)$.

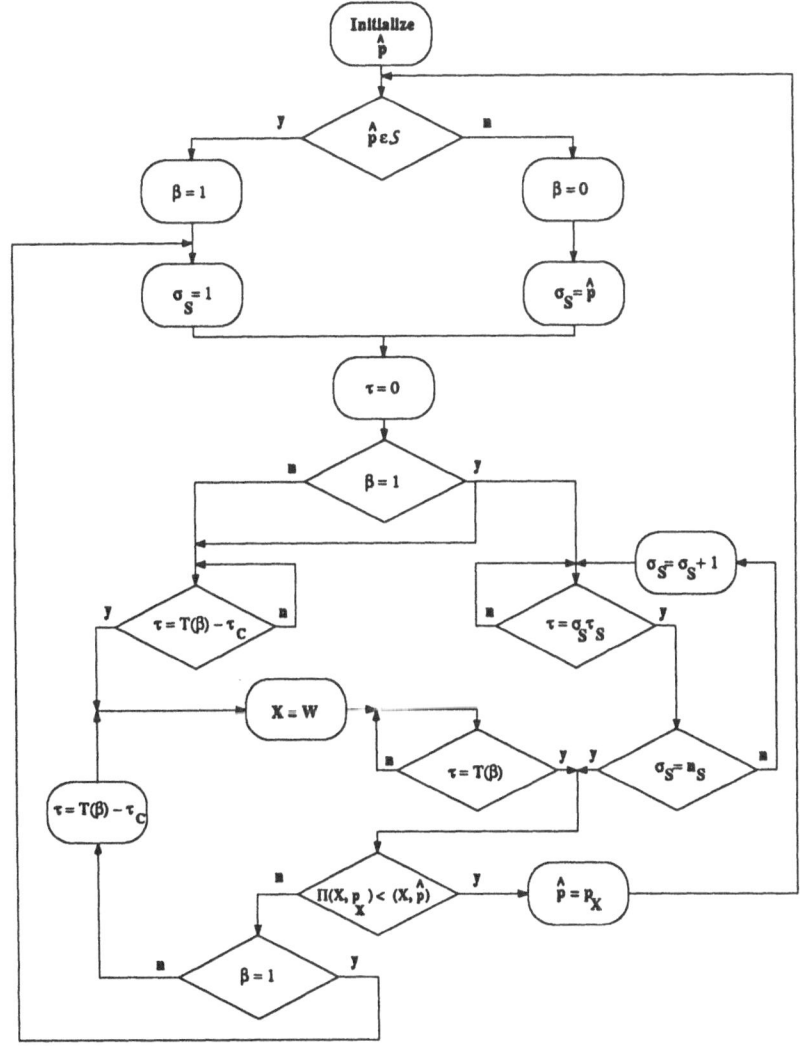

Figure 5: Dwell-Time/Cyclic Switching Logic Σ_{DC}

Because of (13) and the linear dependence of c_p on p {cf. Remark 2}, $\Pi(W,p)$ is a quadratic function of p. This and the assumed convexity of $\mathcal{P}$ imply that the minimization of $\Pi(W,p)$ on $\mathcal{P}$ is a convex, quadratic programming problem.

The functioning of Σ_{DC} can be explained roughly as follows. Suppose that at some time t_0, Σ_S has just changed the value of $\widehat{p}$. Depending on whether $\widehat{p} \in S$ or not, one of two different epochs can occur:

- Suppose $\widehat{p} \notin S$. In this case β is set equal to 0, σ_S is set equal to $\widehat{p}$ and τ is reset to 0. After $\tau_D - \tau_C$ time units have elapsed, W is sampled and X is set equal to this value. During the next τ_C time units, a value $p = p_X$ is computed which minimizes $\Pi(X, p)$. At the end of this period, when $\tau = \tau_D$, if $\Pi(X, p_X)$ is smaller than $\Pi(X, \widehat{p})$, then $\widehat{p}$ is set equal to p_X and the logic goes back to again test whether or not $\widehat{p} \in S$. If, on the other hand, $\Pi(X, \widehat{p})$ is less than or equal to $\Pi(X, p_X)$, τ is reset to $\tau_D - \tau_C$, W is again sampled, X takes on this new sampled value, minimization is again carried out over the next τ_C time units..... and so on.

- Suppose $\widehat{p} \in S$. In this case β is set equal to 1, τ is reset to 0, and two distinct sequences of events occur simultaneously, each lasting $n_S \tau_S$ time units:

  1. At $\tau = 0$, a switching cycle is executed[3].

  2. At $\tau = n_S \tau_S - \tau_C$, W is sampled and X is set equal to this value. During the next τ_C time units, a value $p = p_X$ is computed which minimizes $\Pi(X, p)$. At the end of this period, when $\tau = n_S \tau_S$, if $\Pi(X, p_X)$ is smaller than $\Pi(X, \widehat{p})$, then $\widehat{p}$ is set equal to p_X and the logic goes back to again test whether or not $\widehat{p} \in S$. If, on the other hand, $\Pi(X, \widehat{p})$ is less than or equal to $\Pi(X, p_X)$, τ is reset to 0, another switching cycle is executed ....and so on.

Remark 3 Note that Σ_S is *scale independent* in that its output σ_S remains unchanged if its performance function-weighting matrix pair (Π, W) is replaced by another performance function-weighting matrix pair $(\bar{\Pi}, \bar{W})$ satisfying $\bar{\Pi}(\bar{W}, p) = \theta \Pi(W, p)$, $p \in \mathcal{P}$, where $\theta : [0, \infty) \to \mathbb{R}$ is a positive time function. This is because for any fixed t, the values of p which minimize $\Pi(W(t), p)$ are exactly the same as the values of p which minimize $\theta(t) \Pi(W(t), p)$.

5 Exact Matching

Our aim here is to analyze the closed-loop behavior of a supervisory control system for the case when there is a value $p^* \in \mathcal{P}_P$ for which nominal transfer function ν_{p^*} matches or equals that of process model Σ_P. For the present, we assume that $\lambda = 0$ and that Σ_W and Π are defined by (12) and (13) respectively. Therefore in this case, π_p is the $\mathcal{L}^2$ performance signal

---

[3] The supervisor *executes* a switching cycle at clock time $\tau = 0$ by setting $\sigma_S(t_0 + \tau) = s(\tau)$, $\tau \in [0, n_S \tau_S)$ where t_0 is the actual time τ was reset to 0 and $s : [0, n_S \tau_S) \to \mathcal{I}$ is the piecewise-constant function whose value is i on the subinterval $[(i-1)\tau_S, i\tau_S)$, $i \in \mathcal{I}$.

$$\dot{\pi}_p = e_p^2, \quad p \in \mathcal{P} \tag{15}$$

It is convenient at this point to rewrite the equations describing combined controller (7) and feedback law (9), in alternative form which will prove to be especially useful in the sequel. For each $p, l \in \mathcal{P}$, and $q \in \mathcal{R} \cup \mathcal{I}$, define matrices

$$d_q = g_q b_C + h_C - d_C \tag{16}$$
$$f_{ql} = f_q - g_q c_l \tag{17}$$
$$c_{pl} = c_p - c_l \tag{18}$$
$$A_l = A_C + (d_C - h_C)c_l \tag{19}$$

Fix set-point value r. Let x denote the translated state defined by

$$x = x_C + A_C^{-1} d_C r \tag{20}$$

Using these definitions together with (7) and (9), it is straightforward to verify that

$$\dot{x} = (A_{p\bullet} + b_C f_{\sigma p\bullet})x + d_\sigma e_{p\bullet} \tag{21}$$
$$e_p = c_{pp\bullet} x + e_{p\bullet}, \quad p \in \mathcal{P} \tag{22}$$
$$v = f_{\sigma p\bullet} x + g_\sigma e_{p\bullet} \tag{23}$$
$$\epsilon_T = e_{p\bullet} - c_{p\bullet} x \tag{24}$$

It is important to note that these equations hold for any value of $p^* \in \mathcal{P}_P$ whether or not p^* is actually a point at which exact matching takes place[4]. In other words, equations (21) to (24) have nothing to do with the matching assumption and thus can be used even when the assumption is not made.

The following proposition summarizes the consequences of selecting internal regulator transfer functions so that Property 1 holds and the observation requirement is satisfied.

Proposition 2 *Fix $l \in \mathcal{P}_P$. Then the following are true.*

1. *For each $p \in \mathcal{R}$ the matrix pair $(c_{pl}, A_l + b_C f_{ql})|_{q=p}$ is detectable with stability margin[5] λ_S.*

2. *For each $p \in \mathcal{S}$ there exists a $q \in \mathcal{I}$ such that $(c_{pl}, A_l + b_C f_{ql})$ is detectable with stability margin λ_S.*

---

[4]In deriving these equations, we have used the identities $c_p A_C^{-1} d_C = -1, p \in \mathcal{P}$ and $f_q A_C^{-1} d_C = 0, q \in \mathcal{Q}$; both are direct consequences of (6).

[5]A matrix pair (C, A) is *detectable with stability margin* $\lambda \geq 0$, if $(C, \lambda I + A)$ is a detectable pair.

To understand why the proposition's first claim is so, note from (18) and (19) that there is a vector, namely

$$k_p \triangleq d_C - h_C - b_C g_p$$

such that

$$A_l + b_C f_{pl} + k_p c_{pl} = A_p + b_C f_{pp} \tag{25}$$

Using (3), (6) and (19) it is a simple matter to verify that the characteristic polynomial of $A_p + b_C f_{pp}$ equals

$$\psi_p \omega_{PR}(s\beta_p \rho_p + \alpha_p \gamma_p)$$

where ω_{PR} is the least common multiple of ω_P and ω_R. In view of Property 1 and the definitions of ω_P, ω_R and ψ_p, it must be that $\lambda_S I + A_p + b_C f_{pp}$ is a stability matrix. Thus $\lambda_S I + A_l + b_C f_{pl}$ is an output injection away from a stability matrix which proves that $(c_{pl}, \lambda_S I + A_l + b_C f_{pl})$ is detectable as claimed.

The proof of the second claim of Proposition 2 relies on the following.

Lemma 1 *Fix $l \in \mathcal{P}_P$ and let Λ_{PR} be the root sets of ω_{PR}. For each $p \in S$ and each $q \in \mathcal{I}$ write Λ_q for the the root set of ψ_q, Λ_{plq} for the unobservable spectrum of $(c_{pl}, A_l + b_C f_{ql})$ and Ω_{plq} for the unobservable spectrum of the feedback interconnection of canonical realizations of ν_p, ν_l and g_q shown in Figure 3. Then*

$$\Lambda_{plq} \subset \Omega_{plq} \cup \Lambda_q \cup \Lambda_{PR}$$

The lemma implies that $(c_{pl}, A_l + b_C f_{ql})$ will be detectable with stability margin $-\lambda_S$ if the system in Figure 3 is observable. Therefore, since the set $\{g_q : q \in \mathcal{I}\}$ satisfies the observation requirement, the proposition's second claim is true.

Fix the initial values of the process model state x_P, integrator state u, controller state x_C and supervisor's state variables W and $\hat{p}$. The exact matching assumption made above means that no matter what v is, $e_{p^\bullet}$ must go to zero exponentially fast. This implies that $e_{p^\bullet}$ is bounded on $[0, \infty)$ and has a finite $\mathcal{L}^2[0, \infty)$ norm. Thus because of (15),

$$\lim_{t \to \infty} \pi_{p^\bullet}(t) \triangleq C^* < \infty$$

The limit set [4] of $\hat{p}$, written $\mathcal{L}^*$, can be characterized as follows. If $\hat{p}$ switches at most a finite number of times, define $S^* = \{\bar{p}\} \cap S$ and $\mathcal{R}^* = \{\bar{p}\} \cap (\mathcal{P} - S)$ where $\bar{p}$ is the final value of $\hat{p}$; clearly $\mathcal{L}^* = \mathcal{R}^* \cup S^*$. If $\hat{p}$ does not stop switching, let p_i denote the value of $\hat{p}$ on $[\tilde{t}_{i-1}, \tilde{t}_i)$ where $\tilde{t}_1, \tilde{t}_2, \ldots$ are the times at which $\hat{p}$ switches and $\tilde{t}_0 \triangleq 0$. In this case $\mathcal{L}^*$ is the set of limit points of all convergent infinite subsequences of the sequence $p_1, p_2, \ldots$. For this case define $\mathcal{R}^*$ and $\mathcal{S}^*$ to be the sets of limit points of all infinite, convergent subsequences of $p_1, p_2, \ldots$ which lie in $\mathcal{P} - \mathcal{S}$ and $\mathcal{S}$ respectively. Clearly $\mathcal{R}^* \cup \mathcal{S}^* \subset \mathcal{L}^*$. On the other hand, since the limit point of any convergent infinite sequence s in $\mathcal{P}$ is also the limit point of any convergent infinite subsequence of s, any point in $\mathcal{L}^*$ must be in $\mathcal{R}^*$ or $\mathcal{S}^*$ or both. In other words, $\mathcal{L}^* = \mathcal{R}^* \cup \mathcal{S}^*$ for this case too. Note that whether switching stops or not, $\mathcal{S}^*$ and $\mathcal{R}^*$ are closed bounded subsets of $\mathcal{S}$ and $\mathcal{R}$ respectively and $\hat{p} \to \mathcal{R}^* \cup \mathcal{S}^*$.

Remark 4 The preceding implies that if $\mathcal{R}^*$ is empty, then $\hat{p}$ must eventually enter $\mathcal{S}$ and remain there indefinitely. The preceding also implies that if $\mathcal{R}^*$ is nonempty, then for any number $\epsilon > 0$ there must be a time $\tilde{t}_{i_\epsilon}$ such that for each $t \geq \tilde{t}_{i_\epsilon}$ either $\hat{p}(t) \in \mathcal{S}$ or $|\hat{p}(t) - \bar{p}| \leq \epsilon$ for some $\bar{p} \in \mathcal{R}^*$.

We claim that
$$\int_0^\infty \|e_p\|^2 dt < \infty, \quad p \in (\mathcal{R}^* \cup \mathcal{S}^*)$$

Suppose this inequality fails to hold for some $q \in (\mathcal{R}^* \cup \mathcal{S}^*)$. Then because all π_p are monotone nondecreasing time functions, for fixed $\epsilon > 0$ there would have to be a finite time $\bar{t}$ such that

$$\pi_q(\bar{t}) > C^* + \epsilon \tag{26}$$

Moreover, since q is a limit point, there would have to be a subsequence $p_{i_1}, p_{i_2}, \ldots$ which converges to q. Because of this and the continuous dependence of c_p on p, there would have to be an integer $j \in \{i_1, i_2, \ldots\}$ sufficiently large so that $\tilde{t}_{j-1} > \bar{t} + \max\{\tau_D, n_S\tau_S\}$ and $|\pi_{p_j}(\bar{t}) - \pi_q(\bar{t})| \leq \epsilon$. The latter and (26) would then imply that $\pi_{p_j}(\bar{t}) = \pi_q(\bar{t}) + \pi_{p_j}(\bar{t}) - \pi_q(\bar{t}) \geq \pi_q(\bar{t}) - \epsilon > C^*$. Moreover, since π_{p_j} is monotone, $\pi_{p_j}(t) > C^*$ for all $t \geq \bar{t}$. But $C^* \geq \pi_{p^*}$, so $\pi_{p_j} > \pi_{p^*}$ for all $t \geq \bar{t}$. Therefore $\hat{p}(t) \neq p_j$ for all $t \geq \bar{t} + \max\{\tau_D, n_S\tau_S\}$. In particular, $\hat{p}(\tilde{t}_{j-1}) \neq p_j$ which is a contradiction. Thus the claim is true.

With c_{pq} as in (18), let $\{c_{p_1 p^*}, c_{p_2 p^*}, \ldots, c_{p_m p^*}\}$ be a basis for the span of $\{c_{pp^*} : p \in (\mathcal{R}^* \cup \mathcal{S}^*)\}$. Define $C = [c'_{p_1 p^*} \quad c'_{p_2 p^*} \quad \cdots \quad c'_{p_m p^*}]'$ and

$$\bar{e} = Cx \tag{27}$$

These definitions together with (22) imply that $e_{p_i} - e_{p^*}$ is the ith entry of $\bar{e}$. Since each such entry has a finite $\mathcal{L}^2[0, \infty)$ norm, $\bar{e}$ must have a finite $\mathcal{L}^2[0, \infty)$ norm as well. Note also that the definition of C assures there must be a bounded function $z : (\mathcal{R}^* \cup \mathcal{S}^*) \to \mathbb{R}^{m \times 1}$ for which

$$z(p)C = c_{pp^*}, \ p \in (\mathcal{R}^* \cup \mathcal{S}^*) \tag{28}$$

Remark 5 In view of this and Proposition 2, it must be that

1. for each $q \in \mathcal{R}^*$ the matrix pair $(C, A_{p^*} + b_C f_{qp^*})$ is detectable with stability margin λ_S and

2. there exists a $q \in \mathcal{I}$ such that $(C, A_{p^*} + b_C f_{qp^*})$ is detectable with stability margin λ_S.

Fix $t_0 \geq 0$ and let $\bar{\mathcal{R}}$ be a closed subset of $\mathcal{P}$. Let us agree to call a piecewise-constant function $\sigma : [t_0, \infty) \to \bar{\mathcal{R}} \cup \mathcal{I}$ an *admissible* switching signal if σ is either a constant with value in $\bar{\mathcal{R}}$ or if one of the following conditions is true of each of σ's switching intervals $[t_{i-1}, t_i)$:

• Either

$$\sigma(t) \in \bar{\mathcal{R}}, \ t \in [t_{i-1}, t_i)$$

and $t_i - t_{i-1} \geq \tau_D$ or

•

$$\sigma(t) \in \mathcal{I}, \ t \in [t_{i-1}, t_i)$$

in which case there is set of n_S consecutive switching intervals $\{[t_{j_i}, t_{j_i+1}),$ $[t_{j_i+1}, t_{j_i+2}), \ldots, [t_{j_i+n_S-1}, t_{j_i+n_S})\}$, each of length τ_S, such that

$$[t_{i-1}, t_i) \in \{[t_{j_i}, t_{j_i+1}), [t_{j_i+1}, t_{j_i+2}), \ldots, [t_{j_i+n_S-1}, t_{j_i+n_S})\}$$

and

$$\sigma(t) = q, \ t \in [t_{j_i+q-}, t_{j_i+q}), \ q \in \mathcal{I}$$

Note that because of the way Σ_{DC} has been defined in §4, $\sigma_S : [0, \infty) \to \mathcal{R} \cup \mathcal{I}$ an admissible switching signal as is the restriction of σ_S to any subinterval $[\tilde{t}_i, \infty)$ where $\tilde{t}_i$ is a switching time of $\hat{p}$. Note also that switching cannot occur infinitely fast and thus that existence and uniqueness of solutions to the differential equations involved, is not an issue.

To proceed, let us note that for any time $t^* \geq 0$ and any appropriately sized, bounded, piecewise continuous, matrix-valued function H on $[t^*, \infty)$, (21) can be rewritten as

$$\dot{x} = (A_{p^*} + b_C f_{\sigma s p^*} + HC)x - H\bar{e} + d_{\sigma s} e_{p^*} \tag{29}$$

for $t \geq t^*$. Suppose such a pair (t^*, H) can be shown to exist for which $A_{p^*} + b f_{\sigma s p^*} + HC$ is exponentially stable. Then because $\bar{e}$ and e_{p^*} have finite

$\mathcal{L}^2[t^*, \infty)$ norms, x would have a finite $\mathcal{L}^2[t^*, \infty)$ norm and would in addition tend to zero. Hence x_C would tend to $-A_C^{-1}d_C r$ because of (20). Moreover since e_{p^*} tends to zero and has a finite $\mathcal{L}^2[t^*, \infty)$ norm, (23) and (24) would imply that v and ϵ_T have finite $\mathcal{L}^2[t^*, \infty)$ norms and tend to zero as well. As a consequence x_P and u would have to tend to finite limits; this is because the system defined by (1)-(3) is observable through ϵ_T [6] and r and d are constants. In other words, to show that ϵ_T goes to zero and that x_C, x_P, u tend to finite limits its enough to show that $A_{p^*} + b_C f_{\sigma s p^*} + HC$ is exponentially stable {on $[t^*, \infty)$} for some suitably defined pair (t^*, H).

Consider first the case in which the transfer matrix of (C, A_{p^*}, b_C) is zero. Examination of (28) and (18) reveals that this will be so just in case all of the nominal process model transfer functions in the family $\{\nu_p : p \in (\mathcal{R}^* \cup \mathcal{S}^*)\}$ are the same as ν_{p^*}. Thus for this case

$$(\mathcal{R}^* \cup \mathcal{S}^*) \cap \mathcal{S} = \text{empty set} \tag{30}$$

because of Assumption 1e and

$$f_{pp^*} = f_{p^* p^*}, \; p \in \mathcal{R}^* \tag{31}$$

because of Assumption 3. In view of (30), $\hat{p}$ must tend to $\mathcal{R} - \mathcal{S} \cap \mathcal{R}$ which means that the execution of switching cycles must terminate in finite time. Therefore for all t sufficiently large $\sigma_S(t) = \hat{p}(t)$, so $\sigma_S \to \mathcal{R}^*$. From this, (31) and and the continuity of f_{qp^*} on $\mathcal{R}$ {cf. Remark 2} it follows that $A_{p^*} + b_C f_{\sigma s p^*} \to A_{p^*} + b_C f_{p^* p^*}$ which is a constant stability matrix. Therefore for this case $A_{p^*} + b_C f_{\sigma s p^*}$ is exponentially stable without output injection.

Now consider the case in which the transfer matrix of (C, A_{p^*}, b_C) is nonzero. To deal with this case, we make use of the following result.

Switching Theorem: *Let $\lambda_0 > 0$ be fixed. Let $\bar{\mathcal{R}}$ be a closed, bounded subset of $\mathcal{R}$. Let $(C_{q_0 \times n}, A_{n \times n}, B_{n \times m})$ be a left invertible system. Suppose that $\{F_q : q \in (\mathcal{R} \cup \mathcal{I})\}$ is a closed, bounded subset of matrices in $\mathbb{R}^{m \times n}$ such that*

1. *for each $q \in \bar{\mathcal{R}}$, $(C, A + BF_q)$ is detectable with stability margin no smaller than λ_0 and*

2. *there exists a $q \in \mathcal{I}$ for which $(C, A + BF_q)$ is detectable with stability margin no smaller than λ_0.*

Then there exist a constant $a \geq 0$ and bounded, matrix-valued output injection function $q \longmapsto K_q$ on $\bar{\mathcal{R}} \cup \mathcal{I}$ which, for any $t_0 \geq 0$ and any admissible switching

---

[6]This follows from the standing assumption that the numerator of the transfer function of Σ_P is nonzero at $s = 0$.

signal $\sigma : [t_0, \infty) \rightarrow (\bar{\mathcal{R}} \cup \mathcal{I})$ *causes the state transition matrix of*

$$A + K_\sigma C + BF_\sigma$$

to satisfies

$$|\Phi(t, \mu)| \leq e^{(a - \lambda_0(t - \mu))}, \quad t \geq \mu \geq t_0$$

In view of Remark 5, the switching theorem can be applied to the matrices $A_{p^*}, b_C, C, f_{qp^*}, q \in (\mathcal{R}^* \cup \mathcal{I})$, by identifying $\bar{\mathcal{R}}$ with $\mathcal{R}^*$, A with A_{p^*}, B with b_C, F_q with $f_{qp^*}, q \in \mathcal{R}^* \cup \mathcal{I}$ and λ_0 with λ_S. Therefore by the switching theorem there exists a number $a > 0$ and a bounded output injection matrix, namely K_q, which causes the state transition $\Phi(t, \mu)$ of $A_{p^*} + K_\sigma C + b_C f_{\sigma p^*}$ to satisfy

$$|\Phi(t, \mu)| \leq e^{(a - \lambda_S(t - \mu))}, \quad t \geq \mu \geq t_0 \tag{32}$$

for any $t_0 \geq 0$ and any admissible switching signal $\sigma : [t_0, \infty) \rightarrow \mathcal{R}^* \cup \mathcal{I}$.

Suppose $\mathcal{R}^*$ is empty. This being so, in view of Remark 4 and the definition of Σ_S there must exist a finite time $\bar{t}$ beyond which $\sigma_S \in \mathcal{I}$. Let t^* be the first time beyond $\bar{t}$ at which a switching cycle is initiated. Let $\sigma : [t^*, \infty) \rightarrow \mathcal{I}$ be the restriction of σ_S to $[t^*, \infty)$. Then σ is admissible so (32) holds with $t_0 \triangleq t^*$. Therefore $H \triangleq K_\sigma$ exponentially stabilizes $A_{p^*} + HC + b_C f_{\sigma s p^*}$ on $[t^*, \infty)$ as required.

Now suppose $\mathcal{R}^*$ is non-empty. Pick any positive number

$$\lambda < \lambda_S e^{-a} \tag{33}$$

Let $w : \mathcal{R} \cup \mathcal{I} \rightarrow \mathcal{R}^* \cup \mathcal{I}$ denote a function which assigns to each $p \in \mathcal{R}$ a value of $\bar{p} \in \mathcal{R}^*$ which minimizes $|p - \bar{p}|$ and to each $q \in \mathcal{I}$ the same value q. In view of Remark 4 and the definition of σ_S, it must be that for any number $\epsilon > 0$ there is a switching time $\tilde{t}_{i_\epsilon}$ such that for each $t \geq \tilde{t}_{i_\epsilon}$, either $\sigma_S \in \mathcal{I}$ or $|\sigma_S - w(\sigma_S)| \leq \epsilon$. Because of this and the continuity of f_{qp^*} on $\mathcal{R}$ {cf. Remark 2}, it is possible to pick ϵ so small that for each $t \geq \tilde{t}_{i_\epsilon}$, $|b_C(f_{\sigma s p^*} - f_{w(\sigma_S)p^*})| < \lambda$. Set $t_0 = t^* = \tilde{t}_{i_\epsilon}$ and define $\sigma : [t^*, \infty) \rightarrow \mathcal{R}^* \cup \mathcal{I}$ so that $\sigma = w(\sigma_S)$ for $t \geq t^*$. Then σ is an admissible switching signal and

$$|b_C(f_{\sigma s p^*} - f_{\sigma p^*})| < \lambda \tag{34}$$

Clearly

$$A_{p^*} + K_\sigma C + b_C f_{\sigma s p^*} = (A_{p^*} + K_\sigma C + b_C f_{\sigma p^*}) + b_C(f_{\sigma s p^*} - f_{\sigma p^*})$$

But the state transition matrix of $A_{p^*} + K_\sigma C + b_C f_{\sigma p^*}$ satisfies (32) and $b_C(f_{\sigma sp^*} - f_{\sigma p^*})$ satisfies (34). Therefore by the Bellman-Gronwall Lemma, state transition matrix of $A_{p^*} + K_\sigma C + b_C f_{\sigma sp^*}$ must satisfy

$$|\bar{\Phi}(t, \mu)| \leq e^{(a - (\lambda_S - e^* \lambda)(t - \mu))}, \quad t \geq \mu \geq t^*$$

In view of (33), $A_{p^*} + K_\sigma C + b_C f_{\sigma sp^*}$ is exponentially stable. Therefore $H \triangleq K_\sigma$ exponentially stabilizes $A_{p^*} + HC + b_C f_{\sigma sp^*}$ on $[t^*, \infty)$ as required.

Let Σ denote the closed-loop supervisory control system consisting of process model Σ_P described by (1), tracking error ϵ_T defined by (2), integrating subsystem (3), shared controller Σ_C defined by (7), feedback law (9), switching law (10), performance weight generator Σ_W defined by (12) with $\lambda = 0$, performance function Π given by (13), and the dwell-time/cyclic switching logic Σ_{DC} described in §4. We have proved the following theorem.

Theorem 1 *Suppose that Σ_P's transfer function equals nominal process model transfer function ν_p for some $p = p^* \in \mathcal{P}_P$. Then for each constant set-point value r, each constant disturbance input d, and each initial state $\{x_P(0), u(0), x(0), W(0), X(0), \hat{p}(0), \tau(0), \beta(0), \sigma_S(0)\}$, the tracking error $\epsilon_T \to 0$, and x_P, u, and x_C tend to finite limits as $t \to 0$.*

6   Performance Signals

One of the problems with the preceding is that for $r \neq 0$, the weighting matrix W is generated by (12) with $\lambda = 0$ will typically not remain bounded. There are several ways to remedy this problem [2]. We shall briefly describe one.

Under the exact matching hypothesis, (8) holds. Since ω_{PR} is the minimum polynomial of A_C, all of A_C's eigenvalues have real parts no greater than $-\lambda_S$. Thus there is a non-negative constant C_0 such that $e_{p^*}^2(t) \leq C_0 e^{-\lambda_S t}$. Pick $\lambda \in (0, \lambda_S)$. Let Π and π_p be defined as in (11) and (13) respectively, but rather than using (12) to generate W, use the equation

$$\dot{W} = e^{\lambda t} \begin{bmatrix} x_C \\ y \end{bmatrix} \begin{bmatrix} x_C \\ y \end{bmatrix}' \tag{35}$$

instead. Clearly

$$\dot{\pi}_p = e^{\lambda t} e_p^2$$

As defined, π_p has three crucial properties:

1. For each $p \in \mathcal{P}$, π_p is monotone nondecreasing.

2. $\lim_{t \to \infty} \pi_{p^*} \triangleq C^* \leq \pi_{p^*}(0) + \int_0^\infty C_0 e^{-(\lambda_S - \lambda)t} dt < \infty$

3. If $\mathcal{R}^*$ and $\mathcal{S}^*$ are again defined as before, for each $p \in \mathcal{R}^* \cup \mathcal{S}^*$ e_p has a finite $\mathcal{L}^2[0, \infty)$ norm.

These are precisely the properties needed to define C and $\bar{e}$ as in (28) so that $\bar{e}$ has a finite $\mathcal{L}^2[0, \infty)$ norm and that $(C, A_{p^*} + b_C f_{qp^*})$ has the detectability properties stated in Remark 5. In other words, if one were to use (35) to generate W, then the assertions Theorems 1 would still hold.

Now consider replacing W with the "scaled" weighting matrix

$$\bar{W} \triangleq e^{-\lambda t} W \tag{36}$$

Note that $\Pi(\bar{W}, p) = e^{-\lambda t} \Pi(W, p), \quad p \in \mathcal{P}$. In the light of the scale independence property of Σ_S noted previously in Remark 3, it must be that replacing W with $\bar{W}$ has no effect on σ_S so the claims in Theorem 1 still holds. The key point here is that the weighting matrix $\bar{W}$ defined by (36) can also be generated directly by the stable dynamical system

$$\dot{\bar{W}} = -\lambda \bar{W} + \begin{bmatrix} x_C \\ y \end{bmatrix} \begin{bmatrix} x_C \\ y \end{bmatrix}' \tag{37}$$

Moreover, since Theorem 1 states that x_P and x_C, tend to finite limits, it must be that $\bar{W}$ {and therefore its sampled state $\bar{X}$} tend to finite limits as well. The following corollary summarizes the established properties of the closed-loop supervisory control system considered in this paper.

Corollary 1 *Let λ any number in the interval $[0, \lambda_S)$. Suppose that Σ_P's transfer function equals nominal process model transfer function ν_p for some $p = p^* \in \mathcal{P}_P$. Then the conclusions of Theorem 1 are true. Moreover W and X tend to finite limits.*

7 Concluding Remarks

This paper has shown that the concepts of cyclic switching and dwell-time switching are compatible. A supervisor has been described which utilizes both types of switching in a single hybrid subsystem. As a result, the idea of supervisory control outlined in [2] has been extended to encompass those situations in which the set of nominal models upon which supervisor decision-making is based, contains models which are not stabilizable.

Acknowledgment

The author thanks João Hespanha and Wen Chung Chang for useful discussions contributing to this work.

References

[1] F. M. Pait. *Achieving Tunability in Parameter Adaptive Control.* PhD thesis, Yale University, 1993.

[2] A. S. Morse. Supervisory control of families of linear set-point controllers - part 1: Exact matching. submitted for publication.

[3] F. M. Pait and A. S. Morse. A cyclic switching strategy for parameter-adaptive control. *IEEE Transactions on Automatic Control*, 39(6):1172–1183, jun 1994.

[4] J. P. Lasalle. *The Stability of Dynamical Systems.* Society for Industrial and Applied Mathematics, 1976.

THE UNFALSIFIED CONTROL CONCEPT: A DIRECT PATH FROM EXPERIMENT TO CONTROLLER*

Michael G. Safonov[††] and Tung-Ching Tsao[‡]
Electrical Engineering—Systems
University of Southern California
Los Angeles, CA 90089-2563
USA

Dedicated to George Zames on his 60th birthday.

Abstract

The philosophical issues pertaining to the problem of going from experiment to controller design are discussed. The "unfalsified control" concept is introduced as a framework for determining control laws whose ability to meet given performance specifications is at least not invalidated (i.e., not falsified) by the available data. The approach is "model-free" in the sense that no plant model is required — only plant input-output data. When implemented in real time, the result is an *adaptive* robust controller which modifies itself whenever a new piece of data invalidates the present controller. A simple design example based on fixed-order LTI controllers and an L_2-inequality performance criterion is presented.

The task of science is to stake out the limits of the knowable and to center the consciousness within them.

Rudolf Virchow — Berlin, 1849

1 INTRODUCTION

Problem formulations complicated by unessential assumptions tend to obscure key issues. To get a firm grasp on a problem, consider the following:

*Presented at Conference on Feedback Control, Nonlinear Systems, and Complexity, McGill University, Montreal, Canada, May 6–7, 1994.
[†]Phone 1-213-740-4455. FAX 1-213-740-4449. Email safonov@bode.usc.edu.
[‡]Research supported in part by AFOSR grant F49620-92-J-0014.

Principle (Simple Case First) *Consider first only the very simplest prob-lem — but strive for a representation of the simplest problem which generalizes easily.* □.

Unfortunately in theoretical research, simplicity often gives way to a haze of irrelevant detail which tends to accumulate as each researcher builds upon the work of his predecessors. Consequently, some of the most brilliant theoretical works have been those which cut through the layers of haze left by earlier researchers to produce a general theory in which the underlying simplicity shines through.

An example of such brilliance can be found in the works of George Zames and Irwin Sandberg whose celebrated papers on input-output stability [1, 2, 3, 4, 5] cleared away the accumulated haze of detail associated with state-space stability theory to lay the foundation for a much simpler input/output theory. Features of the approach in its final form in [4, 5] can be summarized by the following paradigm for simplification of problems involving dynamical systems:

- Throw away the state.

- Work with relations. Never mind whether x is input and y is output or vice versa.

- Forget about causality and think of a system as simply a collection of input-output pairs $(x, y) \in \mathcal{X} \times \mathcal{Y}$; i.e., a "graph" consisting of the points (x, y) in the "$\mathcal{X} \times \mathcal{Y}$-plane".

- Think in terms of the simple case in which $\mathcal{X}, \mathcal{Y}$ are the real numbers $\mathbb{R}$, keeping in mind the need to generalize to more general function spaces like the extended L_2 spaces.

While Zames [4, 5] spoke of functional relations in conic sectors contained in function spaces, he drew pictures similar to Figure 1, focusing his readers' attention squarely on the simplest case of conic sectors in the real plane $\mathcal{X} \times \mathcal{Y} = \mathbb{R} \times \mathbb{R}$. As Zames found, the general dynamical case in which signals reside in function spaces more general than $\mathbb{R}$ is relatively simple.

Zames [4, 5] saw that a number of interesting stability theorems reduced to such sector conditions — all easily visualized in the $\mathbb{R} \times \mathbb{R}$-plane as in Figure 1. Some years later Safonov [6, 7, 8] observed that the Zames-Sandberg conic sector stability criteria can be simply interpreted in terms of a topological separation of the graphs of input/output relations — a condition easily visualized in the real plane. This, in turn, led to a general "topological separation" theory of stability providing a unified treatment of both Lyapunov and input-output stability theories. But, the key to obtaining these results was to focus first on the simplest case in which systems are viewed as graphs in $\mathbb{R} \times \mathbb{R}$ and then generalize to appropriate function spaces. That is, these successes can be attributed to the application of the Simple Case First Principle.

An important lesson to be learned from [1]–[8] is that feedback intercon-nections of dynamical systems can be better understood if each subsystem is

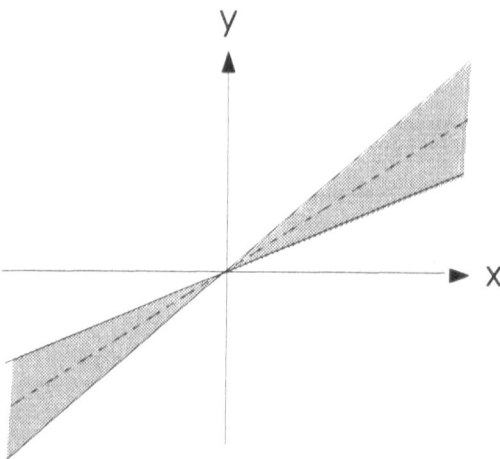

Figure 1: Zames spoke of conic sectors in $L_{2e} \times L_{2e}$ but, drew pictures of simpler conic sectors in $\mathbb{R} \times \mathbb{R}$ such as shown here.

viewed as a set of input/output pairs (x, y) in appropriate function spaces $\mathcal{X}, \mathcal{Y}$, taking note that to gain insight simple function spaces like, say, the reals $\mathbb{R}$ may be substituted for exotic function spaces like the extended L_2 spaces [2, 3, 4, 5]. With the smokescreen of dynamics, causality, and states removed, the effect of nonlinear feedback interconnections on system behavior becomes so very simple that it can be visualized graphically. The "graph" F of points (x, y) can now be plotted in the real plane $\mathbb{R} \times \mathbb{R}$ and visually inspected. In this very simple framework, the solutions to feedback interconnections of two systems become simply the intersection of two graphs in the real plane. The "behavioral system" concept of Willems [9] can be regarded as an application of these principles.

In this paper, the foregoing ideas are applied to help clear the haze of irrelevant detail that has accumulated in research on identification, learning and adaptation. The Zames-Sandberg input-output perspective is adopted and the Simple Case First Principle is applied by initially restricting attention to the very simple case in which there are no dynamics, then generalizing. The "unfalsified control concept" emerges almost trivially from the resultant problem formulation, providing a remarkably clear perspective on the nature of learning in control and paving the way for better robust and adaptive control algorithms.

The present work on unfalsified control was inspired, in part, by recent work on plant model validation (i.e., falsification) [10, 11, 12, 9, 13, 14, 15, 16]. In these works, a hypothesized model with an uncertainty bound is "validated" against input/output data with the hope that robust control methods can be subsequently applied if the uncertainty bound proves to be valid (i.e., unfalsified [9] by the data). In reading these works, we were struck by the observation that

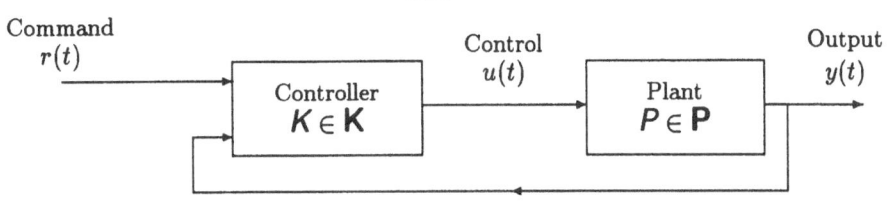

Figure 2: Feedback control system.

it should be just as simple to use input/output data to directly validate control laws with specified performance bounds as to validate plant models with specified uncertainty bounds. And, the result should be a much less conservative, much more direct approach to robust control design. The resultant unfalsified control ideas presented here evolved over the past few years, as reported in [17, 18, 19, 20, 21, 22].

The paper is organized as follows. The unfalsified control concept is developed for the simple nondynamical case in Section 2. The extensions to the general case and the dynamical case are considered in Sections 3 and 4. Practical considerations are discussed in Section 5 and an example of direct, model-free control of a dynamical system using the unfalsified control concept is described in Section 6. In Section 7 generalizations to case where performance inequalities are replaced by cost functions are discussed and issues relating to the creation of efficient recursive learning algorithms are examined. Conclusions are in Section 8.

2 LEARNING: THE SIMPLE CASE

The issues relating to identification and learning in control have for too long been obscured by a haze of superfluous assumptions and detail. Popular misconceptions have evolved concerning what features are essential to the mathematical problem formulation. For example, a popular myth is that plant models are essential to control and that assumptions about the properties of such models are inherently a part of any analysis of learning or adaptive systems. As will be shown below, "model free" control is entirely possible.

Consider the feedback control system in Figure 2. The goal of feedback control theory is to describe a methodology for determining a control law K for a plant P so that the closed-loop system response, say T, satisfies certain given specifications. The need for learning arises when the plant is either unknown or is only partially known and one wishes to extract information from measurements which will be helpful in selecting a suitable control law K. To understand the issues in developing learning control laws, it is helpful to follow the lead of Zames [4, 5] and adopt the input/output perspective. A very simple learning control problem can be formulated as follows:

Problem 1 (Simple Unfalsified Control) *Given*
    a). A data point $(u_0, y_0) \in \mathcal{U} \times \mathcal{Y}$.

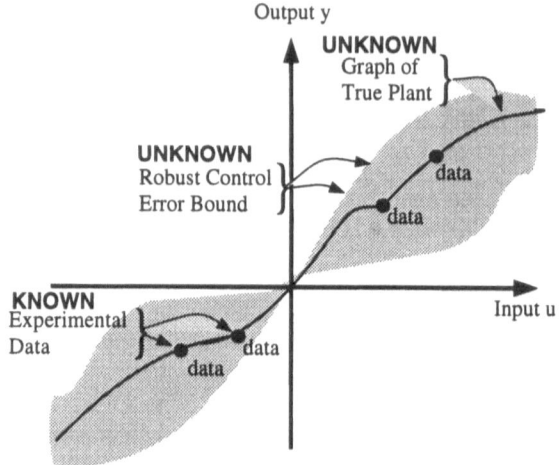

Figure 3: Plant information availability.

b). *A performance specification set*

$$T_{spec} \subset \mathcal{R} \times \mathcal{Y}. \tag{1}$$

c). *A class* **K** *of admissible control laws.*
Determine the subset of $\mathbf{K}_{OK} \subset \mathbf{K}$ *of those control laws* **K** *whose ability to meet the specification* T_{spec} *is not invalidated (i.e., is unfalsified) by the data point* (u_0, y_0). □

The set $\mathbf{K}_{OK}$ is called the *unfalsified controller set* associated with Problem 1.

Like Zames [4, 5], we consider the plant **P** and the controller **K** to be relations in appropriate function spaces:

$$\mathbf{P} \subset \mathcal{U} \times \mathcal{Y} \tag{2}$$

$$\mathbf{K} \subset \mathcal{R} \times \mathcal{Y} \times \mathcal{U} \tag{3}$$

That is, the plant **P** is a locus or "graph" in $\mathcal{U} \times \mathcal{Y}$; see Figure 3. More generally, an uncertain set of possible plants $P \in \mathbf{P}$ might be described by a somewhat larger subset of $\mathcal{U} \times \mathcal{Y}$; such uncertainty sets are used in the formulation of robust control problems [6, 7, 8]. On the other hand, experimental data from a plant corresponds to a but single point $(u_0, y_0) \in \mathbf{P}$.

If all that is known about a plant is the result (u_0, y_0) of an experiment, the question arises "Can Problem 1 be solved?" The answer, it turns out, is affirmative. To see that this is so, apply the Simple Case First Principle and consider the case in which $\mathcal{R}, \mathcal{Y}, \mathcal{U} = \mathbb{R}$ — see Figure 4 . The data is a point (u_0, y_0) in the $\mathcal{U} \times \mathcal{Y}$-plane. The specification T_{spec} is a subset of the $\mathcal{R} \times \mathcal{Y}$-plane. And, each control law **K** is a surface in the space $\mathcal{R} \times \mathcal{Y} \times \mathcal{U}$, say

$$K(r, y, u) = 0. \tag{4}$$

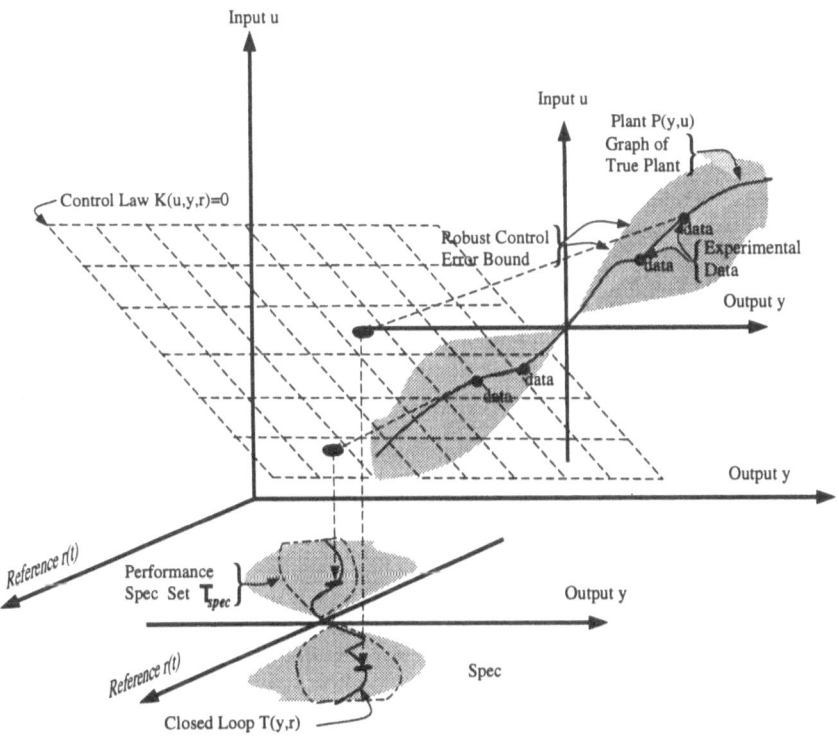

Figure 4: A controller $K(r, y, u) = 0$ is *unfalsified* by the data (u_0, y_0) if and only if $(r_0, y_0) \in T_{spec}$ for every r_0 satisfying $K(r_0, y_0, u_0) = 0$.

We thus have the following theorem.

Theorem 1 (Simple Unfalsified Control) *Consider Problem 1. Let*

$$\tilde{T}_{spec} = \{ \ (r, y, u) \mid (r, y) \in T_{spec} \ \} \subset \mathcal{R} \times \mathcal{Y} \times \mathcal{U}. \tag{5}$$

A control law K *is unfalsified by the measurement data* (u_0, y_0) *if and only if*

$$(K \cap \{ \ (r, y, u) \in \mathcal{R} \times \mathcal{Y} \times \mathcal{U} \mid r \in \mathcal{R}, y = y_0, u = u_0 \ \}) \subset \tilde{T}_{spec} \tag{6}$$

□

Loosely speaking, Theorem 1 says that a controller K's ability to meet the performance specification T_{spec} is unfalsified by the data (u_0, y_0) if and only if the image of (u_0, y_0) in the $\mathcal{R} \times \mathcal{Y}$-plane under constraint $K(r, y, u) = 0$ contains no points (r_0, y_0) which violate the specification T_{spec}. It is really a statement of the obvious — almost a tautology. Yet, simple though it may be, it has far reaching implications:

- Theorem 1 is nonconservative; i.e., it gives "if and only if" conditions on K. It uses all the information in the data — and no more.

- Theorem 1 is "model free"; no plant model is needed to test its conditions.

- Because no plant model is imposed on the problem, there are no gratuitous unverifiable assumptions about the plant. This is in stark contrast to the usual case in the control and identification literature.

- Data (u, y) which invalidates the controller K need not have been generated with the controller K in the feedback loop; it may be open loop data or data generated by some other control law (which need not even be in K).

As we shall see in the following sections, Theorem 1 may be easily generalized to cases where the measurement yields only partial information about the plant input/output pair (u_0, y_0) — including the dynamical case where only the past is revealed by the measurement.

3 LEARNING: PARTIAL OBSERVATIONS

In Problem 1 the assumption is made that data (u_0, y_0) is measured. A more general situation arises when incomplete information about (u_0, y_0) is available. Suppose that, instead of measuring the pair (u_0, y_0) directly, all that can be deduced from the measurement is a set, say M, containing the (u_0, y_0); i.e.,

$$(u_0, y_0) \in M \subset \mathcal{U} \times \mathcal{Y}. \tag{7}$$

Definition (Measurement) *The set* M *is called the* **measurement information**.

□

Additionally, to add flexibility, let us generalize the specification set T_{spec} to include constraints on u as well as on r, y. This leads to the following generalization of Problem 1.

Problem 2 (General Unfalsified Control) *Given*
   a). A measurement information set $M \subset U \times Y$ containing an otherwise unknown plant input/output pair (u_0, y_0).
   b). A performance specification set

$$\tilde{T}_{spec} \subset R \times Y \times U. \tag{8}$$

   c). A class K of admissible control laws.
Determine the subset of $K_{OK} \subset K$ of those control laws K whose ability to meet the specification $\tilde{T}_{spec}$ is not invalidated (i.e., is unfalsified) by the measurement information M. ☐

Definition *The set K_{OK} is called the* **unfalsified set** *and the control laws $K \in K_{OK}$ are said to be* **unfalsified control laws.** ☐

The following theorem is immediate from the definitions.

Theorem 2 (General Unfalsified Control) *Consider Problem 2. A control law K is unfalsified by the knowledge that*

$$(u_0, y_0) \in M \tag{9}$$

if and only if, for some $(u_1, y_1) \in M$,

$$\left(K \cap \{ \ (r, y, u) \in R \times Y \times U \mid r \in R, y = y_1, u = u_1 \ \} \right) \subset \tilde{T}_{spec} \tag{10}$$

☐

Note that in the case where an ensemble of experiments has produced partial knowledge M_α for each element of a family of input/output data pairs (u_α, y_α), the unfalsified controller set K_{OK} would be simply the intersection of the unfalsified controller sets, say $K_{OK\alpha}$, associated with each M_α; i.e.,

$$K_{OK} = \bigcap_\alpha K_{OK\alpha}. \tag{11}$$

4   LEARNING: DYNAMICAL CASE

The dynamical case presents no special conceptual difficulties. It turns out that the dynamical learning control problem is really just a special case of the General Unfalsified Control Problem addressed in the preceding section. The necessary machinery for embedding the dynamical case in the framework of Theorem 2 is provided by the time truncation operator P_τ of input/output stability theory.

Definition (Time Truncation Operator [2, 3, 4, 5]) *For any* $\tau \in \mathbb{R}$, *the* **time truncation operator** P_τ *is a mapping of time signals into time signals defined by*

$$[\mathsf{P}_\tau x](t) \triangleq \begin{cases} x(t), & \text{if } t \leq \tau \\ 0, & \text{if } t > \tau \end{cases} \qquad (12)$$

□

The goal of learning in the dynamical control context is to use *past* plant input/output data to assess the *future* potential of each control law $K \in \mathsf{K}$ to meet the performance specification

$$(r, y, u) \in \tilde{T}_{spec}. \qquad (13)$$

Definition *A control law* $K \in \mathsf{K}$ *is said to be* **falsified** *by past data* $(\mathsf{P}_\tau u_0, \mathsf{P}_\tau y_0)$ *if it can be proved that the control law* K *could not have met the performance specification* $\tilde{T}_{spec}$ *if that control law had been in place when the plant generated the data. Otherwise, the control law* K *is said to be* **unfalsified**.

□

The key observation is that a measurement of *past* input/output data $(\mathsf{P}_\tau u_0, \mathsf{P}_\tau y_0)$ corresponds to *partial* knowledge of a plant input/output pair (u_0, y_0); i.e., it corresponds to the case of Theorem 2 in which the measurement information set is

$$M = \left\{ (u, y) \in \mathcal{U} \times \mathcal{Y} \;\middle|\; \mathsf{P}_\tau \begin{bmatrix} u \\ y \end{bmatrix} = \mathsf{P}_\tau \begin{bmatrix} u_0 \\ y_0 \end{bmatrix} \right\} \qquad (14)$$

Thus, Theorem 2 specializes to the following.

Theorem 3 (Dynamical Unfalsified Control) *Consider Problem 2. A control law* K *is unfalsified by past plant input/output data* $(\mathsf{P}_\tau u_0, \mathsf{P}_\tau y_0)$ *if and only if, for some* $(u_1, y_1) \in M$,

$$(K \cap \{ (r, y, u) \in \mathcal{R} \times \mathcal{Y} \times \mathcal{U} \mid r \in \mathcal{R}, y = y_1, u = u_1 \}) \subset \tilde{T}_{spec} \quad (15)$$

where M *is given by (14).*

□

5 PRACTICAL CONSIDERATIONS

Practical application of Theorem 3 requires that one have characterizations of the sets K and $\tilde{T}_{spec}$ which are both simple and amenable to computations. In the control field, experience has shown that linear equations and quadratic cost functions often lead to tractable problems. A linear parameterization of the set K of admissible control laws is possible by representing each $K \in \mathsf{K}$ as a sum of filters, say $Q_i(s)$, so that the control laws $K \in \mathsf{K}$ are linearly parameterized by an unspecified vector $\theta \in \mathbb{R}^n$; i.e.,

$$K_\theta = \{ (r, y, u) \mid K_\theta(r(s), y(s), u(s)) = 0 \} \qquad (16)$$

where the argument (s) indicates Laplace transformation and

$$K_\theta(r(s), y(s), u(s)) = \sum_{i=1}^{n} \theta_i Q_i(s) \equiv \theta' Q(s) \begin{bmatrix} r(s) \\ y(s) \\ u(s) \end{bmatrix} \quad \forall \theta \in \mathbb{R}^n. \quad (17)$$

Thus,

$$\mathbf{K} = \bigcup_{\theta \in \mathbb{R}^n} K_\theta. \quad (18)$$

The performance specification set $\tilde{T}_{spec}$ might be selected to be a collection of quadratic inequalities, expressed in terms of L_2 inner-products weighted by a given transfer function matrix, say $T_{spec}(s)$; e.g.,

$$\tilde{T}_{spec} = \text{sector}(T_{spec}) \stackrel{\Delta}{=}$$

$$\left\{ (r, y, u) \in L_{2e} \;\middle|\; < z_1, z_2 >_\tau \;\leq 0\; \forall \tau \in [0, \infty), \; \begin{bmatrix} z_1(s) \\ z_2(s) \end{bmatrix} = T_{spec}(s) \begin{bmatrix} r(s) \\ y(s) \\ u(s) \end{bmatrix} \right\}$$

$$(19)$$

where, as above, the argument (s) indicates Laplace transformation and

$$< z_1, z_2 >_\tau \stackrel{\Delta}{=} < \mathsf{P}_\tau z_1, \mathsf{P}_\tau z_2 >_{L_2[0,\infty)} . \quad (20)$$

Note that sets of the type $\text{sector}(T_{spec})$ have been studied by Safonov [6, 7, 8]; such sets are a generalization of the Zames-Sandberg [2, 4, 5] L_{2e} conic sector

$$\text{sector}(a, b) \stackrel{\Delta}{=} \left\{ (u, y) \in L_{2e} \times L_{2e} \;\middle|\; < y - au, y - bu >_\tau \;\leq 0\; \forall \tau \in [0, \infty) \right\}.$$

$$(21)$$

The example in the following section illustrates some of the foregoing ideas.

6 EXAMPLE

In this section we describe a simulation study involving a linear time-invariant robust control design obtained using the unfalsified control concept. Simulation data is generated by the ACC benchmark model [23] with uncertain, time-varying parameters. But, the control design itself is entirely model-free in that the control design proceeds, without any specific information about the benchmark system model, using past input/output data to update the control law until it converges to a robust linear time-invariant control law in the unfalsified set $\mathbf{K}_{OK}$.

The ACC benchmark model consists of two masses connected by a spring as shown in Fig. 5[23]. The nominal value of each mass is 1 and that of the

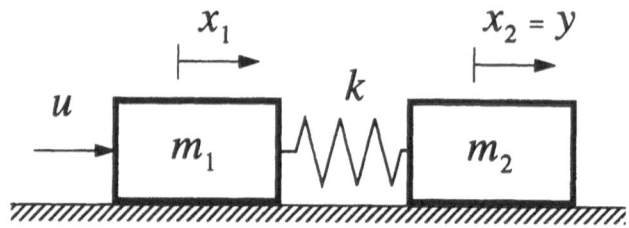

Figure 5: Two-Mass System of ACC Benchmark Problem

spring constant is 1.25. The nominal transfer function from u to y (see Fig. 5) is

$$P(s) = \frac{k}{m_1 s^2 [m_2 s^2 + (1 + \frac{m_2}{m_1})k]} \tag{22}$$

which is of fourth-order. The poles of the nominal plant are $0, 0, \pm 1.5811i$.

To make the problem more challenging, we have added time-varying and nonlinear perturbations for the mass and the spring constant as shown below:

- $m_1(t) = 1 + 0.1 \sin 8t$

- $m_2(t) = 1 + 0.1 \sin 12t$

- $k(t) = 1.25 + 0.1 \sin 10(x_1(t) - x_2(t))$
  where x_1 and x_2 are the positions of the two masses, respectively. We emphasize that the knowledge of the above parameters is for simulation purposes only; it is not used in the construction of the set of unfalsified control laws.

- The performance specification set $\tilde{T}_{spec}$ is taken to be

$$\left\{ (r, y, u) \in L_{2e} \mid \|w_1 * (y - r)\|^2_{L_2[0,\tau]} + \|w_2 * u\|^2_{L_2[0,\tau]} \leq \|r\|^2_{L_2[0,\tau]}, \ \forall \tau \geq 0 \right\}$$

$$\tag{23}$$

which is of the L_{2e} sector type (cf. eqn. (19)). It says that the error signal $r - y$ and the control signal u should be "small" compared to the command signal r; the dynamical "weights" w_1 and w_2 determine what is small. It is essentially the same as the weighted mixed sensitivity performance criterion that is widely used in H^∞ control — cf. [24, 25].

In (23), $w_1(t)$ and $w_2(t)$ are the impulse responses of stable minimum phase weighting transfer functions $W_1(s)$ and $W_2(s)$ respectively, "*" means convolution, r is the input reference signal, y is the plant output signal, and u is the control signal.

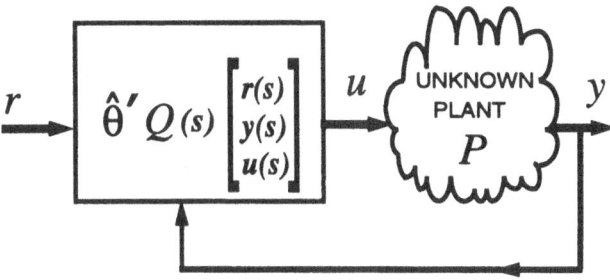

Figure 6: Controller Structure

In keeping with (16)–(17), the set of admissible controllers **K** is chosen to be

$$0 = \theta' Q(s) \begin{bmatrix} r(s) \\ y(s) \\ u(s) \end{bmatrix} \tag{24}$$

where $\theta \in \mathbb{R}^9$ and

$$Q(s) = \begin{bmatrix} 0 & 0 & H_1(s) \\ 0 & 0 & H_2(s) \\ 0 & 0 & H_3(s) \\ 0 & H_1(s) & 0 \\ 0 & H_2(s) & 0 \\ 0 & H_3(s) & 0 \\ 0 & 1 & 0 \\ H_4(s) & 0 & 0 \\ 0 & 0 & -1 \end{bmatrix} \tag{25}$$

and

$$H_1(s) = \frac{1}{s+1}, \quad H_2(s) = \frac{.3}{s+.3}, \quad H_3(s) = \frac{.08}{s+.08}, \quad H_4(s) = (s+1)^3 \tag{26}$$

See Figure 6. Without loss of generality, we *assume*

$$\theta_9 = 1. \tag{27}$$

The simulation was conducted as follows. At each time t a control law in $K_{\hat{\theta}(t)} \in \mathbf{K}$ was connected to the ACC simulation model where $\hat{\theta}(t)$ denotes the value of θ associated with the controller in use at time t. The value of $\hat{\theta}(t)$ was held constant until such time as it is falsified by the past data $(\mathbf{P}_\tau u, \mathbf{P}_\tau y)$, then it is switched to a point near the geometric center of the current unfalsified controller parameter set. The following were used in the simulation:

- $W_1(s) = \frac{s+.5}{s+.1}$

- $W_2(s) = \frac{1}{(s+1)^4}$

- $r(t) = \sin(.0001t) + \sin(.0003t) + \sin(.001t) + \sin(.003t) + \sin(.01t) + \sin(.05t) + 2\sin(.07t) + \sin(.09t) + \sin(.2t) + \sin(2t)$

- Initial conditions are all zero.

- We arbitrarily restricted our search for unfalsified control laws as follows:

$$\hat{\theta}_1^2 + \hat{\theta}_2^2 \leq 250^2, \tag{28}$$

$$|\hat{\theta}_i| \leq 100, \quad \text{for } i = 3, \ldots, 7 \tag{29}$$

$$0.4 \leq \hat{\theta}_i \leq 20, \quad \text{for } i = 8 \tag{30}$$

- We arbitarily initialized

$$\hat{\theta}(0) = [0, 0, 0, 0, 0, 0, 0, 10.2, 1]^T \tag{31}$$

The simulation results are shown in Figure 7. From the plots, one can see that despite the nonlinear and time varying perturbations, the parameter vector converges within finite time to the steady-state value:

$\lim_{t \to \infty} \hat{\theta}(t) =$

$[19.4820, -113.6951, -12.6124, -60.5376, 20.5655, 0.4681, 36.1163, 3.0569, 1]^T$

and the convergent nominal steady-state command-to-error transfer function $T_{r-y,r}$ satisfies $|T_{r-y,r}(j\omega)| < |W_1^{-1}(j\omega)|, \forall\omega$. Although performance specification is achieved after about 740 sec, the ratio

$$\frac{1}{\|r\|_{L_2[0,t]}} \sqrt{\|W_1(y-r)\|_{L_2[0,t]}^2 + \|W_2 u\|_{L_2[0,t]}^2}$$

approaches one rapidly, which means good transient behavior. Further details of the simulation are described in [20].

Discussion of Example

While the foregoing example clearly demonstrates the feasibility of dynamical unfalsified control, it also serves to illustrate some of the computational difficulties which can arise. One such difficulty was that our problem formulation led to the set of unfalsified control laws being described by an increasingly large collection of quadratic inequalities in the parameter vector θ, one inequality for each measurement time, thus preventing an exact recursive implementation of our unfalsified control algorithm. In the computations we sidestepped this difficulty by "gridding" the space $\mathbb{R}^9$ in which θ lies so that, instead of falsifying entire regions of $\mathbb{R}^9$, we could restrict our search for unfalsified control laws $K_\theta(r, y, u) = 0$ to a finite number of grid points. This gridding eliminated the need to remember each of the ever increasing number of quadratic inequalities

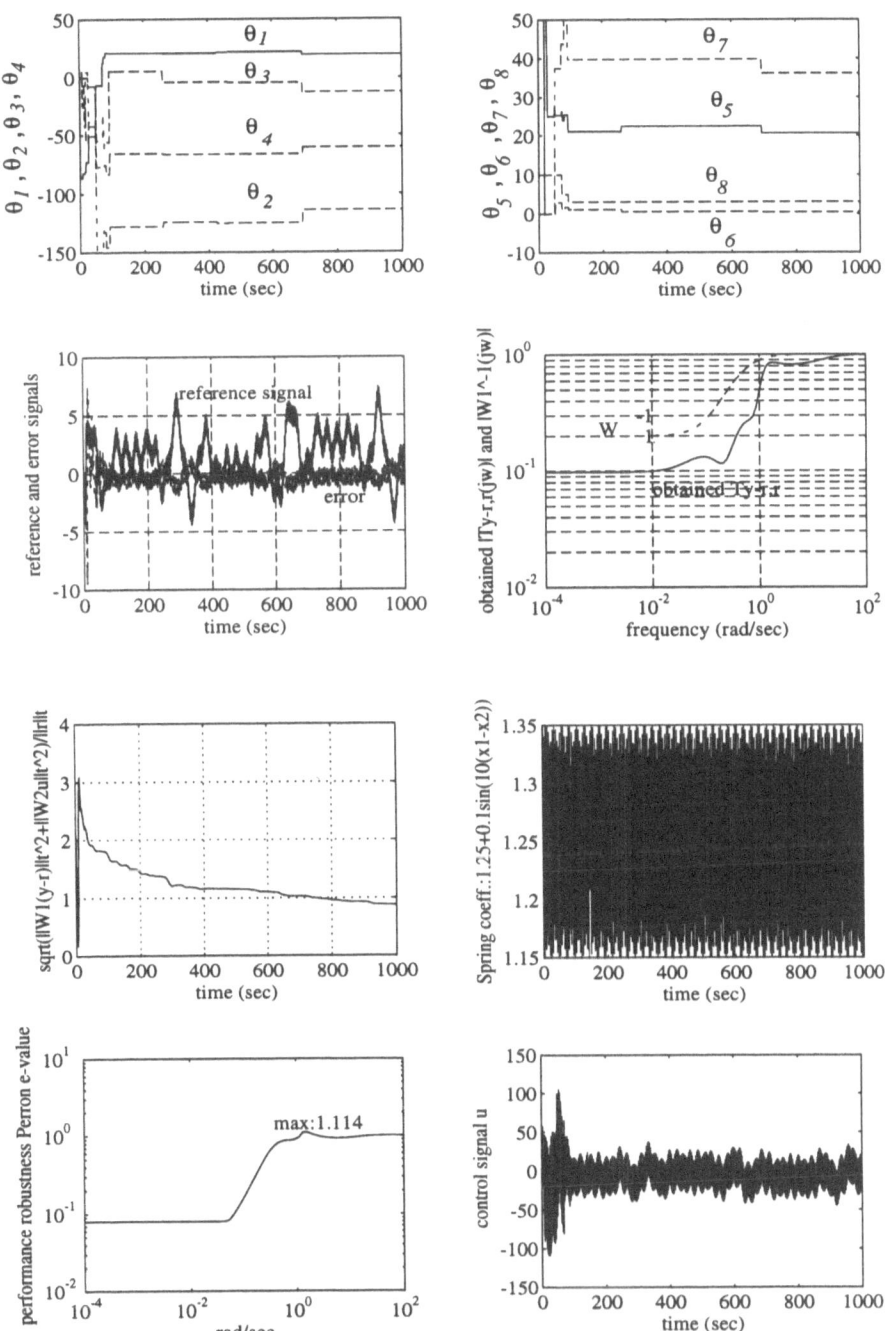

Figure 7: Simulation Results for Benchmark Problem

produced by the measurements, replacing it with a need only to remember the ever decreasing list of as yet unfalsified grid points in $\mathbb{R}^9$.

Our need to "grid" the parameter space $\mathbb{R}^9$ was caused by the fact that the description of the unfalsified set of values of $\theta \in \mathbb{R}^9$, though defined in terms of simple quadratic inequalities, involved an ever increasing number of such inequalities. A related problem was solved without gridding by Schweppe [26]. Schweppe's work concerns an "unknown but bounded noise" Kalman-Bucy filtering problem in which the goal is to track unfalsified portions of the state-space which, it turns out, are defined by a collection of quadratic inequalities — one inequality for each time at which a measurement is taken. In [26] a recursive estimator is produced by using a conservative ellipsoidal bound on the unfalsified set so that only a *single* quadratic inequality need be remembered at any time. Unfortunately, in general the quadratic inequalities in our problem, unlike those of in Schweppe's problem, may be indefinite, leading to non-convex "hyperboloids" not suitable for ellipsoidal bounding. So, perhaps the lesson to be learned from Schweppe's set-theoretic analog of the Kalman-Bucy filter is that simple, optimal recursive algorithms will have to rely on going beyond the unfalsified control concept by replacing the deterministic set-theoretic formulation of the unfalsified control concept with a more flexible probabilistic or, perhaps, fuzzy formulation which is amenable to exact, recursive implementations.

7 BEYOND UNFALSIFIED CONTROL

The unfalsified control concept introduced in this paper is based on a deterministic set-theoretic performance specification (8), viz.

$$(r, y, u) \in \tilde{T}_{spec}.$$

This is essentially a yes/no true/false binary characterization of performance — a point (r, y, u) is either in $\tilde{T}_{spec}$ or it is not. But, there is nothing inherent to prevent extensions to a non-deterministic setting — nothing to prevent extensions to performance specifications defined with respect to more general probabilistic, stochastic or fuzzy settings. Indeed, such extensions may be essential if one is to capture the essence of the issues raised by imprecise measurements and vague performance specifications. And, as explained in Section 6, the additional flexibility provided by a non-binary performance specification might prove useful in creating problem formulations which lead to simple *recursive* computation of the unfalsified controller set. This may be vital if the theory is to become better suited to the task of creating simple real-time adaptive control algorithms.

To achieve the extension of the unfalsified control concept to a non-deterministic setting requires replacing the specification $(r, y, u) \in \tilde{T}_{spec}$ with a suitably chosen real-valued cost function to be optimized, say $\rho_{spec}(r, y, u)$ and, perhaps, replacing the measurement information set M with a real-valued function, say $M(u, y)$. If the measurement function $M(u, y)$ is interpreted as a probability density, then a probabilistic learning result emerges. Alternatively, sticky philosophical issues associated with probabilistic interpretations can be sidestepped by interpreting $M(u, y)$ as a fuzzy set membership function [27]; indeed, the fuzzy/neural "universal controller" concept [28] may be interpreted in this context. Similarly, the value $\rho_{spec}(r, y, u)$ of a cost function $T_{spec} : \mathcal{R} \times \mathcal{Y} \times \mathcal{U} \to \mathbb{R}$ might be selected to be a real number representing the "degree" to which we feel performance requirements are satisfied by each input/output triple (r, y, u). By choosing a cost function $\rho_{spec}(r, y, u)$ which is quadratic in the control law parameters, computational tractability is virtually assured and recursive real-time implementations of adaptive learning control laws should be feasible.

Adaptive Control

It should be noted that the step from unfalsified control to adaptive control is a small one. Simply choosing as the current control law one that is not falsified by the past data produces a control law that is adaptive in the sense that it learns in real time and changes based on what it learns. To accommodate a slowly varying plant, one could either window the data so as to ignore the distant past or use a time-dependent specification set $\tilde{T}_{spec}$ to deemphasize data from the distant past.

8 CONCLUSIONS

Theorem 2 is our main result — with simple case Theorem 1 and the dynamical case Theorem 3 both being special cases of the general result of Theorem 2. Given a performance specification $\tilde{T}_{spec}$ expressible as a constraint to be satisfied by inputs and outputs (r, y, u), these theorems give necessary and sufficient conditions for a control law to be invalidated (i.e., falsified) by experimental data — even if the data was generated open-loop or by another controller. The derivations are simple, even obvious — obtained via a relatively straightforward application of the Simple Case First Principle using the input/output perspective pioneered by Zames and Sandberg [1, 2, 3, 4, 5]. No proofs are provided or needed because the theorems are immediate from the definitions — once the haze of irrelevant detail is cleared away by formulating the problem in the input/output setting.

The results of Theorems 1 through 3 distill the essence of the process of learning and adaptation in a deterministic setting. As time increases, the truncation time τ which demarks the boundary between past and future increases, causing the set M of (14) to evolve with time τ. This, in turn, causes the set $\mathbf{K}_{OK}$ of unfalsified control laws to evolve too. The challenge of adaptive control theory is thus clearly seen to be the task of finding problem formulations which provide a qualitatively correct representation of the evolution of the measurement information set M while allowing efficient, recursive computation of the set $\mathbf{K}_{OK}$ of unfalsified control laws.

A key strength of the results is that they provide a model-free approach to control design — much like the fuzzy/neural "universal controller" problem formulation (cf. [28]) which emerges as a special case of the types of extensions described in Section 7. In stark contrast to nearly all other results in control theory, the results here make no use of plant models or uncertainty bounds — not even for convergence proofs. The only models employed are the candidate controller models $K \in \mathbf{K}$. Experimentally unverifiable prejudices about plant behavior that would be inherent in the use of plant models are not imposed at any step in the derivation or use of the theory. Of course, a plant model or any other a priori information, if available, is welcome and can be accommodated as additional "measurement information" sets M_α à la equation (11).

References

[1] G. Zames. Functional analysis applied to nonlinear feedback systems. *IEEE Trans. on Circuit Theory*, CT-10:392–404, September 1963.

[2] I.W. Sandberg. On the L_2-boundedness of solutions of nonlinear functional equations. *Bell System Technical Journal*, 43(4):1581–1599, July 1964.

[3] I.W. Sandberg. A frequency-domain condition for the stability of feedback systems containing a single time-varying nonlinear element. *Bell System Technical Journal*, 43(4):1601–1608, July 1964.

[4] G. Zames. On the input–output stability of time-varying nonlinear feedback systems — Part I: Conditions derived using concepts of loop gain, conicity, and positivity. *IEEE Trans. on Automatic Control*, AC-15(2):228–238, April 1966.

[5] G. Zames. On the input–output stability of time-varying nonlinear feedback systems — Part II: Conditions involving circles in the frequency plane and sector nonlinearities. *IEEE Trans. on Automatic Control*, AC-15(3):465–467, July 1966.

[6] M. G. Safonov. *Robustness and Stability Aspects of Stochastic Multivariable Feedback System Design*. PhD thesis, Dept. Elec. Eng., MIT, 1977. Supervised by Michael Athans.

[7] M. G. Safonov and M. Athans. On stability theory. In *Proc. 1978 IEEE Conf. on Decision and Control*, pages 301–314, San Diego, CA, January 10–12, 1979. IEEE Press, New York.

[8] M. G. Safonov. *Stability and Robustness of Multivariable Feedback Systems*. MIT Press, Cambridge, MA, 1980.

[9] J. C. Willems. Paradigms and puzzles in the theory of dynamical systems. *IEEE Trans. on Automatic Control*, AC-36:259–294, 1991.

[10] R. L. Kosut. Adaptive uncertainty modeling: On-line robust control design. In *Proc. American Control Conf.*, pages 245–250, New York, June 1987. IEEE. Mineapolis, MN.

[11] R. Smith and J. Doyle. Model invalidation — a connection between robust control and identification. In *Proc. American Control Conf.*, pages 1435–1440, Pittsburgh, PA, June 21-23, 1989. IEEE Press, New York.

[12] J. M. Krause. Stability margins with real paramenter uncertainty: Test data implications. In *Proc. American Control Conf.*, pages 1441–1445, Pittsburgh, PA, June 21-23, 1989. IEEE Press, New York.

[13] R. L. Kosut, M. K. Lau, and S. P. Boyd. Set-membership identification of systems with parametric and nonparametric uncertainty. *IEEE Trans. on Automatic Control*, AC-37(7):929–941, July 1992.

[14] K. Poolla, P. Khargonekar, J. Krause A. Tikku, and K. Nagpal. A time-domain approach to model validation. In *Proc. American Control Conf.*, pages 313–317, New York, June 1992. IEEE. Chicago, IL.

[15] J. J. Krause, G. Stein, and P. P. Khargonekar. Sufficient conditions for robust performance of adaptive controllers with general uncertainty structure. *Automatica*, 28(2):277–288, March 1992.

[16] R. Smith. An informal review of model validation. In R. S. Smith and M. Dahleh, editors, *The Modeling of Uncertainty in Control Systems: Proc. of the 1992 Santa Barbara Workshop*, pages 51–59. Springer-Verlag, New York, 1994.

[17] M. G. Safonov. Thoughts on identification for control. In R. S. Smith and M. Dahleh, editors, *The Modeling of Uncertainty in Control Systems: Proc. of the 1992 Santa Barbara Workshop*, pages 15–17. Springer-Verlag, New York, 1994.

[18] T. C. Tsao and M. G. Safonov. A robust ellipsoidal-bound approach to direct adaptive control. In R. S. Smith and M. Dahleh, editors, *The Modeling of Uncertainty in Control Systems: Proc. of the 1992 Santa Barbara Workshop*, pages 181–196. Springer-Verlag, New York, 1994.

[19] T. C. Tsao and M. G. Safonov. Set theoretic adaptor control systems. In *Proc. American Control Conf.*, pages 3043–3047, San Francisco, CA, June 2-4, 1993. IEEE Press, New York.

[20] T.-C. Tsao and M. G. Safonov. Convex set theoretic adaptor control systems. In *Proc. IEEE Conf. on Decision and Control*, pages 582–584, San Antonio, TX, December 15-17, 1993. IEEE Press, New York.

[21] T. C. Tsao and M. Safonov. Data, consistency and feedback: A new approach to robust direct adaptive control. In *Proc. American Control Conf.*, Baltimore, MD, June 29–July 1, 1994. IEEE Press, New York.

[22] T. C. Tsao. *Set Theoretic Adaptor Systems*. PhD thesis, University of Southern California, May 1994. Supervised by M. G. Safonov.

[23] B. Wie and D. S. Bernstein. Benchmark problems for robust control design. In *Proc. American Control Conf.*, pages 2047–2048, Chicago, IL, June 24–26, 1992. IEEE Press, New York.

[24] M. G. Safonov and R. Y. Chiang. CACSD using the state-space L^∞ theory–A design example. *IEEE Trans. on Automatic Control*, AC-33:477–479, 1988.

[25] R. Y. Chiang and M. G. Safonov. *Robust-Control Toolbox*. Mathworks, South Natick, MA, 1988.

[26] F. Schweppe. Recursive state estimation: Unknown but bounded errors and system inputs. *IEEE Trans. on Automatic Control*, 13:22–29, February 1968.

[27] B. Kosko. *Neural Networks and Fuzzy Systems*. Prentice-Hall, Englewood-Cliffs, NJ, 1992.

[28] B. Kosko and J. Dickerson. Function approximation with additive fuzzy systems. In H. Nguyen, M. Sugeno, R. Tong, and R. Yager, editors, *Theoretical Aspects of Fuzzy Control*, chapter 12. North Holland, New York, 1994. To appear.

[29] R. Virchow. *Der Mensch (On Man)*. Berlin, 1849. In *Disease, Life and Man Selected Essays of Rudolf Virchow* (trans. L. J. Rather), Stanford University Press, Stanford, CA, pp. 67–70, 1958.

State-Space and I/O Stability
for Nonlinear Systems

*Eduardo D. Sontag**
Department of Mathematics
Rutgers University, New Brunswick, NJ 08903[†]

On the occasion of George James' 60th birthday

Abstract

This paper surveys several alternative but equivalent definitions of "input to state stability" (ISS), a property which provides a natural framework in which to formulate notions of stability with respect to input perturbations. Relations to classical Lyapunov as well as operator theoretic approaches, connections to dissipative systems, and applications to stabilization of several cascade structures are mentioned. The particular case of linear systems subject to control saturation is singled-out for stronger results.

1 Introduction

George Zames has long been a proponent of *input/output* approaches to the analysis of control systems. Among his many deep contributions, he pioneered the use of operator techniques for determining the stability of feedback configurations. These techniques focus on the estimation of bounds on solutions, expressed in terms of bounds on forcing functions, and allow powerful tools to be applied, such as small-gain theorems. Conceptually, the main competing variants of the notion of stability are based on *state-space* ideas, which concentrate on the asymptotic stability of equilibria (or of more general attractors) in the absence of —or subject to only small— external "disturbance" inputs. It is the purpose of this paper to briefly survey various links between these two alternative paradigms of stability, through the systematic use of the notion of "input to state stability" (ISS).

Mathematically, the state-space theory is grounded on classical dynamical systems; Lyapunov functions and geometric methods play a central role. In contrast, input/output stability has classically had a more operator-theoretic flavor and developed independently. The latter notion is arguably the most

---

[*]Supported in part by US Air Force Grant AFOSR-91-0346
[†]E-mail: sontag@hilbert.rutgers.edu

useful in many control applications, since it permits the natural quantification of performance bounds and it is well-behaved under operations such as cascading of systems. In addition, i/o stability provides a framework in which to study the classification and parameterization of dynamic controllers. It is also the most natural notion to consider in the context of building observer-based controllers.

Based on linear systems intuition, where all notions coincide, it is perhaps surprising that state-space and i/o stability are not automatically related. Even for feedback linearizable systems, this relation is more subtle than might appear: if one first linearizes a system and then stabilizes the equivalent linearization, in terms of the original system one does not in general obtain a closed-loop system that is input/output stable in any reasonable sense. However, it is always possible to make a choice of a —usually different— feedback law that achieves such stability, in the linearizable case as well as for all other smoothly stabilizable systems. This paper presents a brief and informal survey of such results, and discusses precise definitions of input to state stability, nonlinear gains, and stability margins which lend themselves to useful theoretical analysis.

One important source of inspiration for our approach is the pioneering work of Willems ([27]), who introduced an abstract concept of energy dissipation in order to unify i/o and state space stability, and in particular with the purpose of understanding conceptually the meaning of Kalman-Yakubovich positive-realness (passivity), and frequency-domain stability theorems such as those due to Zames, in a more general nonlinear context. His work was continued by many authors, most notably Hill and Moylan (see e.g. [5, 6]).

However, although extremely close in spirit, technically our work does not make much contact with the existing dissipation literature. Mathematically it is grounded instead in more classical converse Lyapunov arguments in the style of Massera, Kurzweil, and Zubov,

The results reported here regarding equivalences between different notions of input to state stability originate with the paper [14], but the definitive conclusions were obtained in recent work jointly carried out with with Yuan Wang in [20], which in turn built upon research with Wang and Yuandan Lin in [9] and [18]; the input-saturated results are based on joint papers with Wensheng Liu and Yacine Chitour ([10]) as well as Sussmann and Yang ([21]). Some recent and very relevant results by Jiang, Praly, and Teel ([7]) are also mentioned.

In the interest of exposition, the style of presentation in this survey is informal. The reader should consult the references for more details and, in some cases, for precise statements.

Acknowledgements. I wish to especially thank Yuan Wang and Zhong Ping Jiang for a careful reading of this manuscript and many suggestions for its improvement, as well as Andy Teel for suggestions concerning Theorem 2.

Preliminaries

This paper deals with continuous time systems of the standard form

$$\dot{x} = f(x, u), \tag{1}$$

where $x(t) \in \mathbb{R}^n$ and $u(t) \in \mathbb{R}^m$. (Since global asymptotical stability will be of interest, there is no reason to consider systems evolving in more general manifolds than Euclidean space. For undefined terminology from control theory see [17].) It is assumed that $f : \mathbb{R}^n \times \mathbb{R}^m \to \mathbb{R}^n$ is locally Lipschitz and satisfies $f(0,0) = 0$. *Controls* or *inputs* are measurable locally essentially bounded functions $u : \mathbb{R}_{\geq 0} \to \mathbb{R}^m$. The set of all such functions is denoted by $L^m_{\infty,e}$, and one denotes $\|u\|_\infty = (\text{ess}) \sup\{|u(t)|, t \geq 0\} \leq \infty$; when this is finite, one obtains the usual space L^m_∞, endowed with the (essential) supremum norm. (Everywhere, $|\cdot|$ denotes Euclidean norm in the appropriate space of vectors, and $\|\cdot\|$ induced norm for matrices, while $\|\cdot\|_\infty$ is used for sup norm.) For each $x_0 \in \mathbb{R}^n$ and each $u \in L^m_\infty$, $x(t, x_0, u)$ denotes the trajectory of the system (1) with initial state $x(0) = x_0$ and input u. This is a priori defined only on some maximal interval $[0, T_{x_0,u})$, with $T_{x_0,u} \leq +\infty$. If the initial state and input are clear from the context, one writes just $x(\cdot)$ for the ensuing trajectory. The system is *(forward-) complete* if $T_{x_0,u} = +\infty$ for all x_0 and u.

The questions to be studied relate to the "stability," understood in an appropriate sense, of the input to state mapping $(x_0, u(\cdot)) \mapsto x(\cdot)$ (or, in the last section, when an output is also given, of the input to output mapping $\mapsto y(\cdot)$). To appreciate the type of problem that one may encounter, consider the following issue. Suppose that in the absence of inputs the trivial solution $x \equiv 0$ of the differential equation

$$\dot{x} = f_0(x) = f(x, 0) \tag{2}$$

is globally asymptotically stable (for simplicity, in such a situation, we'll simply say that (1), or equivalently the zero-input restricted system (2), is GAS). Then one would like to know if, for solutions of (1) associated to *nonzero* controls, it holds that

$$u(\cdot) \underset{t \to \infty}{\longrightarrow} 0 \quad \Rightarrow \quad x(\cdot) \underset{t \to \infty}{\longrightarrow} 0$$

(the "converging input converging state" property) or that

$$u(\cdot) \text{ bounded} \quad \Rightarrow \quad x(\cdot) \text{ bounded}$$

(the "bounded input bounded state" property). Of course, for linear systems $\dot{x} = Ax + Bu$ these implications are always true. Not only that, but one has explicit estimates

$$|x(t)| \leq \beta(t)|x_0| + \gamma \|u_t\|_\infty$$

where

$$\beta(t) = \|e^{tA}\| \to 0 \quad \text{and} \quad \gamma = \|B\| \int_0^\infty \|e^{sA}\| \, ds$$

for any Hurwitz matrix A, where u_t is the restriction of u to $[0, t]$, though of as a function in L^m_∞ which is zero for $s > t$. From these estimates both properties can be easily deduced.

These implications fail in general for nonlinear systems, however, as has been often pointed out in the literature (see for instance [26]). As a trivial illustration, take the system

$$\dot{x} = -x + (x^2 + 1)u \tag{3}$$

and the control $u(t) = (2t + 2)^{-1/2}$. With $x_0 = \sqrt{2}$ there results the unbounded trajectory $x(t) = (2t + 2)^{1/2}$. This is in spite of the fact that the system is GAS. Thus, the converging input converging state property does not hold. Ever worse, the bounded input $u \equiv 1$ results in a finite-time explosion. This example is not artificial, as it arises from the simplest case of feedback linearization design. Indeed, given the system

$$\dot{x} = x + (x^2 + 1)u,$$

the obvious stabilizing control law (obtained by first cancelling the nonlinearity and then assigning dynamics $\dot{x} = -x$) is

$$u := \frac{-2x}{x^2 + 1} + v$$

where v is the new external input. In terms of this new control (which might be required in order to meet additional design objectives, or may represent the effect of an input disturbance), the closed-loop system is as in (3), and thus is ill-behaved. Observe, however, that if instead of the obvious law just given one would use:

$$u := \frac{-2x}{x^2 + 1} - x + v,$$

then the closed-loop system becomes instead

$$\dot{x} = -2x - x^3 + (x^2 + 1)u.$$

This is still stable when $u \equiv 0$, but in addition it tolerates perturbations far better, since the term $-x^3$ dominates $u(x^2 + 1)$ for bounded u and large x. The behavior with respect to such u is characterized qualitatively by the notion of "ISS" system, to be discussed below. More generally, it is possible to show that up to *feedback equivalence*, GAS always implies (and is hence equivalent) to the ISS property to be defined. This is one of many motivations for the study of the ISS notion, and will be reviewed after the precise definitions have been given.

Besides being mathematically natural and providing the appropriate framework in which to state the above-mentioned feedback equivalence result, there are several other reasons for studying the ISS property, some of which are briefly mentioned in this paper. See for instance the applications to observer design and new small gain theorems in [24], [25], [7], and [12]; the construction of coprime stable factorizations was the main motivation in the original paper [14] which introduced the ISS concept, and the stabilization of cascade systems using these ideas was briefly discussed in [15].

2 The Property ISS

Next, four natural definitions of input to state stability are proposed and separately justified. Later, they turn out to be equivalent. The objective is to express the fact that states remain bounded for bounded controls, with an ultimate bound which is a function of the input magnitude, and in particular that states decay when inputs do.

2.1 From GAS to ISS — A First Pass

The simplest way to introduce the notion of ISS system is as a generalization of GAS, global asymptotic stability of the trivial solution $x \equiv 0$ for (2). The GAS property amounts to the requirements that the system be complete and the following two properties hold:

1. *(Stability)*: the map $x_0 \mapsto x(\cdot)$ is continuous at 0, when seen as a map from $\mathbb{R}^n$ into $C^0([0, +\infty), \mathbb{R}^n)$, and

2. *(Attractivity)*: $\lim_{t \to +\infty} |x(t, x_0)| = 0$.

Note that, under the assumption that 1. holds, the convergence in the second part is automatically uniform with respect to initial states x_0 in any given compact. By analogy, one defines the system (1) to be *input to state stable* (ISS) if the system is complete and the following properties, which now involve nonzero inputs, hold:

1. the map $(x_0, u) \mapsto x(\cdot)$ is continuous at $(0, 0)$ (seen as a map from $\mathbb{R}^n \times L^m_\infty$ to $C^0([0, +\infty), \mathbb{R}^n)$, and

2. there exists a "nonlinear asymptotic gain" $\gamma \in \mathcal{K}$ so that

$$\overline{\lim_{t \to +\infty}} |x(t, x_0, u)| \leq \gamma(\|u\|_\infty) \tag{4}$$

uniformly on x_0 in any compact and all u.

(The class $\mathcal{K}$ consists of all functions $\gamma : \mathbb{R}_{\geq 0} \to \mathbb{R}_{\geq 0}$ which are continuous, strictly increasing, and satisfy $\gamma(0) = 0$. The uniformity requirement means, explicitly: for each r and ε positive, there is a $T > 0$ so that $|x(t, x_0, u)| \leq \varepsilon + \gamma(\|u\|_\infty)$ for all u and all $|x_0| \leq r$ and $t \geq T$.)

In the language of robust control, the inequality (4) is an "ultimate boundedness" condition. Note that this is a direct generalization of attractivity to the case $u \not\equiv 0$; the "lim sup" is now required since the limit need not exist.

2.2 From Lyapunov to Dissipation — A Second Pass

A potentially different concept of input to state stability arises when generalizing classical Lyapunov conditions to certain classes of dissipation inequalities.

A *storage* or *energy* function is a $V : \mathbb{R}^n \to \mathbb{R}_{\geq 0}$ which is continuously differentiable, proper (that is, radially unbounded) and positive definite (that is, $V(0)=0$ and $V(x)>0$ for $x \neq 0$). A (classical) *Lyapunov function* for the zero-input system (2) is a storage function for which there exists some function α of class $\mathcal{K}_\infty$ —that is, of class $\mathcal{K}$ and so that also $\alpha(s) \to +\infty$ as $s \to +\infty$— so that

$$\nabla V(x) \cdot f_0(x) \leq -\alpha(|x|)$$

holds for all $x \in \mathbb{R}^n$. This means that $dV(x(t))/dt \leq -\alpha(|x(t)|)$ along all trajectories.

By analogy, when nonzero inputs must be taken into account, it is sensible to define an ISS-*Lyapunov* function as a storage function for which there exist two class $\mathcal{K}_\infty$ functions α and θ such that

$$\nabla V(x) \cdot f(x, u) \leq \theta(|u|) - \alpha(|x|) \tag{5}$$

for all $x \in \mathbb{R}^n$ and all $u \in \mathbb{R}^m$. Thus, along trajectories one now obtains the inequality $dV(x(t))/dt \leq \theta(|u(t)|) - \alpha(|x(t)|)$.

A *smooth* ISS-Lyapunov function is a V which satisfies these properties and is in addition infinitely differentiable. Smoothness is an extremely useful property in this context, as one may then use iterated derivatives of V along trajetories for various design as well as analysis questions, in particular in so-called "backstepping" design techniques.

In the terminology of [27, 6], (5) is a *dissipation inequality* with storage function V and *supply* function $w(u, x) = \theta(|u|) - \alpha(|x|)$. (In the context of dissipative systems one often postulates the equivalent integral form $V(x(t, x_0, u)) - V(x_0) \leq \int_0^t w(u(s), x(s))ds$, which must hold along all trajectories, and no differentiability is required of V. Moreover, outputs $y=h(x)$ are used instead of states in the estimates, so the present setup corresponds to the case $h(x)=x$.) The estimate (5) is a generalization of the one used by Brockett in [1] when defining "finite gain at the origin;" in that paper, the function θ is restricted to be quadratic, and the concepts are only defined locally, but the ideas are very similar.

2.3 Gain Margins — A Third Pass

Yet another possible approach to formalizing input to state stability is motivated both by the classical concept of total stability and as a generalization of the usual gain margin for linear systems.

In [20], a *(nonlinear) stability margin* for system (1) is defined as any function $\rho \in \mathcal{K}_\infty$ with the following property: for each admissible —possibly nonlinear and/or time-varying— feedback law k bounded by ρ, that is, so that

$$|k(t, x)| \leq \rho(|x|)$$

for all (t, x), the closed-loop system

$$\dot{x} = f(x, k(t, x)) \tag{6}$$

is GAS, uniformly on k. (More precisely, an admissible feedback law is a measurable function $k : \mathbb{R}_{\geq 0} \times \mathbb{R}^n \to \mathbb{R}^m$ for which (6) is well-posed; that is, for each initial state $x(0)$ there is an absolutely continuous solution, defined at least for small times, and any two such solutions coincide on their interval of existence. Uniformity in k means that all limits in the definition of GAS are independent of the particular k, as long as the inequality $|k(t, x)| \leq \rho(|x|)$ holds.) A system is said to be *robustly stable* if there exists some such ρ.

Observe that for arbitrary nonlinear GAS systems, in general only small perturbations can be tolerated (cf. total stability results). The requirement that $\rho \in \mathcal{K}_\infty$ is thus highly nontrivial: it means that for large states relatively large perturbations should not affect stability.

2.4   Estimates — Fourth Pass

A final proposed notion of input to state stability can be introduced by means of an estimate similar to that which holds in the linear case:

$$|x(t, x_0, u)| \leq \|e^{tA}\| |x_0| + \left(\|B\| \int_0^\infty \|e^{sA}\| \, ds\right) \|u_t\|_\infty \ .$$

It is first necessary to review an equivalent —if somewhat less widely known— definition of GAS. This is a characterization in terms of comparison functions. Recall that a function of class $\mathcal{KL}$ is a

$$\beta : \mathbb{R}_{\geq 0} \times \mathbb{R}_{\geq 0} \to \mathbb{R}_{\geq 0}$$

so that $\beta(\cdot, t)$ is of class $\mathcal{K}$ for each fixed $t \geq 0$ and $\beta(s, t)$ decreases to 0 as $t \to \infty$ for each $s \geq 0$ (example of relevance to the linear case: $ce^{-at}s$, with $a > 0$ and a constant c). It is not difficult to prove (this is essentially in [4]; see also [14]) that the system (2) is GAS if and only if there exists a $\beta \in \mathcal{KL}$ so that

$$|x(t, x_0)| \leq \beta(|x_0|, t) \tag{7}$$

for all t, x_0. (Note that sufficiency is trivial, since forward completeness follows from the fact that trajectories stay bounded, the estimate $|x(t, x_0)| \leq \beta(|x_0|, 0)$ provides stability, and $|x(t, x_0)| \leq \beta(|x_0|, t) \to 0$ shows attractivity. The converse is established by formulating and solving a differential inequality for $|x(t, x_0)|$.)

In this context, it is then natural to consider the following "$\beta + \gamma$" property: There exist $\beta \in \mathcal{KL}$ and $\gamma \in \mathcal{K}$ so that, for all initial states and controls, and all $t \geq 0$:

$$|x(t, x_0, u)| \leq \beta(|x_0|, t) + \gamma(\|u_t\|_\infty) \ . \tag{8}$$

(One could use a "max" instead of the sum of the two estimates, but the same concept would result. Also, it makes no difference to write $\|u\|_\infty$ instead of the norm of the restriction $\|u_t\|_\infty$.) This is a direct generalization of both the linear estimate and the characterization of GAS in terms of comparison functions.

2.5 All Are Equivalent

The following result was recently proved by Yuan Wang and the author:

Theorem 1 ([20]) *For any system (1), the following properties:*

1. ISS *(nonlinear asymptotic gain),*

2. *there is an* ISS-*Lyapunov function (dissipativity),*

3. *there is a smooth* ISS-*Lyapunov function,*

4. *there is a nonlinear stability margin (robust stability), and*

5. *there is some $\beta+\gamma$ estimate,*

are all equivalent. ∎

The proof is heavily based on a result obtained by Wang, Lin, and the author in [9], which states essentially that a parametric family of systems $\dot{x} = f(x, d)$, with arbitrary time-varying "disturbances" $d(t)$ taking values on a compact set D, is uniformly globally asymptotically stable if and only if there exists a smooth storage function V and an $\alpha \in \mathcal{K}_\infty$ so that

$$\nabla V(x) \cdot f(x, d) \leq -\alpha(|x|)$$

for all $x \in \mathbb{R}^n$ and values $d \in D$. Note that the construction of a smooth V is not entirely trivial (this subsumes as particular cases several standard converse Lyapunov theorems).

2.6 Checking the ISS Property

Of course, verifying the ISS property is in general very hard —after all, in the particular case of systems with no inputs, this amounts to checking global asymptotic stability. Nonetheless, the dissipation inequality (5) provides in principle a good tool, playing the same role as Lyapunov's direct method for asymptotic stability. Actually, even more useful is the following variant, which is the original definition of "ISS-Lyapunov function" in [14]. Consider a storage function with the property that there exist two class $\mathcal{K}$ functions α and χ so that the implication

$$|x| \geq \chi(|u|) \quad \Rightarrow \quad \nabla V(x) \cdot f(x, u) \leq -\alpha(|x|) \tag{9}$$

holds for each state $x \in \mathbb{R}^n$ and control value $u \in \mathbb{R}^m$. It is shown in [20] that the existence of such a V provides yet another necessary and sufficient characterization of the ISS property. (Other variants are also equivalent, for instance, asking that α be of class $\mathcal{K}_\infty$.)

As an illustration, consider the following system, which will appear again later in the context of an example regarding the stabilization of the angular momentum of a rigid body. The state space is $\mathbb{R}$, the control value space is $\mathbb{R}^2$, and dynamics are given by:

$$\dot{x} = -x^3 + x^2 u_1 - x u_2 + u_1 u_2 . \tag{10}$$

This system is GAS when $u \equiv 0$, and for large states the term $-x^3$ dominates, so it can be expected to be ISS. Indeed, using the storage function $V(x) = x^2/2$ there results

$$\nabla V(x) \cdot f(x, u) \leq -\left(\frac{2}{9}\right) x^4$$

provided that $3|u_1| \leq |x|$ and $3|u_2| \leq x^2$. A sufficient condition for this to hold is that $|u| \leq \nu(|x|)$, where $\nu(r) := \min\{r/3, r^2/3\}$. Thus V is an ISS-Lyapunov function as above, with $\alpha(r) = (2/9)r^4$ and $\chi = \nu^{-1}$.

Another example is as follows. Let SAT $: \mathbb{R} \to \mathbb{R}$ be the standard saturation function: SAT$[r] = r$ if $|r| \leq 1$, and SAT$[r] = \text{sign}(r)$ otherwise. Consider the following one-dimensional one-input system:

$$\dot{x} = -\text{SAT}[x + u] . \tag{11}$$

This is an ISS system, as will be proved next by showing that

$$V(x) := \frac{|x|^3}{3} + \frac{x^2}{2} \tag{12}$$

is an ISS-Lyapunov function. Observe that V is once differentiable, as required. This is a very particular case of a more general result dealing with linear systems with saturated controls, treated in [10]; more will be said later about the general case (which employs a straightforward generalization of this V).

To prove that V satisfies a dissipation inequality, first note that, since $|r - \text{SAT}[r]| \leq r \text{SAT}[r]$ for all r,

$$|x - \text{SAT}[x+u]| \leq |x+u-\text{SAT}[x+u]| + |u| \leq (x+u)\text{SAT}[x+u] + |u| \tag{13}$$

for all values $x \in \mathbb{R}$ and $u \in \mathbb{R}$. It follows that

$$\begin{aligned}
-x\,\text{SAT}[x+u] &= x(-x) + x(x - \text{SAT}[x+u]) \\
&\leq -x^2 + |x|(x+u)\text{SAT}[x+u] + |x||u|
\end{aligned}$$

for all x, u. On the other hand, using that SAT$[r] \leq 1$ for all r,

$$\begin{aligned}
-|x|x\,\text{SAT}[x+u] &= |x|\left[-(x+u)\text{SAT}[x+u] + u\text{SAT}[x+u]\right] \\
&\leq -|x|(x+u)\text{SAT}[x+u] + |x||u| .
\end{aligned}$$

Adding the two inequalities, it holds that

$$-(1+|x|)x\,\text{SAT}[x+u] \leq -x^2 + 2|x||u| \tag{14}$$

so that indeed

$$\nabla V(x) \cdot f(x, u) \leq -\frac{x^2}{2} + 2u^2$$

as desired (note that $\nabla V(x) = x(1 + |x|)$).

2.7 Relations Among Estimates, Zero-State Responses, and Linear Gains

There are many relationships among the various estimates which appear in the alternative characterizations of ISS. Two of them are as follows.

Assume that V is a storage function satisfying the estimates in Equation (5):

$$\nabla V(x) \cdot f(x, u) \leq \alpha_4(|u|) - \alpha_3(|x|) \tag{15}$$

for some $\mathcal{K}_\infty$ functions α_3 and α_4. Since V is proper, continuous, and positive definite, there are as well two other class $\mathcal{K}_\infty$ functions α_1 and α_2 such that

$$\alpha_1(|x|) \leq V(x) \leq \alpha_2(|x|) \tag{16}$$

for all $x \in \mathbb{R}^n$. It then holds that one may pick an asymptotic gain γ in Equation (4) of the form:

$$\gamma = \alpha_1^{-1} \circ \alpha_2 \circ \alpha_3^{-1} \circ \alpha_4 . \tag{17}$$

Moreover, if instead of (15) there holds a slightly stronger estimate of the form

$$\nabla V(x) \cdot f(x, u) \leq \alpha_4(|u|) - \alpha_3(|x|) - \alpha(|x|)$$

where α is any class $\mathcal{K}$ function, then the γ function in the "$\beta + \gamma$" property (8) can also be picked as in Equation (17). These conclusions are implicit in the proofs given in [14] and [20].

For trajectories starting at the particular initial state $x_0 = 0$, for any input function u, and assuming only that V satisfies (15)-(16), it holds that $|x(t, 0, u)| \leq \gamma(\|u\|_\infty)$ for all $t \geq 0$, not merely asymptotically, for the same γ as in (17), that is,

$$\|x(\cdot, 0, u)\|_\infty \leq \gamma(\|u\|_\infty) .$$

Thus the zero-state response has a "nonlinear gain" bounded by this γ.

A particular case of interest is when both of (α_1, α_2) and (α_3, α_4) are *convex estimate pairs* in the following sense: a pair of class $\mathcal{K}$ functions (α, β) is a convex estimate pair if α and β are convex functions and there is some real number $k \geq 1$ such that $\beta(r) \leq k\alpha(r)$ for all $r \geq 0$. Note that for any convex function α in $\mathcal{K}$ and any $k \geq 1$ it holds that $\alpha^{-1}(k\alpha(r)) \leq kr$ for all nonnegative r, from which it follows that $\alpha^{-1}(\beta(r)) \leq kr$ if k is as in this definition. One concludes that if each of (α_1, α_2) and (α_3, α_4) is a convex estimate pair, then the gain γ can be taken to be bounded by a linear function. In other words, the input to state operator, starting from $x_0 = 0$, is bounded as an operator with respect to sup norms:

$$\|x(\cdot, 0, u)\|_\infty \leq g \|u\|_\infty .$$

This is the standard situation in linear systems theory, where V is quadratic (and hence admits estimates in terms of α_1 and α_2 of the form $c_i r^2$, where c_1 and c_2 are respectively the smallest and largest singular values of the associated form) and the supply function can likewise be taken of the form $c_4|u|^2 - c_3|x|^2$.

So finiteness of linear gain, that is, operator boundedness, follows from convexity of the estimation functions. Somewhat surprinsingly, for certain linear systems subject to actuator saturation, convex (but not quadratic) estimates are also possible, and this again leads to finite linear gains. For example, this applies to the function V in Equation (12), as an ISS-Lyapunov function for system (11): there one may pick $\alpha = \alpha_1 = \alpha_2 = r^3/3 + r^2/2$, which is convex since $\alpha''(r) = 2r + 1 > 0$, while α_3 and α_4 can be taken quadratic (cf. Equation (14)).

. As an additional remark, note that, just from the fact that V is nonnegative and $V(0) = 0$, and integrating the dissipation inequality (5), for $x_0 = 0$ there results the inequality $\int_0^{+\infty} \alpha(|x(t, 0, u)|)dt \leq \int_0^{+\infty} \theta(|u(t)|)dt$. In this manner, it is routine to use dissipation inequalities for proving operator boundedness in various pth norms (in particular, when $\alpha(r) = c_1 r^2$ and $\theta(r) = c_2 r^2$ one is estimating "H^∞" norms). But in the current context, more general nonlinearities than powers are being considered.

It is also interesting to note that, if V and α are so that the estimate (9) is satisfied, then there is some θ so that the dissipation estimate (5) also holds, with these same V and α.

3 Interconnections

It is by now well known, and easy to prove, that the cascade of two ISS systems is again ISS (in particular, a cascade of an ISS and a GAS system is GAS). It is interesting to observe that this statement can be understood very intuitively in terms of the dissipation formalism, and it provides further evidence of the naturality of the ISS notion. In addition, proceeding in this manner, one obtains a Lyapunov function (with strictly negative derivative along trajectories) for the cascade.

Theorem 2 *Consider the system in cascade form*

$$\begin{aligned} \dot{z} &= f(z, x) \\ \dot{x} &= g(x, u) \end{aligned}$$

where $f(0,0) = g(0,0) = 0$, the second equation is ISS, and the first equation is ISS when x is seen as an input. Then the composite system is ISS. ∎

The proof can be based on the following argument. First one shows that it is possible to obtain storage functions V_1 and V_2 so that V_1 satisfies a dissipation estimate

$$\nabla V_1(z) \cdot f(z, x) \leq \theta(|x|) - \alpha(|z|)$$

for the first subsystem, while V_2 is a storage function for the x-subsystem so that

$$\nabla V_2(x) \cdot g(x, u) \leq \tilde{\theta}(|u|) - 2\theta(|x|).$$

Then $V_1(z) + V_2(x)$ is a storage function for the composite system, which satisfies the dissipation inequality with derivative bounded by $\tilde{\theta}(|u|) - \theta(|x|) - \alpha(|z|)$.

A beautiful common generalization of both the cascade result and the usual Small-Gain Theorem was recently obtained by Jiang, Teel, and Praly. We write $\tilde{\gamma} \succ \gamma$ for two functions of class $\mathcal{K}$ if there is some $\rho \in \mathcal{K}_\infty$ so that $\tilde{\gamma} = (I + \rho) \circ \gamma$.

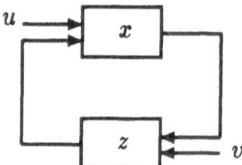

Figure 1: *Composite Feedback Form*

Theorem 3 ([7]) *Consider a system in composite feedback form (cf. Figure 1):*

$$\dot{z} = g(z, x, v)$$
$$\dot{x} = f(x, z, u)$$

where u, v are the inputs to the composite system. Assume:

- *Each of $\dot{x} = f(x, z, u)$ and $\dot{z} = g(z, x, v)$ is an ISS system, when (z, u) and (x, v) are considered as inputs respectively; let γ_1 and γ_2 denote the gains for the x and z subsystems, in the sense of the estimate of type (8).*

- *The following small-gain condition holds: there are $\tilde{\gamma}_1 \succ \gamma_1$ and $\tilde{\gamma}_2 \succ \gamma_2$ so that $(\tilde{\gamma}_1 \circ \tilde{\gamma}_2)(r) \le r$ and $(\tilde{\gamma}_2 \circ \tilde{\gamma}_1)(r) \le r$ for all $r \ge 0$.*

Then, the composite system is ISS. ∎

(Note that in the special case in which the $\gamma_i(r) = g_i r$, the small gain condition is satisfied iff $g_1 g_2 < 1$, thus generalizing the usual case.) It is important to note that the result in [7] is far more general; for instance, it deals with partially observed systems and with "practical stability" notions. Also, the small gain condition can be stated just in terms of the gains with respect to the z and x variables. Related to these results is previous work on small-gain conditions, also relying on comparison functions, in [13, 11].

A different cascade form, with an input feeding into both subsystems, is of interest in the context of stabilization of saturated linear systems (using an approach originally due to Teel, cf. [22]) and in other applications. This provides yet another illustration of the use of ISS ideas. The structure is (cf. Figure 2):

$$\dot{z} = f(z, x, u)$$
$$\dot{x} = g(x, u).$$

First assume that a (locally Lipschitz) feedback law k can be found which makes the system $\dot{z} = f(z, x, k(z))$ GAS *uniformly on x*, that is, $f(0, x, k(0)) = 0$

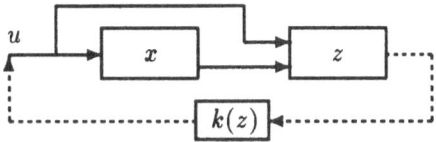

Figure 2: *Special Cascade Configuration*

for all x and an estimate as in (7), $|z(t)| \leq \beta(|z(0)|, t)$ holds, which is independent of $x(t)$. Suppose also that the x subsystem is ISS. Then, the feedback law $u = k(z)$ gives closed-loop equations $\dot{z} = f(z, x, k(z))$, $\dot{x} = g(x, k(z))$; because x is essentially irrelevant in the first equation, these equations behave just as a cascade of a GAS system (the z-system) and an ISS one, so the GAS property results as before. (More precisely, this is because it is still possible to find a Lyapunov function which depends only on z for the z-subsystem, due to the assumed uniformity property; see [9].) The interesting fact is the same global conclusions hold under more local assumptions on the z-subsystem. Assume:

- The z-subsystem is stabilizable with small feedback, uniformly on x small, meaning that for each $0 < \varepsilon \leq \varepsilon_0$ there is a (locally Lipschitz) feedback law k_ε with $|k_\varepsilon(z)| \leq \varepsilon$ for all z so that $\dot{z} = f(z, x, k_\varepsilon(z))$ is GAS uniformly on $|x| \leq \varepsilon_0$; further, under the feedback law $u = k_\varepsilon(z)$ the composite system is forward complete (solutions exist for all $t > 0$).

- The x subsystem is ISS.

(Later we discuss an interesting class of examples where these properties are verified.) Then, the claim is that, for any small enough $\varepsilon > 0$, the composite system under the feedback law $u = k_\varepsilon(z)$ is GAS. Stability is clear: for small x and z, trajectories coincide with those that would result if uniformity would hold globally on x (cf. the previous case). We are left to show that every solution $(x(\cdot), z(\cdot))$ satisfies $x(t) \to 0$ and $z(t) \to 0$ as $t \to +\infty$.

To establish this fact, pick any ε as follows. Let γ be a "nonlinear asymptotic gain" as in Equation (4), so that $\varliminf_{t \to +\infty} |x(t, x_0, u)| \leq \gamma(\|u\|_\infty)$ for all inputs and initial conditions. Now take any $0 < \varepsilon < \varepsilon_0$ so that $\gamma(\varepsilon) < \varepsilon_0$. Pick any k_ε so that $|k_\varepsilon(z)| \leq \varepsilon$ for all z. Consider any solution $(x(\cdot), z(\cdot))$. Seeing $v(t) = k_\varepsilon(z(t))$ as an input to the x-subsystem, with $\|v\|_\infty \leq \varepsilon$, the choice of γ means that for some T, $t \geq T$ implies $|x(t)| < \varepsilon_0$. It follows that $z(t) \to 0$. Now the second equation is an ISS system with an input $v(t) \to 0$, so also $x(t) \to 0$, as required.

4 An Example

As a simple illustration of the use of the ISS concept, we may consider the oft-studied problem of globally stabilizing to zero the angular momentum of a rigid body which is controlled by means of two external torques applied along principal axes, and suggest an alternative way of achieving this objective using

ISS ideas. (This may represent a model of a satellite under the action of a pair of opposing jets.) The components of the state variable $\omega = (\omega_1, \omega_2, \omega_3)$ denote the angular velocity coordinates with respect to a body-fixed reference frame with origin at the center of gravity and consisting of the principal axes. Letting the positive numbers I_1, I_2, I_3 denote the respective principal moments of inertia (positive numbers), this is a system on $\mathbb{R}^3$, with controls in $\mathbb{R}^2$ and equations:

$$I\dot{\omega} = S(\omega)I\omega + Bu , \tag{18}$$

where I is the diagonal matrix with entries I_1, I_2, I_3 and where B is a matrix in $\mathbb{R}^{3 \times 2}$ whose columns describe the axes on which the control torques apply. Since it is being assumed that the two torques act along two principal axes, without loss of generality the columns of B are $(0, 1, 0)'$ and $(0, 0, 1)'$ respectively. The matrix $S(\omega)$ is the rotation matrix

$$S(\omega) = \begin{pmatrix} 0 & \omega_3 & -\omega_2 \\ -\omega_3 & 0 & \omega_1 \\ \omega_2 & -\omega_1 & 0 \end{pmatrix}.$$

Dividing by the I_j's, and applying the obvious feedback and coordinate transformations, there results a system on $\mathbb{R}^3$ of the form:

$$\dot{x}_1 = x_2 x_3$$
$$\dot{x}_2 = u_1$$
$$\dot{x}_3 = u_2$$

where u_1 and u_2 are the controls.

To globally stabilize this system, and following the ideas of [1] for the corresponding local problem, one performs first a change of coordinates into new coordinates (x, z_1, z_2), where $x = x_1$ and

$$x_2 = -x_1 + z_1 , \quad x_3 = x_1^2 + z_2 .$$

The system is now viewed as a cascade of two subsystems. One of these is described by the x variable, with z_1 and z_2 now thought of as inputs, and the second one is the z_1, z_2 subsystem. The first subsystem is precisely the one in example (10), and it is therefore ISS. Since a cascade of an ISS and a GAS system is again GAS, it is only necessary to stabilize the z_1, z_2 subsystem. In other words, looking at the system in the new coordinates:

$$\dot{x} = -x^3 + x^2 z_1 - x z_2 + z_1 z_2$$
$$\dot{z}_1 = u_1 + (-x + z_1)(x^2 + z_2)$$
$$\dot{z}_2 = u_2 - 2x_1(-x + z_1)(x^2 + z_2) ,$$

any feedback that stabilizes the last two equations will also make the composite system GAS. One may therefore use

$$u_1 = -x_1 - x_2 - x_2 x_3 , \quad u_2 = -x_3 + x_1^2 + 2x_1 x_2 x_3 ,$$

which renders the last two equations $\dot{z}_1 = -z_1$ and $\dot{z}_2 = -z_2$. As a remark, note that a conceptually different approach to the same problem can be based upon zero dynamics techniques ([2, 23]). In that context, one uses Lie derivatives of a Lyapunov function for the x-subsystem in building a global feedback law; see the discussion in [17], Section 4.8. For the present rigid body stabilization problem, the feedback stabilizing law obtained using that approach would be as follows ([2]):

$$u_1 = -x_1 - x_2 - x_2x_3 - 2x_1x_3 , \quad u_2 = -x_3 + 3x_1^2 + 2x_1x_2x_3 .$$

5 Linear Systems with Actuator Saturation

For linear systems subject to actuator saturation, more precise results regarding stabilization can be obtained. The objective is to study control problems for plants P that can be described as in Figure 3, where W indicates a linear transfer matrix. For simplicity, we consider here just the state-observation

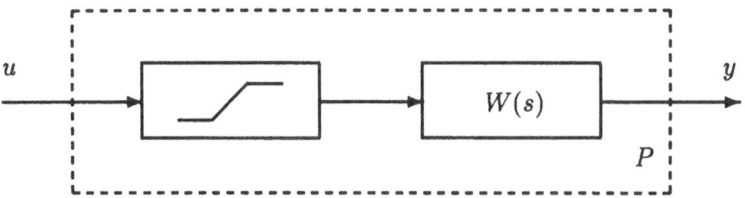

Figure 3: Saturated-Input Linear System

case, that is, systems of the type

$$\dot{x} = Ax + B\text{SAT}[u] . \tag{19}$$

By an L_p-stable system one means that the zero-initial state response induces a bounded operator $L_p \to L_p$. The following result was recently obtained by W. Liu, Y. Chitour, and the author (see also [3] for related results on input to state dependence for such systems):

Theorem 4 ([10]) *Assume that the pair (A, B) is controllable and that A is neutrally stable (i.e., there is some symmetric positive definite Q so that $A^T Q + QA \leq 0$). Then, there exists a matrix F so that the system*

$$\dot{x} = Ax + B\text{SAT}[Fx + u]$$

is L_p-stable for each $1 \leq p \leq \infty$. ∎

The fact that GAS can be achieved for such systems is a well-known and classical application of dissipation ideas, and a quadratic Lyapunov function suffices; obtaining the ISS property, and in particular operator stability, is far harder. Not surprisingly, the proof involves establishing a dissipation inequality involving a suitable storage function. What is perhaps surprising is that the storage function that is used is only of class C^1, in general not smooth: V is of the form $x'Px + |x|^3$, for some positive definite P. One establishes by means of such a V that the system is ISS. Since the used V admits convex estimates (in the sense discussed in Section 2.7), stronger operator stability conclusions can be obtained. The second example given in Section 2.6 (system (11) and storage function (12)) illustrates the detailed calculations in a very simple case.

The hypotheses in Theorem 4 can be relaxed considerably. For instance, controllability can be weakened, and the result is also valid if instead of SAT one uses a more general bounded saturation function σ which satisfies: (1) near the origin, σ is in a sector $[\kappa_1, \kappa_2]$: $0 < \kappa_1 \le \frac{\sigma(r)}{r} \le \kappa_2$ for all $0 < |r| \le 1$, and (2) $\text{sign}(r)\sigma(r) > \kappa > 0$ if $|r| > 1$.

A different line of work concerns linear systems subject to control saturation in the case in which the matrix A is not stable, but still has no eigenvalues with positive real part. This is the case, for instance, if A has a Jordan block of size at least two corresponding to an eigenvalue at the origin (the multiple integrator). In that case, L_p stabilization is not possible, but, since the system is open-loop null-controllable (assuming as in Theorem 4 that the pair (A, B) is controllable, or at least stabilizable as a linear pair), it is realistic to search for a globally stabilizing feedback.

A first result showing that a smooth globally stabilizing feedback always exists was given in work by Sussmann and the author ([19]). A remarkable design in terms of combinations of saturations was supplied by Teel ([22]), for the particular case of single-input multiple integrators, and a general construction based on Teel's ideas was completed recently in work of Sussmann, Yang, and the author ([21]). For simplicity, call a function $\mathbb{R}^n \to \mathbb{R}^m$ each of whose coordinates has the form

$$\varphi_1 x + \alpha_1 \text{SAT}[\varphi_2 x + \alpha_2 \text{SAT}[\ldots \text{SAT}[\varphi_{s-1} x + \alpha_{s-1} \text{SAT}[\varphi_s x]]\ldots]]$$

for some s and some real numbers α_i and linear functionals φ_i a *cascade of saturations*, and one for which coordinates have the form

$$\alpha_1 \text{SAT}[\varphi_1 x] + \alpha_2 \text{SAT}[\varphi_2 x] + \ldots + \alpha_s \text{SAT}[\varphi_s x]$$

a *superposition of saturations*. (In the terminology of artificial neural networks, this last form is a "single hidden layer net.") There are two results, one for each of these controller forms:

Theorem 5 ([21]) *Consider the system (19), where the pair (A, B) is stabilizable and A has no eigenvalues with positive real part. Then there exist a cascade of saturations k and a superposition of saturations ℓ so that $\dot{x} = Ax + B\text{SAT}[k(x)]$ and $\dot{x} = Ax + B\text{SAT}[\ell(x)]$ are both GAS.* ∎

(The coefficients α_i in the second case can be chosen arbitrarily small, which means that the second result could also be stated as stability of $\dot{x} = Ax + B\ell(x)$ since the saturation is then irrelevant.)

For cascades of saturations, this design proceeds in very rough outline as follows (the superposition case is similar). A preliminary step is to bring the original system (19) to the following composite form:

$$\dot{z} = A_1 z + B_1(-Fx + \text{SAT}[u])$$
$$\dot{x} = A_2 x + B_2 \text{SAT}[u],$$

where F is a matrix which has the property that the system $\dot{x} = A_2 x + B_2 \text{SAT}[Fx + u]$ is ISS. (An example of such F is provided by the case $\dot{x} = -\text{SAT}[x + u]$, shown earlier to be ISS, and more generally the case treated in Theorem 4.) Further, it is assumed that for each $\varepsilon > 0$ sufficiently small there is a (locally Lipschitz) feedback law k_ε with $|k_\varepsilon(z)| \leq \varepsilon$ for all z and so that $\dot{z} = A_1 z + B_1 k_\varepsilon(z)$ is GAS. Now the feedback law

$$u = Fx + k_\varepsilon(z)$$

is so that for small x and ε the z-equation is GAS independently of x (in fact, the x variable is completely cancelled out), and hence the discussion given in connection with Figure 2 applies. Thus the composite system is stabilized, assuming only that the z-subsystem can be stabilized with small feedback. Moreover, $Fx + k_\varepsilon(z)$ has a cascade form provided that k_ε be a saturation of a cascade. These assumptions can be in turn obtained inductively, by decomposing the z equation recursively into lower dimensional subsystems. (More precisely, instead of SAT one may use a scaled version with smaller lower bounds, $\text{SAT}_\delta[r] = \delta \text{SAT}[r/\delta]$, and the proof is the same. This provides the small feedback needed in the inductive step.) See [21] for details as well as a far more general result, which allows many other saturation functions σ instead of SAT.

6 Feedback Equivalence

As mentioned earlier, with the concept of ISS, it is possible to prove a general result on feedback equivalence. Consider two systems

$$\dot{x} = f(x, u) \quad \text{and} \quad \dot{x} = g(x, u)$$

with the same state and input value spaces (same n, m). These systems are *feedback equivalent* if there exist a smooth $k : \mathbb{R}^n \rightarrow \mathbb{R}^m$, and an $m \times m$ matrix Γ consisting of smooth functions having $\det \Gamma(x) \neq 0$ for all x, such that

$$g(x, u) = f(x, k(x) + \Gamma(x)u)$$

for all x and u (see Figure 4). The systems are *strongly* feedback equivalent if this holds with $\Gamma = I$ (see Figure 5).

Strong equivalence is the most interesting concept when studying actuator perturbations, while feedback equivalence is a natural concept in feedback linearization and other design techniques.

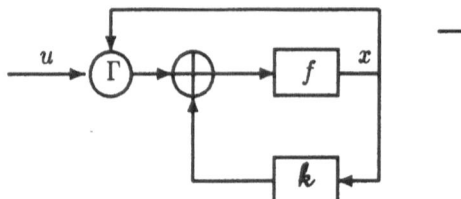 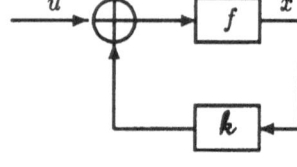

Figure 4: *Feedback Equivalence*   Figure 5: *Strong Feedback Equivalence*

The system (1) is *stabilizable* if there exists a smooth function k (with $k(0)=0$) so that

$$\dot{x} = f(x, k(x))$$

is GAS. Equivalently, the system is strongly feedback equivalent to a GAS system. It is ISS-*stabilizable* if it is feedback equivalent to an ISS system.

Theorem 6 ([14, 16]) *The following properties are equivalent, for any system:*

- *The system is stabilizable.*

- *The system is* ISS-*stabilizable.*

For systems affine in controls u (that is, $f(x, u)$ is affine in u) the above are also equivalent to strong feedback equivalence to ISS. ∎

7 Input/Ouput Stability (IOS)

Until here, only input to state stability was discussed. It is possible to give an analogous definition for input/output operators. This will be done next, and a result will be stated which shows that this property is equivalent to internal stability under suitable reachability and observability conditions, just as with linear systems (cf. for instance Section 6.3 in [17]).

An *i/o operator* is a causal map $F : L_{\infty,e}^m \to L_{\infty,e}^p$. (More generally, partially defined operators can be studied as well, but since only the stable case will be considered, and since stability implies that F is everywhere defined, there is no need to do so here; see [14] for more details.)

The i/o operator F is *input/output stable* (IOS) if there exist two functions $\beta \in \mathcal{KL}$ and $\gamma \in \mathcal{K}$ so that

$$|F(u)(t)| \leq \beta(\|u_t\|_\infty, t - T) + \gamma(\|u^t\|_\infty)$$

for all pairs $0 \leq T \leq t$ (a.e.) and all $u \in L_{\infty,e}^m$. Here u_t denotes as earlier the restriction of the input u to $[0, t]$ and u^t denotes its restriction to $[t, +\infty)$, in both cases seen as elements of $L_{\infty,e}^m$ having zero value outside of the considered range. This notion is well-behaved in various senses; for instance, it is closed under composition (serial connection), and $u \to 0$ implies $F(u) \to 0$.

Consider now a control system $\dot{x} = f(x, u)$ with outputs

$$y = h(x) \tag{20}$$

where $h : \mathbb{R}^n \to \mathbb{R}^p$ is continuous and satisfies $h(0)=0$. With initial state $x_0=0$, this induces an operator

$$F(u)(t) := h(x(t, 0, u))$$

(a priori only partially defined). The system (1)-(20) is called IOS if this operator is.

The system with outputs (1)-(20) is *well-posed observable* ("strongly" observable in [14]) provided that the following property holds: there exist two functions α_1, α_2 of class $\mathcal{K}$ such that, for each triple of state, control, and output functions on $t \geq 0$

$$(x(\cdot), u(\cdot), y(\cdot))$$

satisfying the equations, the norms of these functions necessarily satisfy

$$\|x\|_\infty \leq \alpha_1(\|u\|_\infty) + \alpha_2(\|y\|_\infty). \tag{21}$$

Analogously, one has a notion of a *well-posed reachable* system (1). This is a system for which there is a function α_3 of class $\mathcal{K}$ with the following property: for each $x_0 \in \mathbb{R}^n$ there exists a time $T > 0$ and a control u so that

$$\|u\|_\infty \leq \alpha_3(|x_0|)$$

and so that $x(T, 0, u) = x_0$.

For linear systems, these properties are equivalent to observability and reachability from zero respectively. In general, the first one corresponds to the possibility of reconstructing the state trajectory in a regular fashion —similar notions have been studied under various names such as "algebraic observability" or "topological observability" — and the second models the situation where the energy needed to control from the origin to any given state must be in some sense proportional to how far this state is from the origin. The proof of the following result is a routine argument, and is quite similar to the proofs of analogous results in the linear case as well as in the dissipation literature:

Theorem 7 ([14]) *If (1) is* ISS, *then (1)-(20) is* IOS. *Conversely, if (1)-(20) is* IOS, *well-posed reachable, and well-posed observable, then (1) is* ISS. ∎

Many variants of the notion of IOS are possible, in particular in order to deal with nonzero initial states ([8, 7]), or to study notions of practical stability, in which convergence to a small neighborhood of the origin is desired. Also of interest, is the study of the IOS (or even ISS) property relative to attracting invariant sets, not necessarily the origin and not even necessarily compact; see ([8, 9]) for instance.

References

[1] Brockett, R.W., "Asymptotic stability and feedback stabilization," in *Differential Geometric Control theory* (R.W. Brockett, R.S. Millman, and H.J. Sussmann, eds.), Birkhauser, Boston, 1983, pp. 181-191.

[2] Byrnes, C.I. and A. Isidori, "New results and counterexamples in nonlinear feedback stabilization," *Systems and Control Letters*, **12**(1989): 437-442.

[3] Chitour, Y., W. Liu, and E.D. Sontag, "On the continuity and incremental-gain properties of certain saturated linear feedback loops," *Intern. J. Robust and Nonlinear Control*, to appear.

[4] Hahn, W, *Stability of Motion*, Springer-Verlag, New York, 1967.

[5] Hill, D.J., and P. Moylan, "Dissipative dynamical systems: Basic input-output and state properties," *J. Franklin Institute* **5**(1980): 327-357.

[6] Hill, D.J., "Dissipative nonlinear systems: Basic properties and stability analysis," *Proc. 31st IEEE Conf. Dec. and Control*, Tucson, Arizona, 1992, pp. 3259-3264.

[7] Jiang, Z.-P., A. Teel, and L. Praly, "Small-gain theorem for ISS systems and applications," *Math of Control, Signals, and Systems*, 1994, to appear.

[8] Lin, Y., *Lyapunov Function Techniques for Stabilization*, PhD Thesis, Mathematics Department, Rutgers, The State University of New Jersey, New Brunswick, New Jersey, 1992.

[9] Lin, Y., E.D. Sontag, and Y. Wang, "A smooth converse Lyapunov theorem for robust stability," *SIAM J. Control and Opt.*, to appear. (See also IMA Preprint #1192, Institute for Mathematics and Its Applications, University of Minnesota, November 1993.)

[10] Liu, W., Y. Chitour, and E.D. Sontag, "Remarks on finite gain stabilizability of linear systems subject to input saturation," *Proc. IEEE Conf. Decision and Control, San Antonio, Dec. 1993*, IEEE Publications, 1993, pp. 1808-1813. Journal version to appear in *SIAM J. Control and Opt.*

[11] Mareels, I.M., and D.J. Hill, "Monotone stability of nonlinear feedback systems," *J. Math. Sys, Estimation and Control* **2**(1992): 275-291.

[12] Praly, L., and Z.-P. Jiang, "Stabilization by output feedback for systems with ISS inverse dynamics," *Systems & Control Letters* **21**(1993): 19-34.

[13] Safonov, M.G., *Stability and Robustness of Multivariable Feedback Systems*, MIT Press, Cambridge, MA, 1980.

[14] Sontag, E.D., "Smooth stabilization implies coprime factorization," *IEEE Transactions on Automatic Control* **AC-34**(1989): 435-443.

[15] Sontag, E.D., "Remarks on stabilization and input-to-state stability," *Proc. IEEE Conf. Decision and Control, Tampa, Dec. 1989*, IEEE Publications, 1989, pp. 1376-1378.

[16] Sontag, E.D., "Further facts about input to state stabilization," *IEEE Transactions on Automatic Control* **AC-35**(1990): 473-476.

[17] Sontag E.D., *Mathematical Control Theory, Deterministic Finite Dimensional Systems*, Springer-Verlag, New York, (1990).

[18] Sontag, E.D., and Y. Lin, "Stabilization with respect to noncompact sets: Lyapunov characterizations and effect of bounded inputs," in *Proc. Nonlinear Control Systems Design Symp., Bordeaux, June 1992* (M. Fliess, Ed.), IFAC Publications, pp. 9-14.

[19] Sontag, E.D., and H.J. Sussmann, "Nonlinear output feedback design for linear systems with saturating controls," *Proc. IEEE Conf. Decision and Control, Honolulu, Dec. 1990*, IEEE Publications, 1990: 3414-3416.

[20] Sontag, E.D., and Y. Wang, "On characterizations of the input-to-state stability property," *Systems and Control Letters*, to appear.

[21] Sussmann, H.J., E.D. Sontag, and Y. Yang, "Stabilization of linear systems with bounded controls," *IEEE Trans. Autom. Control*, to appear.

[22] Teel, A.R., "Global stabilization and restricted tracking for multiple integrators with bounded controls," *Systems and Control Letters* **18**(1992): 165-171.

[23] Tsinias, J., "Sufficient Lyapunovlike conditions for stabilization," *Math. of Control, Signals, and Systems* **2**(1989): 343-357.

[24] Tsinias, J., "Versions of Sontag's input to state stability condition and the global stabilizability problem," *SIAM Journal on Control and Optimization* **31**(1993): 928-941.

[25] Tsinias, J., "Sontag's 'input to state stability condition' and global stabilization using state detection," *Systems & Control Letters* **20**(1993): 219-226.

[26] Varaiya, P.P. and R. Liu, "Bounded-input bounded-output stability of nonlinear time-varying differential systems," *SIAM J.Control* **4**(1966): 698-704.

[27] Willems, J.C., "Mechanisms for the stability and instability in feedback systems," *Proc. IEEE* **64** (1976): 24-35.

[28] Zames, G., "On the input-output stability of time-varying nonlinear systems. Part I: Conditions derived using concepts of loop gain, conicity and passivity; Part II: Conditions involving circles in the frequency domain and sector nonlinearities," *IEEE Trans. Autom. Ctr.* **AC-11** (1966): 228-238 and 465-476.

On Optimal Decentralized Control *

Le Yi Wang and Wei Zhan
Department of Electrical and Computer Engineering,
Wayne State University,
Detroit, Michigan 48202, U.S.A.

Abstract

In this paper, decentralized robust stabilization and performance of two-channel interconnected systems are studied. Necessary and sufficient conditions for decentralized robust stability and robust performance are derived.

It is shown that when local systems are not perfectly modelled, there exists a certain irreducible bound on the robustness for uncertainties in interconnecting channels. On the other hand, when local systems are perfectly modelled, it is demonstrated constructively that when the two local plants do not have common unstable poles, robust stabilization can be achieved for any bounded stable uncertainties in interconnecting channels.

1   Introduction

This paper is concerned with the decentralized robust stabilization and robust performance in two-channel interconnected systems. In practical control problems, subsystems are usually coupled via interconnecting channels. While local designs, assuming absence of interactions among subsystems, enjoy the advantages of simplicity in modeling and design, and of available tools for system analysis and synthesis, the overall performance is usually unsatisfactory when control systems are implemented. A common practice in control engineering is to calibrate control designs either manually or by computer-aided search techniques, so that a certain level of subsystem coordinations can be achieved. This approach, however, does not guarantee overall performances of the system.

Alternatively, interconnected systems can be modelled theoretically as multi-input-multi-output systems where all subsystems and interactions are modelled. Assuming accessibility of total information, centralized controllers are designed by using MIMO synthesis methods. The centralized design carries

---

*This research is supported in part by the National Science Foundation under grant ECS-9209001.

large complexity in modelling and design and is often impractical. For example, control systems interconnected by telecommunications have geometrical limitations which make on-line communications among subsystems costly and intractable. This observation constitutes, perhaps, the most important motivation for classical decentralized control in which decentralized controllers are sought for a given MIMO plant (see, e.g., [3] and the references therein).

Even in situations where subsystems are compactly located such as in automotive control problems, decentralized control is often a preference. For instance, modelling of subsystem interconnections is often expensive and difficult. While modelling of subsystems can usually be performed off-line on system components, interactions can only be modelled in normal operating conditions. It is then highly desirable that decentralized control can be synthesized with little modelling in system interactions, or more precisely, interactions are only summarized in certain uncertainty descriptions.

Traditionally, research on decentralized control concentrates on stabilization of (large scale) nominal MIMO plants by decentralized state or output feedback. The reader is referred to [3] [5] [6] [4] and the references therein for a partial list of early work in this research direction. While the early research has resulted in a better understanding of decentralized control, the key issue of decentralized robust control is still not well understood. In particular, decentralized robust control in an input-output setting remains largely unexplored. Exceptions include the work by Zames and Bensoussan [9], who developed a decentralized control strategy for diagonally-dominant systems in the framework of H^∞ sensitivity optimization. Recently, some new results on robust decentralized control have emerged [1] [7]. Especially, the results in [7] demonstrate interesting progress in the development of decentralized H^∞ control for finite-dimensional systems.

In this paper, decentralized robust stabilization and performance of two-channel interconnected systems are studied. The problem was initially motivated by some benchmark control problems in automotive powertrain control systems, such as the coordination of torque and emission control subsystems. Traditionally, in a much simplified version, engine torque T is controlled by adjusting throttle angle θ, and exhaust gas emission E by air/fuel ratio A/F. These two subsystems, however, are intimately interconnected. The torque controller F_T (and the emission controller F_E) designed based on the local torque plant P_T (and the local emission plant P_E) perform poorly when implemented with interconnecting channels P_{TE} and P_{ET} in force. Questions arise: How should decentralized controllers F_T and F_E be modified? Is it necessary to use centralized controllers which are more complex in design and implementation, and less reliable in operation and mantainence? How much modelling, which is costly to carry out, of the interconnecting channels P_{ET} and P_{TE} is required to achieve coordination of subsystems?

Inspired by these questions, we formulate a generic problem of robust stabilization and robust performance of two-channel interconnected systems. It is shown (Example 2) that local optimal design may result in instability when uncertainties in interconnecting channels are ignored. Necessary and sufficient conditions for decentralized robust stability and robust performance (Theorems 1 and 5) are derived. The conditions suggest new optimization problems for decentralized optimal robustness. It is demonstrated (Theorem 2) using Poisson Integration Formula that when local systems are not perfectly modelled, there exists a certain irreducible bound on the robustness (the inverse of uncertainty radius) against uncertainties in interconnecting channels. On the other hand, it is shown (Theorem 3) constructively that when the two local plants are perfectly modelled and do not have common unstable poles, robust stabilization can be achieved for any bounded stable uncertainties in interconnecting channels. In other words, improvement of robustness against interconnecting uncertainties can be achieved, not by better modeling on interconnecting channels, but by better modeling on local systems, revealing a fundamental property of decentralized feedback. Furthermore, it is proven that if local systems have common unstable poles, tolerable interconnecting uncertainties are constrained by certain interpolation conditions (Theorem 4). It is argued by an example (Example 1) that the optimal robustness in this case is not an interpolation problem of Nevanlinna-Pick type and new mathematics tools are needed in this case. A design procedure using outer interpolation and joint H^∞ optimization is developed to achieve suboptimal design in robust stabilization of such systems.

2 Preliminaries

2.1 Spaces and Systems

$\mathbb{R}, \mathbb{C}, \mathbb{Z}$ denote the reals, complex numbers and integers. The absolute value of $x \in \mathbb{C}$ is $|x|$. For a matrix K, $\sigma_m(K)$ will denote its largest singular value.

L^p, $p \geq 1$, denotes the space of all real-valued functions u on $(-\infty, \infty)$ for which

$$\|u\|_p := \begin{cases} (\int_{-\infty}^{\infty} |u(t)|^p dt)^{1/p} < \infty, & 1 \leq p < \infty; \\ \sup_{t \in \mathbb{R}} |u(t)| < \infty, & p = \infty. \end{cases}$$

$\mathbb{H}^\infty$ is the algebra of analytic functions K on the RHP $\{\Re s > 0\}$ satisfying

$$\|K\|_\infty := \sup_\omega |K(j\omega)| < \infty.$$

It is well known that $\mathbb{H}^\infty$ is isometric to the algebra of LTI systems bounded on L^2. $\mathbb{H}^\infty$ will be viewed as a subalgebra of the usual Lebesque space $\mathbb{L}^\infty$ on $\mathbb{R}$. the norm on $\mathbb{L}^\infty$ will still be denoted by $\| \cdot \|_\infty$.

$\mathbb{H}^\infty_e$ will denote the algebra of all causal (possibly unstable) LTI systems. A pair $(\mathbf{N}, \mathbf{D}) \in \mathbb{H}^\infty \times \mathbb{H}^\infty$ is *coprime* in $\mathbb{H}^\infty$ if for some $(\mathbf{X}, \mathbf{Y}) \in \mathbb{H}^\infty \times \mathbb{H}^\infty$

$$(\mathbf{NX} + \mathbf{DY})^{-1} \in \mathbb{H}^\infty. \tag{1}$$

A system $G \in \mathbb{H}_e^\infty$ has a *factorization representation* in $\mathbb{H}^\infty$ if $G = D^{-1}N$, where $(N, D) \in \mathbb{H}^\infty \times \mathbb{H}^\infty$, and $D^{-1} \in \mathbb{H}_e^\infty$. The factorization $D^{-1}N$ is coprime if the corresponding pair (N, D) is coprime in $\mathbb{H}^\infty$.

The interconnection of a feedback F and a plant G in $\mathbb{H}_e^\infty$ is *well-posed* if all elements of the closed-loop mapping

$$\mathcal{K}(G, F) = \begin{pmatrix} K_{11} & K_{12} \\ K_{21} & K_{22} \end{pmatrix} := \begin{pmatrix} (I+FG)^{-1} & G(I+FG)^{-1} \\ F(I+GF)^{-1} & (I+GF)^{-1} \end{pmatrix} \quad (2)$$

are in $\mathbb{H}_e^\infty$, and stable if they are in $\mathbb{H}^\infty$. $F \in \mathbb{H}_e^\infty$ is said to robustly stabilize a subset $\Omega \subseteq \mathbb{H}_e^\infty$ of uncertain plants G if $\mathcal{K}(F, G)$ is stable for all $G \in \Omega$.

2.2 Problem Formulation

The uncertain plant P considered in this paper is represented by the uncertainty set

$$\Omega_P = \left\{ P = \begin{pmatrix} P_{11} & P_{12} \\ P_{21} & P_{22} \end{pmatrix} = \begin{pmatrix} D_1^{-1}(N_1 + W_{11}\Delta_{11}) & W_{12}\Delta_{12} \\ W_{21}\Delta_{21} & D_2^{-1}(N_2 + W_{22}\Delta_{22}) \end{pmatrix} \right\} \quad (3)$$

where

$$\Delta_{ij} \in \mathbb{H}^\infty, \|\Delta_{ij}\|_\infty \leq \epsilon_{ij}, \quad i,j = 1,2.$$

For $i = 1, 2$, $(N_i, D_i) \in \mathbb{H}^\infty \times \mathbb{H}^\infty$ are coprime in $\mathbb{H}^\infty$ and D_i is finite-dimensional. Without loss of generality, we assume D_1 and D_2 are inner, namely, $|D_1(j\omega)| = 1$, $|D_2(j\omega)| = 1$, for all ω. $W_{ij}, W_{ij}^{-1} \in \mathbb{H}^\infty$ are weighting functions for uncertainty descriptions.

Ω_P can be expressed in a form of structured additive uncertainty in left and right factorization forms:

$$\Omega_P = \{ D^{-1}(N + \Delta) = (\tilde{N} + \tilde{\Delta})\tilde{D}^{-1} \} \quad (4)$$

where

$$D = \tilde{D} = \begin{pmatrix} D_1 & 0 \\ 0 & D_2 \end{pmatrix}$$

$$N = \tilde{N} = \begin{pmatrix} N_1 & 0 \\ 0 & N_2 \end{pmatrix}$$

$$\Delta = \begin{pmatrix} W_{11}\Delta_{11} & D_1 W_{12}\Delta_{12} \\ D_2 W_{21}\Delta_{21} & W_{22}\Delta_{22} \end{pmatrix}$$

$$\tilde{\Delta} = \begin{pmatrix} W_{11}\Delta_{11} & D_2 W_{12}\Delta_{12} \\ D_1 W_{21}\Delta_{21} & W_{22}\Delta_{22} \end{pmatrix}$$

P_{11}, P_{22} will be called local systems, and P_{12}, P_{21} interconnecting channels.

We will consider the problems of robust stabilization and robust performance for Ω_P using decentralized controllers.

Problem 1: Robust Stabilization

(1) Under what conditions, can Ω_P be robustly stabilized by a decentralized controller

$$F = \begin{pmatrix} F_1 & 0 \\ 0 & F_2 \end{pmatrix} \tag{5}$$

where $F_1, F_2 \in H_e^\infty$?

(2) How can one construct a decentralized robust controller F for Ω_P when such controllers exist?

Assume that the system performance is measured by the weighted sensitivity function

$$f(P, F) = \|M(1 + PF)^{-1}\|_\infty,$$

where

$$M = \begin{pmatrix} M_1 & 0 \\ 0 & M_2 \end{pmatrix}, \qquad M_1, M_1^{-1}, M_2, M_2^{-1} \in H^\infty$$

is a weighting matrix for disturbances.

Problem 2: Robust Performance

Under what conditions can a decentralized controller F robustly stabilize Ω_P and achieve the robust performance

$$\sup_{P \in \Omega_r} f(P, F) < \mu?$$

The remaining part of the paper is organized as follows. Problem 1 will be discussed in Section 3. First, necessary and sufficient conditions for robust stability are presented in Subsection 3.1. A design procedure for optimal robustness is given in Subsection 3.2, which employs the idea of outer interpolation and joint H^∞ optimization. Finally, in Section 4, necessary and sufficient conditions for robust performance are presented.

3 Robust Stabilization

3.1 Robust Stability

We start with the problem of stability analysis for uncertain plants P. Without loss of generality, we assume that D_i are inner rational functions in H^∞. By (4), $P \in \Omega_P$ can be expressed in a left and a right factorization representations:

$$P = D^{-1}(N + \Delta) = (\tilde{N} + \tilde{\Delta})\tilde{D}^{-1}$$

By the coprimeness of (N_1, D_1) and (N_2, D_2), there exist $X_1, Y_1, X_2, Y_2 \in H^\infty$ such that

$$N_i X_i + D_i Y_i = 1, \qquad i = 1, 2. \tag{6}$$

Apparently, by the uncertainty description, F_i must robustly stabilize P_{ii} in H^∞ ($i = 1, 2$). Namely, local stabilization must first be achieved. By Youla's parametrization, all stabilizing controllers for the local nominal systems can be represented by

$$F_i = \frac{X_i - D_i Q_i}{Y_i + N_i Q_i} \qquad i = 1, 2$$

where $Q_i \in \mathbb{H}^\infty$, subject only to $(Y_i + N_i Q_i)^{-1} \in \mathbb{H}_e^\infty$. As a result, robustness of stabilization against uncertainties in local and interconnecting channels is to be achieved by suitable design of Q_i in $\mathbb{H}^\infty$.

For a given $Q_i \in \mathbb{H}^\infty$, $i = 1, 2$, we have

$$F_i = \frac{X_i - D_i Q_i}{Y_i + N_i Q_i} \qquad i = 1, 2.$$

Then F can be expressed in a left and a right coprime factorization representations:

$$F = XY^{-1} = Y^{-1}X$$

where

$$X = \begin{pmatrix} X_1 - D_1 Q_1 & 0 \\ 0 & X_2 - D_2 Q_2 \end{pmatrix}$$

$$Y = \begin{pmatrix} Y_1 + N_1 Q_1 & 0 \\ 0 & Y_2 + N_2 Q_2 \end{pmatrix}.$$

Theorem 1 *(1) F is robustly stabilizing for Ω_p if and only if*
(a)

$$\|W_{11}(X_1 - D_1 Q_1)\|_\infty < \frac{1}{\epsilon_{11}}, \qquad \|W_{22}(X_2 - D_2 Q_2)\|_\infty < \frac{1}{\epsilon_{22}},$$

(b)

$$\left\| \frac{W_{12}W_{21}(X_1 - D_1 Q_1)(X_2 - D_2 Q_2)}{(1 - \epsilon_{11}|W_{11}(X_1 - D_1 Q_1)|)(1 - \epsilon_{22}|W_{22}(X_2 - D_2 Q_2)|)} \right\|_\infty < \frac{1}{\epsilon_{12}\epsilon_{21}}. \quad (7)$$

(2) In particular, if $\epsilon_{11} = \epsilon_{22} = 0$, then F is robustly stabilizing for Ω_P if and only if

$$\|W_{12}W_{21}(X_1 - D_1 Q_1)(X_2 - D_2 Q_2)\|_\infty < \frac{1}{\epsilon_{12}\epsilon_{21}}. \quad (8)$$

Proof: (1) By the algebraic system theory, F is robustly stabilizing for Ω_P if and only if for all $\|\Delta_{ij}\|_\infty \le \epsilon_{ij}$, $i, j = 1, 2$, the left and right resolvents

$$\begin{aligned} R_l &= (N + \Delta)X + DY \\ &= \begin{pmatrix} 1 + W_{11}\Delta_{11}(X_1 - D_1 Q_1) & D_1 W_{12}\Delta_{12}(X_2 - D_2 Q_2) \\ D_2 W_{21}\Delta_{21}(X_1 - D_1 Q_1) & 1 + W_{22}\Delta_{22}(X_2 - D_2 Q_2) \end{pmatrix} \end{aligned}$$

and

$$\begin{aligned} R_r &= X(\tilde{N} + \tilde{\Delta}) + Y\tilde{D} \\ &= \begin{pmatrix} 1 + W_{11}\Delta_{11}(X_1 - D_1 Q_1) & D_2 W_{12}\Delta_{12}(X_1 - D_1 Q_1) \\ D_1 W_{21}\Delta_{21}(X_2 - D_2 Q_2) & 1 + W_{22}\Delta_{22}(X_2 - D_2 Q_2) \end{pmatrix} \end{aligned}$$

have inverses in $\mathbb{H}^\infty$. By the determinant test, the condition $\mathbf{R}_l^{-1}, \mathbf{R}_r^{-1} \in \mathbb{H}^\infty$ is equivalent to

$$[(1 + \mathbf{W}_{11}\Delta_{11}(\mathbf{X}_1 - \mathbf{D}_1\mathbf{Q}_1))(1 + \mathbf{W}_{22}\Delta_{22}(\mathbf{X}_2 - \mathbf{D}_2\mathbf{Q}_2))$$
$$-\mathbf{W}_{12}\mathbf{W}_{21}\mathbf{D}_1\mathbf{D}_2\Delta_{12}\Delta_{21}(\mathbf{X}_1 - \mathbf{D}_1\mathbf{Q}_1)(\mathbf{X}_2 - \mathbf{D}_2\mathbf{Q}_2)](s) \neq 0, \qquad (9)$$

for all $s \in \{\Re s \geq 0\}$.

Now, if (a) and (b) are satisfied, we have the bound

$$|[(1 + \mathbf{W}_{11}\Delta_{11}(\mathbf{X}_1 - \mathbf{D}_1\mathbf{Q}_1))(1 + \mathbf{W}_{22}\Delta_{22}(\mathbf{X}_2 - \mathbf{D}_2\mathbf{Q}_2))$$
$$-\mathbf{W}_{12}\mathbf{W}_{21}\mathbf{D}_1\mathbf{D}_2\Delta_{12}\Delta_{21}(\mathbf{X}_1 - \mathbf{D}_1\mathbf{Q}_1)(\mathbf{X}_2 - \mathbf{D}_2\mathbf{Q}_2)](s)|$$
$$\geq (1 - \varepsilon_{11}|\mathbf{W}_{11}(\mathbf{X}_1 - \mathbf{D}_1\mathbf{Q}_1)|)(1 - \varepsilon_{22}|\mathbf{W}_{22}(\mathbf{X}_2 - \mathbf{D}_2\mathbf{Q}_2)|)$$
$$-\varepsilon_{12}\varepsilon_{21}|\mathbf{W}_{12}\mathbf{W}_{21}(\mathbf{X}_1 - \mathbf{D}_1\mathbf{Q}_1)(\mathbf{X}_2 - \mathbf{D}_2\mathbf{Q}_2)|$$
$$> 0.$$

Hence, (9) holds, implying that (a) and (b) are sufficient for robust stability.

Conversely, by substituting $\Delta_{12} = \Delta_{21} = 0$ in (9), we verify that (a) is necessary for robust stability. Furthermore, if (b) is violated, then by continuity there exists ω_0 such that

$$[(1 - \varepsilon_{11}|\mathbf{W}_{11}(\mathbf{X}_1 - \mathbf{D}_1\mathbf{Q}_1)|)(1 - \varepsilon_{22}|\mathbf{W}_{22}(\mathbf{X}_2 - \mathbf{D}_2\mathbf{Q}_2)|)$$
$$-\varepsilon_{12}\varepsilon_{21}|\mathbf{W}_{12}\mathbf{W}_{21}(\mathbf{X}_1 - \mathbf{D}_1\mathbf{Q}_1)(\mathbf{X}_2 - \mathbf{D}_2\mathbf{Q}_2)|](j\omega_0)$$
$$= 0.$$

Consequently, there exist $|\Delta_{ij}| \leq \varepsilon_{ij}$ such that (9) is violated. Therefore, (a) and (b) are also necessary for robust stability.

(2) It follows immediately from (1) after substituting $\varepsilon_{11} = \varepsilon_{22} = 0$. $\qquad \square$

3.2 Optimal Robust Stabilization

By Theorem 1, the optimal robustness is achieved by optimizing

$$\mu = \inf_{\mathbf{Q}_1 \in \mathbb{H}^\infty} \inf_{\mathbf{Q}_2 \in \mathbb{H}^\infty} \left\| \frac{\mathbf{W}_{12}\mathbf{W}_{21}(\mathbf{X}_1 - \mathbf{D}_1\mathbf{Q}_1)(\mathbf{X}_2 - \mathbf{D}_2\mathbf{Q}_2)}{(1 - \varepsilon_{11}|\mathbf{W}_{11}(\mathbf{X}_1 - \mathbf{D}_1\mathbf{Q}_1)|)(1 - \varepsilon_{22}|\mathbf{W}_{22}(\mathbf{X}_2 - \mathbf{D}_2\mathbf{Q}_2)|)} \right\|_\infty . \tag{10}$$

Except for the trivial case $\mathbf{D}_1 = 1$ or $\mathbf{D}_2 = 1$ (then obviously $\mu = 0$), we will show that when $\varepsilon_{11} > 0$ and $\varepsilon_{22} > 0$, we have

$$\mu \geq f(\varepsilon_{11}, \varepsilon_{22}) > 0$$

where $f(\varepsilon_{11}, \varepsilon_{22})$ is a monotone function of ε_{11} and ε_{22} with $f(0,0) = 0$. In other words, there exists an irreducible margin of robustness against interconnecting uncertainties whenever local systems are not perfectly modeled. The better the local modelling (i.e., the smaller the modelling errors $\varepsilon_{11}, \varepsilon_{22}$), the better the robustness margin. On the other hand, if $\varepsilon_{11} = \varepsilon_{22} = 0$, i.e., the case of perfect local modelling, it will be shown that $\mu = 0$ whenever $\mathbf{D}_1$ and $\mathbf{D}_2$ do not have common unstable zeros.

3.2.1 Optimal Robustness: Lower Bounds

We shall start with a simple situation where $\mu > 0$, independent of local modelling error bounds ε_{11} and ε_{22}.

Let $A_1 = \{z_1^1, \cdots, z_n^1\}$ and $A_2 = \{z_1^2, \cdots, z_m^2\}$ be the sets of all zeros of $\mathbf{D}_1$ and $\mathbf{D}_2$ in the closed RHP, respectively. Assume $A_1 \cap A_2 \neq \emptyset$. Define

$$\mu_c = \inf \left\{ \|\mathbf{K}\|_\infty : \mathbf{K} \in \mathbb{H}^\infty, \mathbf{K}(p) = \frac{\mathbf{W}_{12}(p)\mathbf{W}_{21}(p)}{\mathbf{N}_1(p)\mathbf{N}_2(p)}, \quad p \in A_1 \cap A_2 \right\}.$$

Proposition 1 provides a lower bound on μ.

Proposition 1 *If $A_1 \cap A_2$ is not empty, then*

$$\mu \geq \mu_c.$$

Proof: Apparently,

$$\mu \geq \mu' := \inf_{\mathbf{Q}_1 \in \mathbb{H}^\infty} \inf_{\mathbf{Q}_2 \in \mathbb{H}^\infty} \|\mathbf{W}_{12}\mathbf{W}_{21}(\mathbf{X}_1 - \mathbf{D}_1\mathbf{Q}_1)(\mathbf{X}_2 - \mathbf{D}_2\mathbf{Q}_2)\|_\infty$$

We will show that

$$\mu' \geq \mu_c.$$

Since for any choice of $\mathbf{Q}_1, \mathbf{Q}_2 \in \mathbb{H}^\infty$,

$$\mathbf{K} := \mathbf{W}_{12}\mathbf{W}_{21}(\mathbf{X}_1 - \mathbf{D}_1\mathbf{Q}_1)(\mathbf{X}_2 - \mathbf{D}_2\mathbf{Q}_2) \in \mathbb{H}^\infty,$$

and

$$\mathbf{K}(p) = \mathbf{W}_{12}(p)\mathbf{W}_{21}(p)\mathbf{X}_1(p)\mathbf{X}_2(p) = \frac{\mathbf{W}_{12}(p)\mathbf{W}_{21}(p)}{\mathbf{N}_1(p)\mathbf{N}_2(p)}, \quad p \in A_1 \cap A_2$$

we have

$$\mu' \geq \inf \left\{ \|\mathbf{K}\|_\infty : \mathbf{K} \in \mathbb{H}^\infty, \mathbf{K}(p) = \frac{\mathbf{W}_{12}(p)\mathbf{W}_{21}(p)}{\mathbf{N}_1(p)\mathbf{N}_2(p)}, \quad p \in A_1 \cap A_2 \right\}$$
$$= \mu_c.$$

$\square$

Hence, in the following discussions, we will assume $A_1 \cap A_2 = \emptyset$.

Theorem 2 *Suppose $A_1 \cap A_2 = \emptyset$, $A_1 \neq \emptyset$ and $A_2 \neq \emptyset$. To avoid triviality, assume $0 < \varepsilon_{ii} < 1, i = 1, 2$. Let $p_1 = \sigma_1 + j\omega_1 \in A_1$, $p_2 = \sigma_2 + j\omega_2 \in A_2$, and*

$$a_1 = \frac{\mathbf{W}_{11}(p_1)}{\mathbf{N}_1(p_1)} \neq 0, \qquad a_2 = \frac{\mathbf{W}_{22}(p_2)}{\mathbf{N}_2(p_2)} \neq 0$$

Then,

$$\mu \geq \max\{\beta_1 \varepsilon_{11}^{\alpha_1}, \beta_2 \varepsilon_{22}^{\alpha_2}\} \tag{11}$$

where $\alpha_1, \alpha_2, \beta_1, \beta_2 > 0$ are constants,

$$
\begin{aligned}
\alpha_i &= -\log_2 |a_i \epsilon_{ii}| > 0, \qquad i = 1, 2, \qquad (\textit{note: } |a_i \epsilon_{ii}| < 1) \\
\beta_1 &= \frac{\min\{|a_1|^{1/\delta_1}, |a_1|\}|a_2|}{2m} > 0, \\
\beta_2 &= \frac{\min\{|a_2|^{1/\delta_2}, |a_2|\}|a_1|}{2m} > 0, \\
\delta_i &= (1 - \log_2 |a_i| \epsilon_{ii})^{-1}, \qquad i = 1, 2 \\
m &= \left\| \frac{W_{12} W_{21}}{W_{11} W_{22}} \right\|_\infty.
\end{aligned}
$$

Proof: In Appendix.

3.2.2 Optimal Robustness: Interconnecting Uncertainties

In the special case of $\epsilon_{11} = \epsilon_{22} = 0$, the robustness optimization problem becomes

$$
\mu = \inf_{Q_1 \in H^\infty} \inf_{Q_2 \in H^\infty} \| W_{12} W_{21} (X_1 - D_1 Q_1)(X_2 - D_2 Q_2) \|_\infty \qquad (12)
$$

Recall that $A_1 = \{z_1^1, \cdots, z_n^1\}$ and $A_2 = \{z_1^2, \cdots, z_m^2\}$ are the sets of all zeros of D_1 and D_2 in the closed RHP, respectively.

Theorem 3 *If $A_1 \cap A_2 = \emptyset$, namely, D_1 and D_2 have no common zeros in the closed RHP $\{\Re s \geq 0\}$, then $\mu = 0$.*

Proof: In Appendix.

Theorem 3 claims that if the two subsystems do not have common unstable poles, any bounded uncertainties in the interconnecting channels can be tolerated, provided the decentralized controller is appropriately designed.

Theorem 3 seems to suggest that disjoint zeros of D_1 and D_2 do not constrain the achievable robustness. Together with Proposition 1, it is then tempting to conjecture that $\mu = \mu_c$. Example 1 shows that in general it is not the case. A direct consequence is that in general the joint optimization problem (12) is not a Nevanlinna-Pick interpolation problem, namely not a standard H^∞ optimization problem.

Example 1: Suppose $A_1 \cap A_2 = \{z_0\}$. Then the unique optimal interpolant for

$$
\inf \left\{ \|K\|_\infty : \quad K \in H^\infty, \quad K(z_0) = \frac{W_{12}(z_0) W_{21}(z_0)}{N_1(z_0) N_2(z_0)} \right\}
$$

is

$$
K = \frac{W_{12}(z_0) W_{21}(z_0)}{N_1(z_0) N_2(z_0)}.
$$

The question becomes: Can one find $Q_1, Q_2 \in H^\infty$ such that

$$K = W_{12}W_{21}(X_1 - D_1Q_1)(X_2 - D_2Q_2)? \qquad (13)$$

Since K is a constant, both $X_1 - D_1Q_1$ and $X_2 - D_2Q_2$ must be outer functions, i.e., no zeros in the RHP. However, if there exist real positive $z_1, z_2 \in A_1$ such that $N_1(z_1)N_1(z_2) < 0$, by [8] for any $Q_1 \in H^\infty$, $X_1 - D_1Q_1$ is not an outer function. As a result, the problem (13) has no solution for $Q_1, Q_2 \in H^\infty$. For instance,

$$N_1 = \frac{s-2}{(s+1)^2}, \quad D_1 = \frac{(s-1)(s-3)}{(s+1)^2}; \quad N_2 = 1, \quad D_2 = \frac{s-1}{s+1}$$

is such an example.

A pair $(L, B) \in H^\infty \times H^\infty$ is said to be *outer interpolable in H^∞* if there exists $Q \in H^\infty$ such that $L - BQ$ is outer in H^∞. By [8], (L, B) is outer interpolable if and only if the signs of L on the real RHP zeros of B are same.

Theorem 4 *If $A_1 \cap A_2 \neq \emptyset$ and (X_1, D_1) is outer interpolable, then*

$$\mu \leq \mu_0$$

where

$$\mu_0 = \inf\{\|K\|_\infty : K \in H^\infty \text{ and } K(z_j^2) = b_j, z_j^2 \in A_2\}$$

and

$$b_j = \begin{cases} \dfrac{W_{12}(z_j^2)W_{21}(z_j^2)}{N_1(z_j^2)N_2(z_j^2)}, & z_j^2 \in A_1 \cap A_2 \\[2mm] 0, & z_j^2 \in A_2/A_1. \end{cases}$$

Remark: When (X_1, D_1) is outer interpolable, we have

$$\mu_c \leq \mu \leq \mu_0.$$

Proof: In Appendix.

3.3   Examples

Example 2: Suppose a two-channel plant has the nominal part

$$N_1 = \frac{s-3}{s+1}, \quad D_1 = \frac{s-1}{s+1}; \quad N_2 = \frac{s-4}{s+2}, \quad D_2 = \frac{s-2}{s+2}.$$

The weighting functions are

$$W_{11} = \frac{s+4}{s+3}, \quad W_{22} = \frac{s+6}{s+5}, \quad W_{12} = W_{21} = 1.$$

Let $\varepsilon_0 := \varepsilon_{12}\varepsilon_{21} > \frac{7}{20}$.

Local stabilizing controllers for the nominal part:
It is easy to verify that one solution to

$$N_1 X_1 + D_1 Y_1 = 1$$

is

$$X_1 = -1, \quad Y_1 = 2.$$

Similarly, a solution to

$$N_2 X_2 + D_2 Y_2 = 1$$

is

$$X_2 = -2, \quad Y_2 = 3.$$

Locally optimal controllers:
The controllers achieving optimal radii of robustness for the local uncertainties are obtained by optimizing

$$\inf_{Q_i \in H^\infty} \|W_{ii}(X_i - D_i Q_i)\|_\infty, \qquad i = 1, 2.$$

By Nevanlinna-Pick interpolation, the optimal interpolants in this case are

$$K_1 = W_{11}(X_1 - D_1 Q_1) = -\frac{5}{4}, \qquad K_2 = W_{22}(X_2 - D_2 Q_2) = -\frac{16}{7}.$$

Correspondingly,

$$X_1 - D_1 Q_1 = \frac{K_1}{W_{11}} = -\frac{5}{4}\left(\frac{s+3}{s+4}\right),$$

$$X_2 - D_2 Q_2 = \frac{K_2}{W_{22}} = -\frac{16}{7}\left(\frac{s+5}{s+6}\right),$$

Now, since

$$\varepsilon_{12}\varepsilon_{21}|W_{12}W_{21}(X_1 - D_1 Q_1)(X_2 - D_2 Q_2)| = \varepsilon_0\frac{20}{7}\left|\left(\frac{s+3}{s+4}\right)\left(\frac{s+5}{s+6}\right)\right|,$$

we have

$$\left\|\varepsilon_0\frac{20}{7}\left(\frac{s+3}{s+4}\right)\left(\frac{s+5}{s+6}\right)\right\|_\infty = \varepsilon_0\frac{20}{7} > 1$$

whenever

$$\varepsilon_0 > \frac{7}{20}.$$

Therefore, the condition (7) is always violated. It follows that the interconnected system is not robustly stable.

Modified design:

When ϵ_{11} and ϵ_{22} are small, we would like to redesign the controller such that the inequality (8) is satisfied. This can be done by optimizing

$$\mu = \inf_{Q_1 \in H^\infty} \inf_{Q_2 \in H^\infty} \|(X_1 - D_1 Q_1)(X_2 - D_2 Q_2)\|_\infty.$$

Following the construction steps in the proof of Theorem 2, we first solve the outer interpolation problem:
Find $K, K^{-1} \in H^\infty$ *such that*

$$K(1) = \epsilon, \qquad K(2) = \frac{1}{\epsilon},$$

where ϵ is selected such that

$$\epsilon < \frac{1}{\sqrt{2\epsilon_0}}$$

Since $\epsilon > 0$, $\frac{1}{\epsilon} > 0$, such K exists by [8].

The construction of K can be accomplished as follows[1]: It is known that outer functions $K \in H^\infty$ can be represented by

$$K = e^F,$$

where $F \in H^\infty$. Since $F = \ln K$, we shall solve the interpolation problem:

$$F \in H^\infty: \qquad F(1) = \ln \epsilon, \quad F(2) = \ln \frac{1}{\epsilon} = -\ln \epsilon.$$

A solution to this interpolation problem is

$$F(s) = \alpha \frac{s - \beta}{s + \beta},$$

where α and β are solved from

$$\alpha \frac{1 - \beta}{1 + \beta} = \ln \epsilon, \qquad \alpha \frac{2 - \beta}{2 + \beta} = -\ln \epsilon.$$

As a result,

$$\beta = \sqrt{2}, \qquad \alpha = -\frac{\sqrt{2} + 1}{\sqrt{2} - 1} \ln \epsilon = -(3 + 2\sqrt{2}) \ln \epsilon.$$

It follows that

$$K = e^{\alpha \frac{s - \beta}{s + \beta}}.$$

It can be verified that

$$\|K\|_\infty = \|K^{-1}\|_\infty = e^\alpha = \epsilon^{-(3 + 2\sqrt{2})}.$$

---

[1] For the purpose of demonstration, we do not constrain K to be rational. However, rational solutions are possible, as shown in several examples in [8].

Now,

$$
\begin{aligned}
\mu &= \inf_{Q_1 \in H^\infty} \inf_{Q_2 \in H^\infty} \|K(X_1 - D_1 Q_1)K^{-1}(X_2 - D_2 Q_2)\|_\infty \\
&\le \inf_{Q_1 \in H^\infty} \|K(X_1 - D_1 Q_1)\|_\infty \inf_{Q_2 \in H^\infty} \|K^{-1}(X_2 - D_2 Q_2)\|_\infty \\
&= \varepsilon^2
\end{aligned}
$$

The optimal interpolants are easily obtained as

$$
K(X_1 - D_1 Q_1) = -\varepsilon;
$$

$$
K^{-1}(X_2 - D_2 Q_2) = -2\varepsilon.
$$

Since

$$
2\varepsilon^2 < \frac{1}{\varepsilon_0} = \frac{1}{\varepsilon_{12}\varepsilon_{21}}
$$

by Theorem 1, the interconnected system is robustly stable (when $\varepsilon_{11} = \varepsilon_{22} = 0$).

Note that in this example,

$$
(X_1 - D_1 Q_1) = -\varepsilon K^{-1};
$$

$$
(X_2 - D_2 Q_2) = -2\varepsilon K.
$$

Hence, the local radii of robustness are

$$
\varepsilon_{11} < (\varepsilon \|W_{11} K^{-1}\|_\infty)^{-1} \approx \left(\varepsilon \varepsilon^{-(3+2\sqrt{2})}\right)^{-1} = \varepsilon^{2(1+\sqrt{2})};
$$

$$
\varepsilon_{22} < (2\varepsilon \|W_{22} K\|_\infty)^{-1} \approx \frac{1}{2}\varepsilon^{2(1+\sqrt{2})}.
$$

3.4 A Comparison to the Case of Unstructured Uncertainty

It will be of interest to compare the results of the previors section with the optimal robustness achieved by decentralized controllers when uncertainties are unstructured. More precisely, let the uncertainty set of the plant be described by

$$
\Omega_p^U = \{D^{-1}(N + \Delta): \quad \|\Delta\|_\infty \le \varepsilon\},
$$

where

$$
D = \begin{pmatrix} D_1 & 0 \\ 0 & D_2 \end{pmatrix}
$$

$$
N = \begin{pmatrix} N_1 & 0 \\ 0 & N_2 \end{pmatrix}
$$

The decentralized stabilizing controllers for the nominal plant $\mathbf{D}^{-1}\mathbf{N}$ can be easily obtained by solving the local Bezout equations:

$$\mathbf{N}_i\mathbf{X}_i + \mathbf{D}_i\mathbf{Y}_i = 1, \qquad i = 1, 2,$$

with $\mathbf{X}_i, \mathbf{Y}_i \in \mathbb{H}^\infty$; and

$$\mathbf{F}_i = \frac{\mathbf{X}_i - \mathbf{D}_i\mathbf{Q}_i}{\mathbf{Y}_i + \mathbf{N}_i\mathbf{Q}_i}, \qquad \mathbf{Q}_i \in \mathbb{H}^\infty.$$

For unstructured uncertainties Ω_p^U, it is well known that the (decentralized) controllers robustly stabilize Ω_p^U if and only if

$$\inf_{\mathbf{Q}_1,\mathbf{Q}_2\in\mathbb{H}^\infty} \left\| \begin{pmatrix} \mathbf{X}_1 - \mathbf{D}_1\mathbf{Q}_1 & 0 \\ 0 & \mathbf{X}_2 - \mathbf{D}_2\mathbf{Q}_2 \end{pmatrix} \right\|_\infty < \frac{1}{\varepsilon},$$

namely,

$$\mu = \max\{\mu_1, \mu_2\}$$

where

$$\mu_i = \inf_{\mathbf{Q}_i\in\mathbb{H}^\infty} \|\mathbf{X}_i - \mathbf{D}_i\mathbf{Q}_i\|_\infty, \qquad i = 1, 2.$$

For instance, for the plant of Example 2, it is easily computed that

$$\mu_1 = 1, \qquad \mu_2 = 2.$$

therefore, $\mu = 3$ and the optimal robustness radius is

$$\frac{1}{\mu} = \frac{1}{2}.$$

Since arbitrarily large interconnecting uncertainties can be tolerated in Example 2, it is apparent that structural information is of critical importance in robust design. In particular, in decentralized control problems where structural constraints are imposed on controllers, neglection of available structural information on uncertainties can significantly limit achievable robustness.

4   Robust Performance

We are considering in this section the problem of robust performance. Here, the assumption $\varepsilon_{11} = \varepsilon_{22} = 0$ will be imposed. In this case,

$$\Omega_P = \left\{ \begin{pmatrix} \mathbf{D}_1^{-1}\mathbf{N}_1 & \mathbf{W}_{12}\Delta_{12} \\ \mathbf{W}_{21}\Delta_{21} & \mathbf{D}_2^{-1}\mathbf{N}_2 \end{pmatrix}, \quad \|\Delta_{12}\|_\infty \leq \varepsilon_{12}, \|\Delta_{21}\|_\infty \leq \varepsilon_{21} \right\}.$$

where $\mathbf{D}_1, \mathbf{D}_2$ are assumed, without loss of generality, to be inner. Define the weighting matrix $\mathbf{M}$ as

$$\mathbf{M} = \begin{pmatrix} \mathbf{M}_1 & 0 \\ 0 & \mathbf{M}_2 \end{pmatrix}, \qquad \mathbf{M}_1, \mathbf{M}_1^{-1}, \mathbf{M}_2, \mathbf{M}_2^{-1} \in \mathbb{H}^\infty.$$

The performance measure is the weighted sensitivity

$$f(\mathbf{P}, \mathbf{F}) = \left\| \mathbf{M}(1 + \mathbf{PF})^{-1} \right\|_{\infty}.$$

We would like to derive conditions for robust performance

$$\sup_{\mathbf{P} \in \Omega_{\gamma}} f(\mathbf{P}, \mathbf{F}) < \mu, \tag{14}$$

via decentralized control.

Let $\mathbf{X}_1, \mathbf{Y}_1, \mathbf{X}_2, \mathbf{Y}_2$ be sulotions to the Bezout equations (6). In this section we shall denote

$$
\begin{aligned}
\mathbf{S}_1 &= \mathbf{M}_1 \mathbf{Y}_1 \\
\mathbf{S}_2 &= \mathbf{M}_2 \mathbf{Y}_2 \\
\mathbf{V}_1 &= -\mathbf{M}_1 \mathbf{D}_1 \mathbf{Y}_1 \mathbf{X}_2 \mathbf{W}_{12} = -\mathbf{S}_1 \mathbf{D}_1 \mathbf{X}_2 \mathbf{W}_{12} \\
\mathbf{V}_2 &= -\mathbf{M}_2 \mathbf{D}_2 \mathbf{Y}_2 \mathbf{X}_1 \mathbf{W}_{21} = -\mathbf{S}_2 \mathbf{D}_2 \mathbf{X}_1 \mathbf{W}_{21} \\
\mathbf{V} &= \mathbf{D}_1 \mathbf{D}_2 \mathbf{W}_{12} \mathbf{W}_{21} \mathbf{X}_1 \mathbf{X}_2
\end{aligned}
$$

Observe that

$$\mathbf{V}_1 \mathbf{V}_2 = \mathbf{S}_1 \mathbf{S}_2 \mathbf{V}. \tag{15}$$

Define

$$\mathbf{K}_\mu = \frac{1}{\sqrt{2}\mu} \sqrt{b_0 + \sqrt{b_0^2 - 4|\mathbf{S}_1 \mathbf{S}_2|^2 (1 - |\mathbf{V}|\epsilon_{12}\epsilon_{21})} + |\mathbf{V}|\epsilon_{12}\epsilon_{21}}$$

where

$$b_0 = |\mathbf{S}_1|^2 + |\mathbf{S}_2|^2 + \epsilon_{12}|\mathbf{V}_1|^2 + \epsilon_{21}|\mathbf{V}_2|^2.$$

Theorem 5 *The robust performance*

$$\sup_{\mathbf{P} \in \Omega_{\gamma}} f(\mathbf{P}, \mathbf{F}) < \mu$$

is achieved if and only if

$$\|\mathbf{K}_\mu\|_\infty < 1.$$

Proof: In Appendix.

In the special case of stable local systems, we have $\mathbf{D}_1 = \mathbf{D}_2 = 1$, and

$$
\begin{aligned}
\mathbf{X}_1 &= \mathbf{Q}_1, & \mathbf{Y}_1 &= 1 - \mathbf{N}_1 \mathbf{Q}_1, & \mathbf{Q}_1 &\in \mathbf{H}^\infty \\
\mathbf{X}_2 &= \mathbf{Q}_2, & \mathbf{Y}_2 &= 1 - \mathbf{N}_2 \mathbf{Q}_2, & \mathbf{Q}_2 &\in \mathbf{H}^\infty.
\end{aligned}
$$

Then,

$$
\begin{aligned}
\mathbf{S}_1 &= \mathbf{M}_1 (1 - \mathbf{N}_1 \mathbf{Q}_1) \\
\mathbf{S}_2 &= \mathbf{M}_2 (1 - \mathbf{N}_2 \mathbf{Q}_2) \\
\mathbf{V}_1 &= -\mathbf{M}_1 (1 - \mathbf{N}_1 \mathbf{Q}_1) \mathbf{Q}_2 \mathbf{W}_{12} \\
\mathbf{V}_2 &= -\mathbf{M}_2 (1 - \mathbf{N}_2 \mathbf{Q}_2) \mathbf{Q}_1 \mathbf{W}_{21} \\
\mathbf{V} &= \mathbf{W}_{12} \mathbf{W}_{21} \mathbf{Q}_1 \mathbf{Q}_2
\end{aligned}
$$

Consequently, the optimal robust performance is reduced to the following iterative procedure:

Step1: Choose the upper bound μ_u and lower bound μ_l:

$$\mu_u = \max\{\|M_1\|_\infty, \|M_2\|_\infty\}, \quad \mu_l = 0.$$

Step 2: Let $\mu = \frac{1}{2}(\mu_u + \mu_l)$. Compute

$$q(\mu) = \inf_{Q_1, Q_2 \in H^\infty} \|K_\mu\|_\infty. \tag{16}$$

Step 3: If $q(\mu) < 1$, then assign a new value to μ_l:

$$\mu \to \mu_l,$$

and go to Step 2.

If $q(\mu) > 1$, then assign a new value to μ_u:

$$\mu \to \mu_u,$$

and go to Step 2.

This binary search will result in

$$\mu_u \to \mu_l,$$

and the limit is the optimal robust performance achievable by decentralized controllers.

The closed-form solutions to the optimization problem (16) remain a challenging open question. Since this problem is more complicated than the standard H^∞ robust performance problem for which closed-form solutions are still not available, numerical solutions to (16) via, say, convex optimization, should be pursued in the future.

5 Concluding Remarks

In this paper, decentralized robust stabilization and performance of two-channel interconnected systems are studied, in the framework of structured uncertainty and $\mathbb{H}^\infty$ optimization. Necessary and sufficient conditions are derived for robust stability and robust performance. It is demonstrated constructively that when the two local plants do not have common unstable poles, robust stabilization can be achieved for any bounded stable uncertainties in interconnecting channels. This result reveals an important relationship between modeling of local systems and achievable robustness against interconnecting uncertainties.

While the problems considered in this paper are for two-channel interconnected systems, they are of importance in two aspects: (a) many practical decentralized control problems, such as those occuring in automotive control problems, are represented in this structure; (b) a deep understanding of decentralized control can be potentially gained from a comprefensive study of these problems.

Extensions of the results to more complicated MIMO systems are being investigated.

6 Appendix: Proofs of Theorems

Proof of Theorem 2: Since

$$\mu \geq \mu' := \inf_{Q_1 \in H^\infty} \inf_{Q_2 \in H^\infty} \|W_{12}W_{21}(X_1 - D_1Q_1)(X_2 - D_2Q_2)\|_\infty$$

we will show that

$$\mu' \geq \max\{\beta_1 \epsilon_{11}^{\alpha_1}, \beta_2 \epsilon_{22}^{\alpha_2}\}. \tag{17}$$

Denote

$$
\begin{aligned}
L_1 &= W_{11}X_1, & K_1 &= L_1 - D_1Q_1 \\
L_2 &= W_{22}X_2, & K_2 &= L_2 - D_2Q_2 \\
L &= \frac{W_{12}W_{21}}{W_{11}W_{22}}
\end{aligned}
$$

Note that $L, L^{-1} \in H^\infty$. Let $m := \|L^{-1}\|_\infty$.

To achieve local robust stability, K_i $(i = 1, 2,)$ must satisfy

$$K_i \in H^\infty, \qquad \mu_i = \|K_i\|_\infty < \frac{1}{\epsilon_{ii}} \qquad i = 1, 2.$$

Also, by interpolation constraints,

$$K_i(p_i) = L_i(p_i) = \frac{W_{ii}(p_i)}{N_i(p_i)} := a_i \neq 0, \quad i = 1, 2.$$

Therefore, by the maximum modulus principle

$$\mu_i \geq |a_i|.$$

Given any choice of $Q_1, Q_2 \in H^\infty$, denote

$$K = LK_1K_2 \in H^\infty, \quad \text{and} \quad \rho = \|K\|_\infty.$$

We only need to show that

$$\rho \geq \max\{\beta_1 \epsilon_{11}^{\alpha_1}, \beta_2 \epsilon_{22}^{\alpha_2}\}.$$

Let

$$\Omega_i = \{\omega : \frac{|a_i|}{2} \leq |K_i(j\omega)| \leq \mu_i\}$$

and $\Omega_i' = \mathbb{R}/\Omega_i$. Define the following constants

$$
\begin{aligned}
c_1 &= \frac{1}{\pi} \int_{\Omega_1} \frac{\sigma_1}{\sigma_1^2 + (\omega - \omega_1)^2} d\omega \\
d_1 &= \frac{1}{\pi} \int_{\Omega_1'} \frac{\sigma_1}{\sigma_1^2 + (\omega - \omega_1)^2} d\omega \\
c_2 &= \frac{1}{\pi} \int_{\Omega_2} \frac{\sigma_2}{\sigma_2^2 + (\omega - \omega_2)^2} d\omega \\
d_2 &= \frac{1}{\pi} \int_{\Omega_2'} \frac{\sigma_2}{\sigma_2^2 + (\omega - \omega_2)^2} d\omega
\end{aligned}
$$

Since

$$\frac{1}{\pi}\int_{\mathbb{R}}\frac{\sigma_i}{\sigma_i^2+(\omega-\omega_i)^2}d\omega=1,\qquad i=1,2,$$

all the constants are bounded by 1 and $d_i=1-c_i$. We will show that c_i has a uniform lower bound. Indeed, suppose $K_i=B_iO_i$ is an inner-outer factorization of K_i in $\mathbb{H}^\infty$. Then,

$$O_i(p_i)=\frac{K_i(p_i)}{B_i(p_i)}=\frac{a_i}{B_i(p_i)}.\tag{18}$$

Observe that $|B_i(p_i)|\le 1$.

Now, by Poisson Integration Formula,

$$\begin{aligned}\log|O_i(p_i)| &= \frac{1}{\pi}\int_{-\infty}^{\infty}\log|K_i(j\omega)|\frac{\sigma_i}{\sigma_i^2+(\omega-\omega_i)^2}d\omega\\ &= \frac{1}{\pi}\int_{\Omega_i}\log|K_i(j\omega)|\frac{\sigma_i}{\sigma_i^2+(\omega-\omega_i)^2}d\omega\\ &\quad +\frac{1}{\pi}\int_{\Omega_i'}\log|K_i(j\omega)|\frac{\sigma_i}{\sigma_i^2+(\omega-\omega_i)^2}d\omega\\ &\le c_i\log\frac{1}{\varepsilon_{ii}}+d_i\log\frac{|a_i|}{2}.\end{aligned}$$

On the other hand, by (18)

$$\log|O_i(p_i)|=\log|a_i|-\log|B_i(p_i)|\ge\log|a_i|$$

which implies that

$$c_i\log\frac{1}{\varepsilon_{ii}}+d_i\log\frac{|a_i|}{2}\ge\log|a_i|.$$

Namely, noting $d_i=1-c_i$,

$$c_i\log\frac{2}{|a_i|\varepsilon_{ii}}\ge\log|a_i|-\log\frac{|a_i|}{2}=\log 2.$$

As a result,

$$c_i\ge\delta_i:=\frac{\log 2}{\log\frac{2}{|a_i|\varepsilon_{ii}}}=\frac{1}{1-\log_2|a_i\varepsilon_{ii}|}.$$

To achieve $\|K\|_\infty\le\rho$, we must have

$$|K_2(j\omega)|\le\frac{\rho}{|L(j\omega)||K_1(j\omega)|}\le\frac{2m\rho}{|a_1|},\qquad\forall\omega\in\Omega_1.\tag{19}$$

On the other hand, by Poisson Integration Formula,

$$\log|O_2(p_2)|=\frac{1}{\pi}\int_{-\infty}^{\infty}\log|K_2(j\omega)|\frac{\sigma_2}{\sigma_2^2+(\omega-\omega_2)^2}d\omega.$$

Therefore,

$$
\begin{aligned}
\log|a_2| - \log|B_2(p_2)| &= \frac{1}{\pi}\int_{\Omega_1}\log|K_2(j\omega)|\frac{\sigma_2}{\sigma_2^2+(\omega-\omega_2)^2}d\omega \\
&\quad + \frac{1}{\pi}\int_{\Omega_2}\log|K_2(j\omega)|\frac{\sigma_2}{\sigma_2^2+(\omega-\omega_2)^2}d\omega \\
&\le c_2\log\frac{2m\rho}{|a_1|}+(1-c_2)\log\frac{1}{\varepsilon_{22}}.
\end{aligned}
$$

It follows that

$$
c_2\log\frac{2m\rho}{|a_1|}+(1-c_2)\log\frac{1}{\varepsilon_{22}}\ge\log|a_2|,
$$

or

$$
\begin{aligned}
\rho &\ge \frac{\varepsilon_{22}^{\frac{1-c_2}{c_2}}|a_2|^{1/c_2}|a_1|}{2m} \\
&\ge \frac{\varepsilon_{22}^{\frac{1-\delta_2}{\delta_2}}\min\{|a_2|^{1/\delta_2},|a_2|\}|a_1|}{2m} \qquad \text{since } c_2\ge\delta_2, \varepsilon_{22}\le1 \\
&= \beta_2\varepsilon_{22}^{\alpha_2}
\end{aligned}
$$

with

$$
\begin{aligned}
\beta_2 &= \frac{\min\{|a_2|^{1/\delta_2},|a_2|\}|a_1|}{2m}, \\
\alpha_2 &= \frac{1-\delta_2}{\delta_2} = -\log_2|a_2\varepsilon_{22}|.
\end{aligned}
$$

Similarly, we can get

$$
\rho\ge\beta_1\varepsilon_{11}^{\alpha_1}
$$

with

$$
\begin{aligned}
\beta_1 &= \frac{\min\{|a_1|^{1/\delta_1},|a_1|\}|a_2|}{2m}, \\
\alpha_1 &= \frac{1-\delta_1}{\delta_1} = -\log_2|a_1\varepsilon_{11}|.
\end{aligned}
$$

Therefore,

$$
\rho\ge\max\{\beta_1\varepsilon_{11}^{\alpha_1},\beta_2\varepsilon_{22}^{\alpha_2}\},
$$

and (11) follows. $\qquad\square$

Proof of Theorem 3: We will prove by construction. Suppose

$$
\begin{aligned}
\tilde{a}_i &= W_{12}(z_i^1)W_{21}(z_i^1)X_1(z_i^1), \qquad i=1,\cdots,n \\
\tilde{b}_j &= X_2(z_j^2), \qquad j=1,\cdots,m.
\end{aligned}
$$

Note that $\tilde{a}_i \neq 0$, $\tilde{b}_j \neq 0$ since $W_{12}^{-1}, W_{21}^{-1} \in \mathbb{H}^\infty$ by assumption, and $X_1(z_i^1) = \frac{1}{N_1(z_i^1)} \neq 0$, $X_2(z_i^2) = \frac{1}{N_2(z_i^2)} \neq 0$.

For any $\varepsilon > 0$, define

$$a_i = \begin{cases} \varepsilon/\tilde{a}_i, & \text{if } z_i^1 \text{ is complex (not real);} \\[2mm] \varepsilon/|\tilde{a}_i|, & \text{if } z_i^1 \text{ is real.} \end{cases} \qquad i = 1, \cdots, n$$

$$b_j = \begin{cases} \tilde{b}_j/\varepsilon, & \text{if } z_j^2 \text{ is complex (not real);} \\[2mm] |\tilde{b}_j|/\varepsilon, & \text{if } z_j^2 \text{ is real.} \end{cases} \qquad j = 1, \cdots, m$$

Since A_1 and A_2 are disjoint, a_i and b_j are well-defined.

We would like to solve the following outer interpolation problem:

Find $K, K^{-1} \in \mathbb{H}^\infty$ *such that*

$$\begin{aligned} K(z_i^1) &= a_i, & z_i^1 &\in A_1, & i &= 1, \cdots, n \\ K(z_j^2) &= b_j, & z_j^2 &\in A_2, & j &= 1, \cdots, m \end{aligned}$$

By the definition of a_i and b_j, for all real z_i^1, z_j^2, the corresponding a_i and b_j are real and positive. As a result, the solution to the outer interpolation exists and can be constructed via, e.g., the algorithms in [8].

Now,

$$\begin{aligned} \mu &= \inf_{Q_1 \in \mathbb{H}^\infty} \inf_{Q_2 \in \mathbb{H}^\infty} \|W_{12} W_{21} K K^{-1} (X_1 - D_1 Q_1)(X_2 - D_2 Q_2)\|_\infty \\ &\leq \inf_{Q_1 \in \mathbb{H}^\infty} \|W_{12} W_{21} K (X_1 - D_1 Q_1)\|_\infty \inf_{Q_2 \in \mathbb{H}^\infty} \|K^{-1}(X_2 - D_2 Q_2)\|_\infty \\ &= \mu_1 \mu_2. \end{aligned}$$

By optimal interpolation theory, μ_1 and μ_2 are solutions to the optimal interpolation problems:

$$\mu_1 = \inf\{\|K_1\|_\infty : K_1 \in \mathbb{H}^\infty, K_1(z_i^1) = W_{12}(z_i^1) W_{21}(z_i^1) K(z_i^1) X(z_i^1), z_i^1 \in A_1\}$$

It is easy to verify that

$$W_{12}(z_i^1) W_{21}(z_i^1) K(z_i^1) X(z_i^1) = c_i^1 \varepsilon,$$

where

$$c_i^1 = \begin{cases} 1, & \text{if } z_i^1 \text{ is complex (not real);} \\[2mm] sgn(\tilde{a}_i), & \text{if } z_i^1 \text{ is real.} \end{cases}$$

It follows that, if

$$\rho_1 = \inf\{\|K_1'\|_\infty : \quad K_1' \in \mathbb{H}^\infty, K_1'(z_i^1) = c_i^1, \quad i = 1, \cdots, n\}$$

then

$$\mu_1 = \rho_1 \varepsilon.$$

Similarly,

$$\mu_2 = \rho_2 \varepsilon,$$

where

$$\rho_2 = \inf\{\|K_2'\|_\infty : \quad K_2' \in \mathbb{H}^\infty, K_2'(z_j^2) = c_j^2, \quad j = 1, \cdots, m\},$$

and

$$c_i^2 = \begin{cases} 1, & \text{if } z_j^2 \text{ is complex (not real)}; \\ sgn(\tilde{a}_j), & \text{if } z_j^2 \text{ is real}. \end{cases}$$

As a result,

$$\mu \le \mu_1 \mu_2 = \rho_1 \rho_2 \varepsilon^2.$$

Since $\varepsilon > 0$ is arbitrary, we conclude that

$$\mu = 0.$$

$\square$

Proof of Theorem 4: Suppose $A_i, i = 1, 2$ is the set of all RHP zeros of D_i and A_i^r its real subset. Since (X_1, D_1) is outer interpolable, all $X_1(a), a \in A_1^r$ have the same sign. Without loss of generality, assume $X_1(a) > 0, a \in A_1^r$. Let $\varepsilon > 0$. We formulate the following outer interpolation problem, depending on ε:

Find an outer function $K \in \mathbb{H}^\infty$ *such that*

$$K(z_i) = b_i, \quad z_i \in A_1 \cup A_2$$

where

$$b_i = \begin{cases} X_1(z_i), & z_i \in A_1 \\ \varepsilon, & z_i \in A_2/A_1. \end{cases}$$

The problem is solvable since b_i are all positive for $z_i \in A_1^r \cup A_2^r$.

Since $K(z_i^1) = X_1(z_i^1), z_i^1 \in A_1$, K is an interpolant of $X_1 - D_1 Q_1$. As a result, there exists $Q_1 \in \mathbb{H}^\infty$ such that

$$K = X_1 - D_1 Q_1.$$

It follows that

$$\begin{aligned} \mu &\le \inf_{Q_2 \in \mathbb{H}^\infty} \|W_{12} W_{21} K(X_2 - D_2 Q_2)\|_\infty \\ &= \inf_{Q_2 \in \mathbb{H}^\infty} \|W_{12} W_{21} K X_2 - D_2 Q_2)\|_\infty \\ &\quad \text{since } W_{12} W_{21} K \text{ is outer} \\ &:= \mu_\varepsilon \end{aligned}$$

Let $\mathbf{L} = \mathbf{W}_{12}\mathbf{W}_{21}\mathbf{K}\mathbf{X}_2$. Observe that for $a \in A_1 \cap A_2$

$$
\begin{aligned}
\mathbf{L}(a) &= \mathbf{W}_{12}(a)\mathbf{W}_{21}(a)\mathbf{K}(a)\mathbf{X}_2(a) \\
&= \mathbf{W}_{12}(a)\mathbf{W}_{21}(a)\mathbf{X}_1(a)\mathbf{X}_2(a) \\
&= \frac{\mathbf{W}_{12}(a)\mathbf{W}_{21}(a)}{\mathbf{N}_1(a)\mathbf{N}_2(a)}
\end{aligned}
$$

and for $a \in A_2/A_1$

$$
\mathbf{L}(a) = \varepsilon\mathbf{W}_{12}(a)\mathbf{W}_{21}(a)\mathbf{X}_2(a).
$$

As a result,

$$
\mu_\varepsilon = \inf\{\|\mathbf{M}\|_\infty : \mathbf{M}(a) = \mathbf{L}(a), a \in A_2\}. \tag{20}
$$

Since this is valid for all $\varepsilon > 0$, we obtain

$$
\mu \le \lim_{\varepsilon \to 0} \mu_\varepsilon = \mu_0, \tag{21}
$$

the last equality being valid from the continuity of Nevanlinna-Pick interpolation. This completes the proof. □

Proof of Theorem 5: For $\mathbf{P} \in \Omega_P$, it is easy to derive that

$$
f(\mathbf{P}, \mathbf{F}) = \left\| \begin{pmatrix} \mathbf{M}_1\mathbf{Y}_1 & 0 \\ 0 & \mathbf{M}_2\mathbf{Y}_2 \end{pmatrix} \right.
$$

$$
\left[1 + \begin{pmatrix} 0 & \mathbf{D}_1\mathbf{X}_2\mathbf{W}_{12}\Delta_{12} \\ \mathbf{D}_2\mathbf{X}_1\mathbf{W}_{21}\Delta_{21} & 0 \end{pmatrix} \right]^{-1} \left. \begin{pmatrix} \mathbf{D}_1 & 0 \\ 0 & \mathbf{D}_2 \end{pmatrix} \right\|_\infty
$$

$$
= \left\| \begin{pmatrix} \mathbf{M}_1\mathbf{Y}_1 & 0 \\ 0 & \mathbf{M}_2\mathbf{Y}_2 \end{pmatrix} \begin{pmatrix} 1 & \mathbf{D}_2\mathbf{X}_2\mathbf{W}_{12}\Delta_{12} \\ \mathbf{D}_1\mathbf{X}_1\mathbf{W}_{21}\Delta_{21} & 1 \end{pmatrix}^{-1} \right\|_\infty \tag{22}
$$

since $\mathbf{D}_1, \mathbf{D}_2$ are inner. By Theorem 1, the inverse in (22) exists in $\mathbb{H}^\infty$ if and only if

$$
\|\mathbf{V}\|_\infty < \frac{1}{\varepsilon_{12}\varepsilon_{21}}. \tag{23}
$$

To simplify notation, we will omit ω in all expressions, such as $\mathbf{D}_1(\omega)$ ect.. in the following derivations.

Under the condition (23), denote

$$
\begin{aligned}
g(\omega) &:= \sigma_m(\mathbf{M}(1 + \mathbf{P}\mathbf{F})^{-1}(\omega)) \\
&= \frac{\sigma_m \begin{pmatrix} \mathbf{M}_1\mathbf{Y}_1 & -\mathbf{M}_1\mathbf{D}_1\mathbf{Y}_1\mathbf{X}_2\mathbf{W}_{12}\Delta_{12} \\ -\mathbf{M}_2\mathbf{D}_2\mathbf{Y}_2\mathbf{X}_1\mathbf{W}_{21}\Delta_{21} & \mathbf{M}_2\mathbf{Y}_2 \end{pmatrix}}{|1 - \mathbf{D}_1\mathbf{D}_2\mathbf{W}_{12}\mathbf{W}_{21}\mathbf{X}_1\mathbf{X}_2\Delta_{12}\Delta_{21}|} \\
&= \frac{\sigma_m \begin{pmatrix} \mathbf{S}_1 & \mathbf{V}_1\Delta_{12} \\ \mathbf{V}_2\Delta_{21} & \mathbf{S}_2 \end{pmatrix}}{|1 - \mathbf{V}\Delta_{12}\Delta_{21}|}
\end{aligned}
$$

Then,

$$f(\mathbf{P}, \mathbf{F}) = \sup_{\omega} g(\omega).$$

Since $\sigma_m(K) = \lambda_m^{1/2}(K^*K)$, where λ is the largest eigenvalue, we have

$$g(\omega) = \frac{\lambda_m^{1/2}\left[\begin{pmatrix} S_1 & V_1\Delta_{12} \\ V_2\Delta_{21} & S_2 \end{pmatrix}^*\begin{pmatrix} S_1 & V_1\Delta_{12} \\ V_2\Delta_{21} & S_2 \end{pmatrix}\right]}{|1 - V\Delta_{12}\Delta_{21}|}.$$

Here,

$$\begin{pmatrix} S_1 & V_1\Delta_{12} \\ V_2\Delta_{21} & S_2 \end{pmatrix}^*\begin{pmatrix} S_1 & V_1\Delta_{12} \\ V_2\Delta_{21} & S_2 \end{pmatrix} = \begin{pmatrix} a_{11} & a_{12} \\ a_{21} & a_{22} \end{pmatrix}$$

where

$$\begin{aligned}
a_{11} &= |S_1|^2 + |V_2|^2|\Delta_{21}|^2 \\
a_{12} &= S_1^*V_1\Delta_{12} + V_2^*\Delta_{21}^*S_2 \\
a_{21} &= V_1^*\Delta_{12}^*S_1 + S_2^*V_2\Delta_{21} \\
a_{22} &= |V_1|^2|\Delta_{12}|^2 + |S_2|^2
\end{aligned}$$

The corresponding characteristic polynomial is

$$\lambda^2 - b\lambda + c$$

where

$$\begin{aligned}
b &= a_{11} + a_{22} \\
&= |S_1|^2 + |V_2|^2|\Delta_{21}|^2 + |V_1|^2|\Delta_{12}|^2 + |S_2|^2 \\
c &= a_{11}a_{22} - a_{12}a_{21} \\
&= (|S_1|^2 + |V_2|^2|\Delta_{21}|^2)(|V_1|^2|\Delta_{12}|^2 + |S_2|^2) \\
&\quad -(S_1^*V_1\Delta_{12} + V_2^*\Delta_{21}^*S_2)(V_1^*\Delta_{12}^*S_1 + S_2^*V_2\Delta_{21}) \\
&= (S_1S_2 - V_1V_2\Delta_{12}\Delta_{21})^*(S_1S_2 - V_1V_2\Delta_{12}\Delta_{21}) \\
&= (S_1S_2 - S_1S_2V\Delta_{12}\Delta_{21})^*(S_1S_2 - S_1S_2V\Delta_{12}\Delta_{21}), \quad \text{by (15)} \\
&= |S_1S_2|^2|1 - V\Delta_{12}\Delta_{21}|^2.
\end{aligned}$$

Apparently, $b \geq 0$, $c \geq 0$.

Now, λ_m can be computed by solving

$$\lambda^2 - b\lambda + c = 0$$

which implies

$$\lambda_m = \frac{b + \sqrt{b^2 - 4c}}{2}, \quad \text{and} \quad \lambda_m^{1/2} = \sqrt{\frac{b + \sqrt{b^2 - 4c}}{2}}.$$

As a result,

$$g(\omega) = \frac{\sqrt{b + \sqrt{b^2 - 4c}}}{\sqrt{2}|1 - V\Delta_{12}\Delta_{21}|}$$

$$= \frac{\sqrt{b + \sqrt{b^2 - 4|S_1S_2|^2|1 - V\Delta_{12}\Delta_{21}|^2}}}{\sqrt{2}|1 - V\Delta_{12}\Delta_{21}|}$$

Since for any $\|\Delta_{12}\|_\infty \leq \epsilon_{12}, \|\Delta_{21}\|_\infty \leq \epsilon_{21}$,

$$b \leq |S_1|^2 + |V_2|^2\epsilon_{21}^2 + |V_1|^2\epsilon_{12}^2 + |S_2|^2 = b_0,$$

$$|1 - V\Delta_{12}\Delta_{21}| \geq 1 - |V|\epsilon_{12}\epsilon_{21},$$

we obtain

$$g(\omega) \leq \frac{\sqrt{b_0 + \sqrt{b_0^2 - 4|S_1S_2|^2(1 - |V|\epsilon_{12}\epsilon_{21})^2}}}{\sqrt{2}(1 - |V|\epsilon_{12}\epsilon_{21})}.$$

On the other hand, if $V(\omega) = |V(\omega)|e^{j\theta}$, then for $\Delta_{12}(\omega) = \epsilon_{12}, \Delta_{21}(\omega) = \epsilon_{21}e^{-j\theta}$, $g(\omega)$ is

$$g(\omega) = \frac{\sqrt{b_0 + \sqrt{b_0^2 - 4|S_1S_2|^2(1 - |V|\epsilon_{12}\epsilon_{21})^2}}}{\sqrt{2}(1 - |V|\epsilon_{12}\epsilon_{21})}.$$

Therefore, the worst case $g(\omega)$ satisfies

$$g_\Omega(\omega) = \sup_{\Delta_{12},\Delta_{21}} g(\omega)$$

$$= \frac{\sqrt{b_0 + \sqrt{b_0^2 - 4|S_1S_2|^2(1 - |V|\epsilon_{12}\epsilon_{21})^2}}}{\sqrt{2}(1 - |V|\epsilon_{12}\epsilon_{21})} \tag{24}$$

It follows that the robust performance (14) is achieved if and only if

$$g_\Omega(\omega) < \mu, \qquad \forall \omega$$

or equivalently

$$\sqrt{b_0 + \sqrt{b_0^2 - 4|S_1S_2|^2(1 - |V|\epsilon_{12}\epsilon_{21})^2}} < \sqrt{2}\mu(1 - |V|\epsilon_{12}\epsilon_{21}), \qquad \forall \omega$$

That is,

$$\frac{1}{\sqrt{2}\mu}\sqrt{b_0 + \sqrt{b_0^2 - 4|S_1S_2|^2(1 - |V|\epsilon_{12}\epsilon_{21})^2}} + |V|\epsilon_{12}\epsilon_{21} < 1.$$

The theorem follows after taking $\sup_\omega$. $\qquad\qquad\square$

References

[1] Y.H. Chen, Decentralized robust control for large-scale uncertain systems: A design based on the bound of uncertainty, *J. Dynamic Systems, Measurementt and Control*, Vol. 114, pp. 1-9, 1992.

[2] J.C. Doyle, B.A. Francis and A. Tannenbaum, Feedback control theory, Macmillan, 1992.

[3] N. Gundes and C.A. Desoer, Algebraic theory of linear feedback systems with full and decentralized compensators, Springer-Verlag, Berlin, 1990.

[4] L.Trave, A. Titli and A. Tarras, Large scale systems: decentralization, structure constraints and fixed modes, *Lecture Notes in Control and Information Sciences*, Springe-Verlag, Berlin, 1989.

[5] U. Ozguner and E.J. Davison, Sampling and decentralized fixed modes, *Proc. 1985 ACC Conference*, pp. 257-262, 1985.

[6] S.H. Wang and E.J. Davison, On the stabilization of decentralized control systems, *IEEE Trans. Automat. Control*, Vol. AC-18, pp. 473-478, 1973.

[7] L. Xie, Y. Wang, C.E. De Souza, Decentralized output feedback control of discrete-time interconnected uncertain systems, *Proc. IEEE 32nd CDC Conference*, pp. 3762-3767, 1993.

[8] D.C. Youla, J.J. Bongiorno, Jr., and C.N. Lu, Single-loop feedback stabilization of linear multivariable dynamic plants, *Automatica*, Vol. 10, pp. 159-173, 1974.

[9] G. Zames and D. Bensoussan, Multivariable feedback, sensitivity, and decentralized control, *IEEE Trans. Automat. Control*, Vol. AC-28, pp. 1030-1035, 1983.

Control as Interconnection

Jan C. Willems
Mathematics Institute
University of Groningen
P.O. Box 800
9700 AV Groningen
The Netherlands
email: J.C.Willems@math.rug.nl

Abstract

This paper puts forward the idea of system interconnection as the central idea of control. It contrasts this with intelligent control, which refers to the usual measurement-to-control-action feedback type of control. As illustrations, stabilization and linear-quadratic control are treated from this vantage point.

1 Introduction.

The purpose of this essay is to question the universal appropriateness of the usual *signal flow graph*, input/output structure which is invariably taken as the starting point in control. We will argue that it is much more reasonable and pragmatic to view instead *interconnection* as the basic idea in control.

This paper is written in honor of George Zames at the occasion of his sixtieth birthday. I first met George in the mid-sixties when I was a beginning graduate student at MIT. My doctoral dissertation, which appeared in 1968, later expanded into the monograph [1], dealt with input/output stability. It was greatly influenced by Zames' seminal papers [2] and built on the ideas of $\mathcal{L}_2$-stability, the small loop gain theorem, and the positive operator theorem which he had laid out in [2]. One of the things which George's work taught me was to appreciate the importance of clear and elegant problem formulations (as $\mathcal{L}_2$-stability) and of general principles (as the small loop gain theorem). The present paper is written in this spirit.

The usual approach to thinking about control design takes the feedback loop shown in figure 1 as its starting point. The control inputs drive the actuators, while the sensors produce the measured outputs. In this view, the aim of control theory is to design a feedback processor, i.e., a device that processes the observed outputs in order to compute the control inputs. Very powerful control design principles have emerged from this paradigm, combining identification and parameter estimation, observers and state reconstructors, motion planning enhanced with set point servo control and gain scheduling, etc.

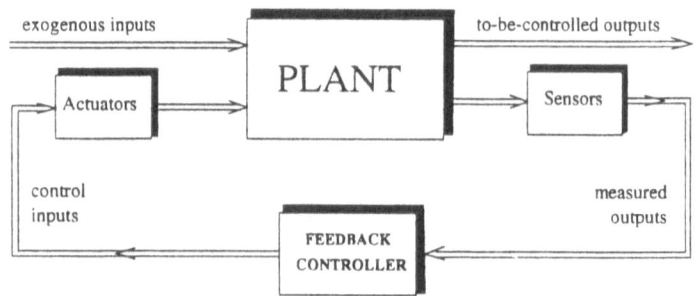

Figure 1: Intelligent control

We will refer to the control scheme of figure 1 as an *intelligent controller*. Before elaborating, we would like to make a short comment on the use of the word *intelligent*. We use it because the controller of figure 1 involves *observation* (through the sensors), *analysis* (for example in the form of state and parameter estimation), *decision making* for example, in the form of an (adaptive) control algorithm), and *action* (through the actuators). Of course, this scheme could involve a very low level of intelligence (for example in a room thermostat). Without wanting to engage in the AI debate, we feel that for the purposes of this paper the use of the term *intelligent* is an appropriate one for clarifying the issues which will be discussed. In fact we use *intelligent* control in contrast to *passive* control.

2 Passive control.

Many, if not most, control devices used in engineering practice do not function as intelligent controllers. In many such devices it is unnatural to view the controller action as feedback, and to think of it in terms of a signal flow graph.

Examples of such situations abound: mechanical dampers to attenuate vibrations, commercial devices for temperature, flow and pressure control, passive elements introduced in electrical signal processing devices in order to improve frequency transfer characteristics, etc. When one analyzes such devices it is often impossible to regard one variable as being measured and used in order to decide what value another variable should take on. When a resistor is added in a circuit in order to improve the response characteristics, it makes no sense to view the current as input and the voltage as output, or vice-versa. To think of a damper, for example a damper in a car, as a device that *measures* the position (or velocity) of the body and the chassis, and *decides* from there what force to exert on the body of the car, is an anthropomorphic caricature (and an unnecessary one at that). It makes just a much (or better just as little) sense to turn the situation around and to view a damper as a device that measures the force exerted on the piston and decides how fast to make the piston move. In other examples, say a simple expansion value with a spring for pressure control, it would perhaps be possible to *think* of the pressure as the measured input and the extension of the spring as the control output. However, it is obvious that this signal flow is merely in our minds (and could hence be useful for enhancing our understanding of the situation), not in nature.

Our point of view will be the following. When a controller is attached to a plant there are some variables which interface between the controller and the plant. Before the controller is attached, these variables have to obey only the laws imposed by the plant. After the controller is put in place, they have to obey *both* the laws of the plant and of the controller. Through this intervention, the dynamic behavior of the plant variables is adjusted. This adjustment effects not only the interfacing variables, but, through these, also the other plant variables. This is important when the to-be-controlled variables are not available for interconnection. By properly choosing the controller one can achieve in this way desired dynamic characteristics for the interconnected system.

The plant and the controller may or may not have an input/output structure when viewed from the interfacing variables. When this input/output structure is present, then the intelligent control paradigm is suitable. When this input/output structure is not present, then it is not. The aim of this paper is to put forward a framework for control design which does not take the signal flow input/output structure as its starting point, but treats it as a special case.

In order to avoid misunderstandings, we would again like to emphasize that there are many situations where the intelligent control paradigm is eminently suitable. It is a deep and attractive paradigm which is basically a must whenever logic devices are involved in the controller. Thus it will undoubtedly gain in importance as logic devices become cheaper and more reliable and more intelligence can be incorporated in controllers. Nevertheless, it remains puzzling why control theory textbooks, those with a practical as well as those with a theoretical outlook, have, ever since the subject was formalized, chosen to work in an input/output framework. By regarding control this way, our subject can be viewed as a part of signal processing, instead of making it a part of integrated system design, of properly designing subsystems.

3 Control as interconnection.

The behavioral framework [3,4,5] provides a suitable setting for our purposes. One of the important features of this approach is that it does not take the usual input/output structure as its starting point. Instead, it lays out a framework for discussing dynamical systems in which all external system variables are treated on equal footing. However, it recognizes *ab initio* the importance in modelling of internal variables, auxiliary variables which unavoidably need to be introduced in the modelling process and which are different from the variables whose behavior the model aims at. These internal variables are called *latent variables*, while the variables which are being modelled are called *manifest* variables.

In control applications, it is natural to distinguish between those variables which are available for control and those which are not. Using again the above language we will call the variables available for control, *manifest* variables, and the remaining plant variables, *latent* variables. We can think of the manifest variables as *control variables*, and of the latent variables as *to-be-controlled* variables. This leads to the setting for control shown in figure 2.

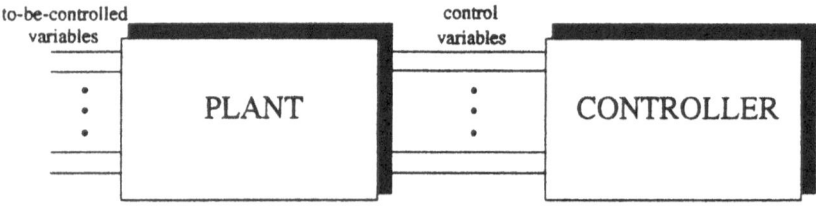

Figure 2: Control as interconnection

However, for the sake of exposition we will first assume that *all* the plant variables are available to the controller. This leads to the situation shown in figure 3. This will now be formalized.

Let $\Sigma_p = (T, W, \mathfrak{B}_p)$ be the *plant*, viewed as a dynamical system (in the sense explained in [5]). Thus $T \subseteq \mathbf{R}$ denotes the time-axis. For the purposes of this

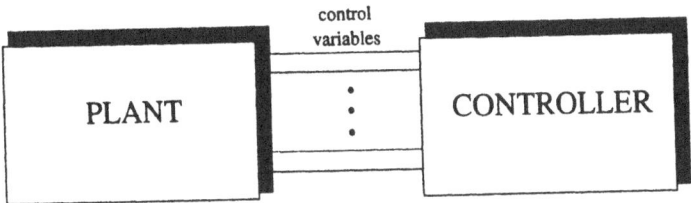

Figure 3: Control as interconnection

paper, in which we will consider for simplicity continuous-time systems, think $T = \mathbf{R}$. W denotes the space of plant variables. Think $W = \mathbf{R}^q$. $\mathfrak{B}_p$ is a subset of W^T, i.e., a subset of all maps from T to W. The (model of the) plant imposes certain laws on the plant variables. This results in the fact that some time function $w : T \to W$ (namely those in $\mathfrak{B}_p$) are declared possible by the plant model, while the others (those not in $\mathfrak{B}_p$) are declared impossible. $\mathfrak{B}_p$ is called the *behavior* of Σ_p.

The *controller* $\Sigma_c = (T, W, \mathfrak{B}_c)$ is, as Σ_p, a dynamical system. It imposes new laws on the variables: the functions $w : T \to W$ are now required to lie in $\mathfrak{B}_c$.

The *controlled* system Σ is the interconnection of Σ_p and Σ_c. This is denoted as $\Sigma = \Sigma_p \wedge \Sigma_c$. The *interconnection* is defined as

$$\Sigma = \Sigma_p \wedge \Sigma_c := (T, W, \mathfrak{B}_p \cap \mathfrak{B}_c) \tag{1}$$

In other words, in the interconnected system, the realizable trajectories $w : T \to W$ will have to obey both the laws imposed by the plant and by the controller. The control design problem is thus:
Given the plant Σ_p and given a family of admissible controllers $\mathfrak{C}$, find a $\Sigma_c \in \mathfrak{C}$ such that $\Sigma_p \wedge \Sigma_c$ has desirable properties.

4 Stabilization.

We will now explain, as an example, how the stabilization question for linear time-invariant systems can be formulated in this setting. Let $\mathcal{L}^q$ denote the family of linear differential systems with q variables. Thus each element of $\mathcal{L}^q$ is described through a polynomial matrix $R \in \mathbf{R}^{\bullet \times q}[\xi]$ by the differential equation

$$R(\frac{d}{dt})w = 0 \tag{2}$$

(2) yields the dynamical system $\Sigma = (\mathbf{R}, \mathbf{R}^q, \mathfrak{B}) \in \mathcal{L}^q$, where $\mathfrak{B}$ consists of all $w : \mathbf{R} \to \mathbf{R}^q$ satisfying (2). The precise meaning of what it signifies that w satisfies a system of differential equations as (2), is not important for the purposes of this paper.

Call $\Sigma = (\mathbf{R}, \mathbf{R}^q, \mathfrak{B}) \in \mathcal{L}^q$ *controllable* [5] if for all $w_1, w_2 \in \mathfrak{B}$ there exists $t' > 0$ and $w \in \mathfrak{B}$ such that $w(t) = w_1(t)$ for $t < 0$ and $w(t + t') = w_2(t)$ for $t \geq 0$. Call it *stabilizable* if for all $w \in \mathfrak{B}$ there exists $w' \in \mathfrak{B}$ such that $w(t) = w'(t)$ for $t < 0$ and such that $\lim_{t \to \infty} w'(t) = 0$. It can be shown that (2) is controllable [5] if and only if $rank(R(\lambda)) = rank(R)$ for all $\lambda \in \mathbf{C}$ and stabilizable if and only if this holds for all $\lambda \in \mathbf{C}^+ := \{\lambda \in \mathbf{C} | Re(\lambda) \geq 0\}$.

Next, call Σ *autonomous* if $w_1, w_2 \in \mathfrak{B}$ and $w_1(t) = w_2(t)$ for $t < 0$ imply $w_1 = w_2$. It can be shown that (2) is autonomous if and only if $rank(R) = q$. Finally, define Σ to be *stable* if $w \in \mathfrak{B}$ implies $\lim_{t \to \infty} w(t) = 0$. It is easy to see that stability requires autonomy. In describing systems $\Sigma \in \mathcal{L}^q$ by differential equations (2) (many R's will generate the same $\mathfrak{B}$!) we can always assume that R has q rows. If the number of rows in the original R is less than q, then this is trivial: simply add zero rows to R. If it is more than q then this fact requires an (easy) proof. Let $\Sigma \in \mathcal{L}^q$ and let it be described by (2), with $R \in \mathbf{R}^{q \times q}[\xi]$. Now define χ_Σ, the *characteristic polynomial* of Σ, by $\chi_\Sigma := det(R)$. Thus Σ is autonomous if and only if $\chi_\Sigma \neq 0$, and stable if and only if χ_Σ is Hurwitz.

Now consider a plant $\Sigma_p \in \mathcal{L}^q$. Assume that it is described by

$$P(\frac{d}{dt})w = 0 \tag{3}$$

Take as controllers elements of $\mathcal{L}^q$. A typical element is thus described by

$$C(\frac{d}{dt})w = 0 \tag{4}$$

Now restrict the class of admissible controllers to those for which

$$rank(\begin{bmatrix} P \\ C \end{bmatrix}) = rank(P) + rank(C) = q \tag{5}$$

Thus in this case $\mathfrak{C} = \{\Sigma_c \in \mathcal{L}^q | \Sigma_c$ can be described by (4), and (5) will be satisfied$\}$

We will return to condition (5) later. For the moment, treat it as a technical requirement. As far as stabilization and pole placement is concerned, the following results can be obtained:

1. *Let $\Sigma_p \in \mathcal{L}^q$. Then there exists $\Sigma_c \in \mathfrak{C}$ such that $\Sigma_p \wedge \Sigma_c$ is stable if and only if Σ_p is stabilizable.*

2. *Let $\Sigma_p \in \mathcal{L}^q$. Then the following conditions are equivalent:*

   (i) Σ_p is controllable
   (ii) For each monic polynomial $r \in \mathbf{R}[\xi]$, there exists a $\Sigma_c \in \mathfrak{C}$ such that $\chi_{\Sigma_p \wedge \Sigma_c} = r$

5 Implementation of controllers.

In intelligent control, it is usually taken for granted that every controller which processes the sensor measurements and delivers the actuator inputs is admissible. From an applications point of view, it is somewhat surprising that this very broad feasibility and implementability of intelligent controllers has not been questioned more often. In many practical circumstances it is difficult to see why and how such sophisticated devices should be used. Computer control is important, but it is not the whole picture.

In addition, there is usually some reference to causality which is sometimes formulated as implying the absence of a differentiating action. This last condition is then justified by referring to noise amplification which differentiating controllers may cause.

There are, however, many control devices which, if we insist in viewing them as input/output processors, will act as differentiators, but which cause no trouble at all. Take as an example a controller consisting of a mass/spring/damper combination which is attached to a given mass and serves to hold this given mass in a particular equilibrium, while achieving a gentle transient response. A traditional door closing mechanism is an example of a device which functions exactly in this way. Our point of view is that it is much better in such examples not to insist on an input/output interpretation, but simply take the interconnection point of view.

Nevertheless, the implementation issue is an important one. In our approach, it can simply not be avoided and should be incorporated in the specifications of $\mathfrak{C}$. In particular, the following question occurs: *if $\Sigma_p \in \mathcal{L}^q$ and we want to control it by means of a controller $\Sigma_c \in \mathcal{L}^q$, how would we achieve this?*

For example:

(i) Assume that the control terminals correspond to electrical terminals or terminals of a mechanical system, can the controller be realized using passive components?

(ii) Can the controller be implemented by means of an input/output device? In other words, can we choose some of the control variables and make them act as inputs to the controller, while the other variables act as outputs?

(iii) Is (ii) possible while making the transfer function of the controller proper? In this case the controller could in principle be implemented as an intelligent controller using logic devices.

(i) is a research question, for which we have obtained some partial results. We will not report them here. We can give complete answers to (ii) and (iii). However, before doing this, we need to introduce some integer invariants of $\mathcal{L}^q$. Define three integer valued maps on $\mathcal{L}^q$ as follows:

$$
\begin{aligned}
m : & \quad \mathcal{L}^q \to \{0, 1, 2, \ldots, q\} \\
p : & \quad \mathcal{L}^q \to \{0, 1, 2, \ldots, q\} \\
n : & \quad \mathcal{L}^q \to \{0, 1, 2, \ldots\} = \mathbb{Z}_+
\end{aligned}
$$

Take $\Sigma \in \mathcal{L}^q$, let it be represented by (2), and assume (without loss of generality - see [5]) that R has full row rank. Define

$$
\begin{aligned}
m(\Sigma) : \ & = \ q - rowdimension(R) \\
p(\Sigma) : \ & = \ rowdimension(R) \\
n(\Sigma) : \ & = \ McMillandegree(R)
\end{aligned}
$$

Recall that the *McMillan degree* of a full row rank polynomial matrix is the largest degree of its maximal size minors. In terms of minimal input/state/output representations, $m(\Sigma), p(\Sigma)$ and $n(\Sigma)$ can be given a very concrete interpretation. They correspond respectively to the number of input variables, the number of output variables, and the number of state variables of Σ [5].

Let $\Sigma_p \in \mathcal{L}^q$ and $\Sigma_c \in \mathcal{L}^q$, and assume that $\Sigma_p \wedge \Sigma_c$ is autonomous. Assume that

$$
p(\Sigma_p) + p(\Sigma_c) = p(\Sigma_p \wedge \Sigma_c) = q \tag{6}
$$

(note that condition (5) is precisely equivalent to this). Then it can be shown that the control variables can be partitioned in such a way that Σ_p and Σ_c have a complementary input/output structure, and with the transfer function in Σ_p proper, but that of Σ_c in general not proper. If we want the transfer function of Σ_c to be proper and the feedback system to be well posed, then we need also

$$
n(\Sigma_p) + n(\Sigma_c) = n(\Sigma_p \wedge \Sigma_c) \tag{7}
$$

For a precise statement see [6]. Generalizations to the case that $\Sigma_p \wedge \Sigma_c$ is not autonomous are also possible.

6   LQ-control.

Let $\Sigma \in \mathcal{L}^q = (\mathbf{R}, \mathbf{R}^q, \mathfrak{B})$ and assume for simplicity that it is controllable. Let (2) be a representation of it. Assume, to avoid smoothness difficulties, that $\mathfrak{B}$ consists of the C^∞ solutions of (2). As in [5] we will call this a *kernel representation* of Σ. Consider also the two-variable polynomial matrix $L \in \mathbf{R}^{q \times q}[\zeta, \eta]$ and assume that it is symmetric, i.e., $L(\zeta, \eta) = L^T(\eta, \zeta)$. L induces the quadratic differential form $Q_L : C^\infty(\mathbf{R}, \mathbf{R}^q) \to C^\infty(\mathbf{R}, \mathbf{R})$ defined by

$$Q_L(w) := \sum_{k,\ell} \left(\frac{d^k w}{dt^k}\right)^T L_{k\ell} \left(\frac{d^\ell w}{dt^\ell}\right) \tag{8}$$

where

$$L(\zeta, \eta) =: \sum_{k,\ell} L_{k\ell} \zeta^k \eta^\ell \tag{9}$$

Consider now the following optimization problem: *Determine $\mathfrak{B}^* \subseteq \mathfrak{B}$ such that $w^* \in \mathfrak{B}^*$ implies*

(i) *(stability)* $\lim_{t \to \infty} w^*(t) = 0$

(ii) *(optimality) for all $\Delta \in \mathfrak{B}$ of compact support, there should hold:*

$$\int_{-\infty}^{+\infty} (Q_L(w^* + \Delta) - Q_L(w^*))dt \geq 0 \tag{10}$$

View (10) as a question of determining the optimal trajectories of the system (2) with the cost functional $\int Q_L(w)dt$. Note that this formulation of the *LQ*-problem departs radically from the classical formulation: we do not start from a state model, there are no specified initial conditions, no inputs, the cost functional may contain higher order derivatives.

The optimal behavior $\mathfrak{B}^*$ can be characterized as follows. It is non-empty (equivalently, $0 \in \mathfrak{B}^*$) if and only if there exists a polynomial matrix $X \in \mathbf{R}^{q \times \bullet}[\xi]$ such that

$$L(-i\omega, i\omega) + X^T(-i\omega)R(i\omega) + R^T(-i\omega)X(i\omega) \geq 0 \tag{11}$$

for all $\omega \in \mathbf{R}$. Assume that this is the case with the left hand side of (11) of rank $m(\Sigma)$ (regard this as a technical condition, similar to observability in the classical formulation). Then $\mathfrak{B}^*$ can be computed as follows: find $X \in \mathbf{R}^{q \times \bullet}[\xi]$ and $C \in \mathbf{R}^{m(\Sigma) \times q}$ such that

$$L(-\xi, \xi) + X^T(-\xi)R(\xi) + R^T(-\xi)X(\xi) = C^T(-\xi)C(\xi) \tag{12}$$

and such that

$$det(\begin{bmatrix} R \\ C \end{bmatrix})\text{is Hurwitz} \tag{13}$$

Proofs and further details will be given in [7]. This way of obtaining $\mathcal{B}^*$ is reminiscent of the spectral factorization approach to LQ control as explained for example in [8, section 26]. Equation (12) is a complete generalization of the Riccati equation.

Note that our way of thinking about optimal control *comes up with $\mathcal{B}^*$, the optimal behavior, not with the optimal controller* ! The question thus remains how to implement the closed loop system $\Sigma^* = (\mathbf{R}, \mathbf{R}^q, \mathcal{B}^*)$. We will take this up in the next section.

7   Implementation of controlled behavior.

Let $\Sigma_p = (\mathbf{R}, \mathbf{R}^q, \mathcal{B}_p) \in \mathcal{L}^q$ be a plant. A system $\Sigma = (\mathbf{R}, \mathbf{R}^q, \mathcal{B}) \in \mathcal{L}^q$ is said to be a *subsystem* of Σ_p if $\mathcal{B} \le \mathcal{B}_p$. A controller $\Sigma_c = (\mathbf{R}, \mathbf{R}^q, \mathcal{B}_c)$ is said to *implement* Σ if

$$\Sigma = \Sigma_p \wedge \Sigma_c \tag{14}$$

This implementation problem is very akin to what, particularly in the Russian literature, is called a *synthesis*. Note that if we do not restrict Σ_c, then it is trivial to solve this problem: take $\Sigma_c = \Sigma$. The problem becomes interesting when Σ_c is further constrained, for example to be passive, or such that in addition to (14), (6), or (6) and (7) hold.

The one case where we have rather complete results is specifying when (14) is implementable with a controller satisfying (6). The result says that if $\Sigma \in \mathcal{L}^q$ is *any* subsystem of a controllable system $\Sigma_p \in \mathcal{L}^q$, then there exists a $\Sigma_c \in \mathcal{L}^q$ such that (14) and (6) hold. Using the results mentioned in section 5, this implies that Σ can be implemented using a controller with an input/output structure which is complementary (in the sense that the input/output structure of the plant and the controller go in opposite directions) to that of the input/output structure of the plant but with a transfer function which may not be proper. Thus any subsystem of a controllable system is implementable using an improper controller of feedback type.

Let us now return to the optimal LQ controller of section 6. The algorithm given there yields the optimal behavior $\mathcal{B}^*$, i.e., the family of all optimal trajectories. The question is how to implement $\mathcal{B}^*$ by means of a controller. Let $\Sigma^* :=$ $(\mathbf{R}, \mathbf{R}^q, \mathcal{B}^*)$. Obviously $\Sigma^* \in \mathcal{L}^q$. Then every time someone proposes a controller Σ_c, we can check whether $\Sigma^* = \Sigma_p \wedge \Sigma_c$. Note that this departs from the usual situation in LQ control, since, while our $\mathfrak{B}^*$ is unique, there will be many $\Sigma_c \in \mathcal{L}^q$ such that

$$\Sigma_p \wedge \Sigma_c = \Sigma^* \tag{15}$$

In particular it can be shown that since Σ_p is controllable, there exists a Σ_c such (15) and (6) hold. In addition, it can be shown that if the 2-variable polynomial matrix L is of degree 0 (i.e., if it is a constant), then Σ_c can be taken to be a memoryless state controller.

We close this section by making the remark that the non-uniqueness of the controller Σ_c which achieves (15) has important practical consequences, since it shows that optimal control can be achieved by means of controllers which in the traditional point of view would not be admissible controllers.

8 Control of latent variables.

We will now return to the situation of figure 2, in which not all the plant variables are available to the controller. Assume again that the plant is described by a constant coefficient linear differential equation. This yields

$$R(\frac{d}{dt})w = M(\frac{d}{dt})\ell \tag{16}$$

as the plant model, with $w = col(w_1, \ldots, w_q)$ the manifest variables (the variables available to the controller) and $\ell = col(\ell_1, \ldots, \ell_d)$ the latent variables (the to-be-controlled variables). The question which we will discuss is what controlled behavior can be achieved by a controller

$$C(\frac{d}{dt})w = 0 \tag{17}$$

which acts on the manifest variables only.

We will call (16) *observable* [5] if ℓ can be deduced from w, i.e., if whenever (w, ℓ_1) and (w, ℓ_2) satisfy (16), then $\ell_1 = \ell_2$ must hold. It can be shown [5] that (16) is observable if and only if the complex matrix $M(\lambda)$ has full column rank for all $\lambda \in \mathbf{C}$. In this case (16) can equivalently, in the sense that (16) and (18) have the same solutions (w, ℓ), be described by a system of differential equations of the form

$$\ell = M'(\frac{d}{dt})w \tag{18a}$$

$$R'(\frac{d}{dt})w = 0 \tag{18b}$$

for suitable polynomial matrices R', M'.

Now, assume that a controller is designed for (16) which disregards the fact that only the w's are available for interconnection:

$$C_1(\frac{d}{dt})w + C_2(\frac{d}{dt})\ell = 0 \tag{19}$$

Then, using (18a), it easily follows that the alternative controller

$$(C_1 + C_2M')(\frac{d}{dt})w = 0 \tag{20}$$

will achieve the same controlled behavior. In other words, the solutions of (16, 19) will be the same as those of (16, 20). It follows that *for an observable system* any controlled behavior which can be achieved by a controller using the (w, ℓ)'s as control variables, can also be achieved by a controller using only the w's as control variables. We remark, however, that the observability requirement is more severe than it might appear at first sight. For example, the usual situation of additive noise in the measurements already obstructs observability.

9   Epilogue.

In this paper we have put forward a theory of control in which system *interconnection* is the central idea. This in contrast with what we have called *intelligent* control, where we can view a controller as a signal processor, processing the measurements in order to compute the control action. This last type of controller is actually a special type of interconnection. In fact, the issue of what interconnections can be implemented by means of an intelligent controller comes up naturally and was briefly discussed in the present paper.

Ever since control theory was established as a scientific discipline, it has chosen to formalize its questions in an input/output setting. This can be observed in older and in more recent texts alike [8, 9,10,11]. Also mathematical system theory, which can be seen as an outgrowth of control theory, has invariably adopted the input/output framework. This may be seen, for example, from the attempts at axiomatization in [1,12].

One thing is clear: as a basic structure for modelling dynamical systems, the input/output framework is unsuitable. In an off-the-shelf modelling package, modules (standard elements) and interconnections (standard ways of interconnecting standard elements) are the key components: modelling proceeds by *tearing* (examining the interconnections) and *zooming* (examining the subsystems) in a hierarchical fashion. Physical components (resistors, capacitors, transformers, masses, spring, dampers, etc, etc.) will be specified by giving their parameter values. The specification of the system architecture will tell how the components are inserted in the system. As such, it is unnecessary, awkward, and illogical to specify the components in input/output form. It may, but need not be the case that the overall system needs to be specified in input/output form, for example because of the presence of logic devices in the system. This input/output structure will have its effect on the components and may translate into an input/output structure on a particular component. However in what input/output structure this particular component will function will depend on the architecture. It is because of such considerations that modelling packages based on simple interconnection ideas (as SPICE) are bound to be much more usable that those based on input/output thinking (as MATLAB's SIMULINK).

True, many of the issues underlying the behavioral framework have been touched on before. The limitations of input/output thinking is very clearly addressed in books on circuit theory [13,14]. Also Zames' original papers [2] on $\mathcal{L}_2$-stability work with input/output relations, rather than with maps (as in [1]). This input/output *relation* point of view was even more strongly emphasized by Safonov [15], where, however, the for-that-time-radical-step of dropping inputs and outputs altogether was not taken. The need to work with both external and internal variables is one of the key ingredients in the state space description of dynamical systems. The feeling that the state is but a limited implementation of this need for involving internal variables in models is what led Rosenbrock to introduce (always in an input/output setting) the *partial state* [16]. We may view this as a precursor of our latent variables. Also descriptor system formulations of state models (implicit systems, singular systems) point to discomfort with the usual input/state/output approach. Indeed, anyone who examines the suitability of input/output structures for models obtained from first principles will discover that it is unreasonable to take $\frac{dx}{dt} = f(x, u); y = f(x, u)$ as the starting point for dynamics.

The fact that in optimal control, it is sometimes convenient to first define the optimal behavior and the proceed to find the optimal controller is implicit in Brockett's approach [8, section 26].

Finally, the fact that it is the solution set of equations and not the equations themselves that are important in modelling underlies the system representation questions, from co-prime factorizations to Hankel matrices to the state space isomorphism theorem.

Whenever an axiomatic framework, as the behavioral setting, is put forward, whenever its effectiveness is argued and compared to an existing framework, as the input/output setting, it is unavoidable that there will be countless links with older work. It is unavoidable that many researchers will recognize their own discomforts. The merit of the behavioral framework has been in bringing the appropriate framework to the foreground explicitly. Mathematics is mainly discovery of existing structures, and inventing the language to discuss them in.

References.

[1] J.C. Willems, *The Analysis of Feedback Systems*, The MIT Press, 1971.

[2] G. Zames, "On the input-output stability of time-varying nonlinear feedback systems. Part I: Conditions derived using concepts of loop gain, conicity, and positivity; Part II: Conditions involving circles in the frequency planeand sector nonlinearities", *IEEE Transactions on Automatic Control*, 11, pp. 228–238, 465–476, 1966.

[3] J.C. Willems "From Time Series to Linear System. Part I: Finite Dimensional Linear Time Invariant Systems; Part II: Exact modelling; Part III: Approximate modelling", *Automatica*, 22, pp. 561-580, (1986); pp. 675-694, (1986); 23, pp. 87-115, 1987.

[4] J.C. Willems, "Models for Dynamics", *Dynamics Reported*, 2, pp. 171-269, 1989.

[5] J.C. Willems, "Paradigms and puzzles in the theory of dynamical systems", *IEEE Transactions on Automatic Control*, 36, pp. 259–294, 1991.

[6] M. Kuijper, "Why do stabilizing controllers stabilize?", *Automatica*, to appear.

[7] J.C. Willems, "On interconnections, control, and feedback", *IEEE Transactions on Automatic Control*, submitted,

[8] R.W. Brockett, *Finite Dimensional Linear Systems*, Wiley, 1970.

[9] J.G. Truxal, *Automatic Feedback Control System Synthesis*, McGraw-Hill, 1955.

[10] J.-C. Gille, P. Decauline, and M. Pellegrin, *Théorie et Technique des Asservissements*, Dunod, 1958.

[11] J.C. Doyle, B.A. Francis, and A.R. Tannenbaum, *Feedback Control Theory*, MacMillan, 1992.

[12] R.E. Kalman, P.L. Falb, and M.A. Arbib, *Topics in Mathematical Systems Theory*, McGraw-Hill, 1969.

[13] V. Belevitch, *Classical Network Theory*, Holden-Day, 1968.

[14] R.W. Newcomb, *Linear Multiport Synthesis*, McGraw-Hill, 1966.

[15] M.G.Safonov, *Stability and Robustness of Multivariable Feedback Systems*, MIT Press, 1980.

[16] H.H. Rosenbrock, *State-space and Multivariable Theory*, Nelson, 1970.

Lecture Notes in Control and Information Sciences

Edited by M. Thoma

1992–1994 Published Titles:

Vol. 182: Hagenauer, J. (Ed.)
Advanced Methods for Satellite and Deep
Space Communications. Proceedings of an
International Seminar Organized by Deutsche
Forschungsanstalt für Luft-und Raumfahrt
(DLR), Bonn, Germany, September 1992.
196 pp. 1992 [3-540-55851-9]

Vol. 183: Hosoe, S. (Ed.)
Robust Control. Proceesings of a Workshop
held in Tokyo, Japan, June 23-24, 1991.
225 pp. 1992 [3-540-55961-2]

Vol. 184: Duncan, T.E.; Pasik-Duncan, B.
(Eds)
Stochastic Theory and Adaptive Control.
Proceedings of a Workshop held in Lawrence,
Kansas, September 26-28, 1991.
500 pages. 1992 [3-540-55962-0]

Vol. 185: Curtain, R.F. (Ed.); Bensoussan, A.;
Lions, J.L.(Honorary Eds)
Analysis and Optimization of Systems: State
and Frequency Domain Approaches for Infinite-
Dimensional Systems. Proceedings of the 10th
International Conference, Sophia-Antipolis,
France, June 9-12, 1992.
648 pp. 1993 [3-540-56155-2]

Vol. 186: Sreenath, N.
Systems Representation of Global Climate
Change Models. Foundation for a Systems
Science Approach.
288 pp. 1993 [3-540-19824-5]

Vol. 187: Morecki, A.; Bianchi, G.;
Jaworeck, K. (Eds)
RoManSy 9: Proceedings of the Ninth
CISM-IFToMM Symposium on Theory and
Practice of Robots and Manipulators.
476 pp. 1993 [3-540-19834-2]

Vol. 188: Naidu, D. Subbaram
Aeroassisted Orbital Transfer: Guidance and
Control Strategies.
192 pp. 1993 [3-540-19819-9]

Vol. 189: Ilchmann, A.
Non-Identifier-Based High-Gain Adaptive
Control.
220 pp. 1993 [3-540-19845-8]

Vol. 190: Chatila, R.; Hirzinger, G. (Eds)
Experimental Robotics II: The 2nd International
Symposium, Toulouse, France, June 25-27
1991.
580 pp. 1993 [3-540-19851-2]

Vol. 191: Blondel, V.
Simultaneous Stabilization of Linear Systems.
212 pp. 1993 [3-540-19862-8]

Vol. 192: Smith, R.S.; Dahleh, M. (Eds)
The Modeling of Uncertainty in Control
Systems.
412 pp. 1993 [3-540-19870-9]

Vol. 193: Zinober, A.S.I. (Ed.)
Variable Structure and Lyapunov Control
428 pp. 1993 [3-540-19869-5]

Vol. 194: Cao, Xi-Ren
Realization Probabilities: The Dynamics of
Queuing Systems
336 pp. 1993 [3-540-19872-5]

Vol. 195: Liu, D.; Michel, A.N.
Dynamical Systems with Saturation
Nonlinearities: Analysis and Design
212 pp. 1994 [3-540-19888-1]

Vol. 196: Battilotti, S.
Noninteracting Control with Stability for
Nonlinear Systems
196 pp. 1994 [3-540-19891-1]

Vol. 197: Henry, J.; Yvon, J.P. (Eds)
System Modelling and Optimization
975 pp approx. 1994 [3-540-19893-8]

Vol. 198: Winter, H.; Nüßer, H.-G. (Eds)
Advanced Technologies for Air Traffic Flow
Management
225 pp approx. 1994 [3-540-19895-4]

Vol. 199: Cohen, G.; Quadrat, J.-P. (Eds)
11th International Conference on Analysis and
Optimization of Systems – Discrete Event
Systems: Sophia-Antipolis, June 15–16–17,
1994
648 pp. 1994 [3-540-19896-2]

Vol. 200: Yoshikawa, T.; Miyazaki, F. (Eds)
Experimental Robotics III: The 3rd International
Symposium, Kyoto, Japan, October 28-30,
1993
624 pp. 1994 [3-540-19905-5]

Vol. 201: Kogan, J.
Robust Stability and Convexity
192 pp. 1994 [3-540-19919-5]